.

Handbook of Milk Proteins

Handbook of Milk Proteins

Edited by **Caroline Gardner**

R CALLISTO REFERENCE

New York

Published by Callisto Reference,
106 Park Avenue, Suite 200,
New York, NY 10016, USA
www.callistoreference.com

Handbook of Milk Proteins
Edited by Caroline Gardner

International Standard Book Number: 978-1-63239-404-0 (Hardback)

Contents

Preface

A detailed account based on milk proteins has been highlighted in this profound book. It provides comprehensive information on a wide range of topics related to milk proteins. The content provided within will prove to be valuable to those interested in dairy foods, milk chemistry, human health, neonatal growth, lactation and mammary gland biology, milk proteins production and other related topics. This book delivers an insight into a range of topics related to milk proteins, including bioactivities of milk proteins and the peptides generated from those proteins, new functions assigned to some milk proteins, impact of processing of milk on milk proteins, allergies associated with consumption of milk, genetic variation of milk proteins, application of genomic technologies for exploring expression of proteins during milk synthesis and production of milk and milk proteins as affected by environmental factors.

The researches compiled throughout the book are authentic and of high quality, combining several disciplines and from very diverse regions from around the world. Drawing on the contributions of many researchers from diverse countries, the book's objective is to provide the readers with the latest achievements in the area of research. This book will surely be a source of knowledge to all interested and researching the field.

In the end, I would like to express my deep sense of gratitude to all the authors for meeting the set deadlines in completing and submitting their research chapters. I would also like to thank the publisher for the support offered to us throughout the course of the book. Finally, I extend my sincere thanks to my family for being a constant source of inspiration and encouragement.

Editor

Bioactivity of Milk Peptides

Milk Derived Peptides with Immune Stimulating Antiviral Properties

Haiyan Sun and Håvard Jenssen

Additional information is available at the end of the chapter

1. Introduction

Milk is thought to be the main source of biologically active compounds for infants, providing antibacterial and antiviral activities, facilitating nutrient absorption, promoting bone growth, enhancing immunological protection and supporting the development of host immune competence. In milk, the main categories of compounds related to antiviral activity through immune stimulation and suppression of host immune inflammation are the casein proteins, whey proteins and their derived peptides [1-3].

Casein proteins, as well as casein fragments, function as antiviral and immune regulatory factors by regulating the innate immune response both through up-regulation to enhance killing of viruses, and down-regulation to reduce detrimental conditions such as sepsis [1, 3-7]. Additionaly, caseins link the innate immune system to the adaptive immune system by activating and/or enhancing B- and T-cell mediated functions. The whey protein lactoferrin, and pepsin derived peptide fragments of this protein (e.g. lactoferricin) have been studied extensively for its antiviral properties [8-10] i.e. its direct interaction with the virus particle, interaction with cellular receptors on the target cells, and lately more complex antiviral mechanisms involving stimulation and regulation of the immune system have been discovered [2, 11-16]. Similarly, peptides tailored on specific protein fragments of casein and α-lactalbumin have also been investigated for their antiviral and immunomodulatory properties. Many of these studies have identified biologically active peptides that can prevent a viral infection, as well as regulate the immune status of the host [1, 2, 17-21]. Currently, some of these peptides are being investigated in clinical trials, like human lactoferrin fragment 1-11 (AM Pharma, Bunnvik, The Netherlands) [22] and LTX-302 (Lytix Biopharma, Tromsø, Norway) [20]. Moreover, another promising class of synthetic peptides with therapeutic potential is a group of innate defence regulator peptides, which exhibit immune protection by enhancing or suppressing the host immune response [23-26]. It is tremendously encouraging that many of these proteins and peptides have pharmaceutical potential within antiviral and anticancer therapy, as vaccine adjuvants, as immunosuppressants for the treatment of autoimmune diseases and in conjunction with organ transplantation, etc. [20, 23-25, 27-29].

Studies have demonstrated that active milk protein and peptide compounds can be extracted from a variety of species including humans, bovine, porcine, mice and camel. The main focus in this paper is the antiviral and immune regulating properties (Table 1) of milk proteins (Table 2) and their peptide derived fragments (Table 3). The vast majority of the discussed studies deal with proteins and peptides of bovine origin, and these will be referenced with their protein names, while proteins and peptides from other origins will be explicitly specified with the species name.

Virus	Protein or peptide	Model of antiviral function	Reference
Enveloped Virus			
Herpes simplex virus 1	human and bovine lactoferrin and lactoferricin, lactoperoxidase chemically modified milk proteins e.g. serum albumin, α-lactalbumin, β-lactoglobulin	Binding to both virus particle and cellular receptors (heparan sulphate) to prevent viral adsorption and entry; Interference with intracellular replication events or synthesis of progeny viral components	[9, 35, 55, 77-79, 81-83] [89, 93, 99, 100]
Herpes simplex virus 2	human and bovine lactoferrin	Binding to virus receptor of non-GAG nature	[10]
	β-lactoglobuline	Binding to virus particle	[93]
Hepatitis C virus	lactoferrin	Binding to viral envelope protein E1 and E2	[52, 63, 71, 220]
Hepatitis B virus	iron- or zinc-saturated lactoferrin	Binding to cellular molecules interfering with viral attachment/entry	[47, 59]
Hepatitis G virus	lactoferrin	Unknown	[63]
Respiratory syncytial virus	lactoferrin, lactoperoxidase	Binding to F1 subunit of RSV F protein to inhibit viral absorption	[35, 44-46]
Human immunodeficiency virus	human and bovine lactoferrin, lactoperoxidase	Binding to cellular receptor to inhibit viral absorption and replication	[8, 18, 19, 34, 48, 49, 62, 68, 70]
	chemically modified milk proteins like serum albumin, α-lactalbumin, β-lactoglobulin		[9, 89-92]
Influenza virus (H3N2, H1N1 and H5N1)	lactoferrin, κ-casein, glycomacropeptide, lactoperoxidase	Binding to hemagglutinin of virus	[36, 53, 135]
	modified human serum		[94, 101-103]

Virus	Protein or peptide	Model of antiviral function	Reference
	albumin and β-lactoglobulin, α-lactalbumin, lactoferrin		
Human cytomegalovirus	lactoferrin and lactoferricin	Interfere with virus target cells; up-regulation of killer cells; synergistic antiviral effect with cidofovir	[12-16]
	chemically modified milk proteins like serum albumin, α-lactalbumin, β-lactoglobulin	Binding to virus particle	[12, 93, 96, 221]
Feline herpes virus 1	human and bovine lactoferrin	Binding to cellular molecules	[54]
Canine herpes virus	human and bovine lactoferrin (apo- and holo-)	Binding to virus particle and cellular receptor on target cell	[74]
Hantavirus	lactoferrin	Binding to cellular molecules; synergistic effect with Ribavirin on inhibiting viral replication	[72, 122]
Vesicular stomatitis virus	lactoferrin	Induction interferon-α/β expression to inhibit viral replication	[11]
Friend virus complex	human lactoferrin	Regulation on the myelopoiesis; synergistic effect with interferon-γ	[11, 56, 57, 64]
Human papillomavirus	human and bovine lactoferrin, human and bovine lactoferricin	Binding to heparan sulphate cell receptor	[65, 121]
Alphavirus heparan sulphate- adapted sindbis virus and semliki forest virus	human lactoferrin, charge-modified human serum albumin	Binding to heparan sulphate cell receptor	[73]
Severe acute respiratory syndrome coronavirus	lactoferrin	Binding to heparan sulphate cell receptor	[120]

Non-enveloped virus

Virus	Protein or peptide	Model of antiviral function	Reference
Rotavirus	human lactoferrin (apo-/holo-), α-lactalbumin, β-lactoglobulin	Binding to viral particles to prevent both rotavirus haemagglutination and viral binding to receptors on susceptible cells	[30, 40]
	human lactadherin	Binding to structural protein of rotavirus and inhibits virus replication	[37, 38]

Virus	Protein or peptide	Model of antiviral function	Reference
	high molecular glycoprotein (e.g. mucin)	Inhibitor for viral-cell binding to prevent productive virus infection	[107, 108]
	immune globulin	In vivo effect on inhibition of viral replication	[107]
Poliovirus	lactoferrin, modified bovine β-lactoglobulin	Binding to viral receptor on target cell	[50, 58, 104]
Coxsackie virus	modified bovine β-lactoglobulin	Binding to viral receptor on target cell	[104]
Adenovirus	lactoferrin	Binding to viral protein III and IIIa; competition with virus for common membrane receptors	[60, 61, 69]
Enterovirus (71, echovirus 6)	lactoferrin	Binding to both cellular receptors and the viral surface protein VP1	[51, 66, 67]
Felin calicivirus	lactoferrin	Binding to cell receptor	[50]
Echovirus	lactoferrin, lactoperoxidase	Binding to cell receptor and viral structural proteins	[35, 75, 95]

Table 1. Models of antiviral proteins & peptides from milk proteins

2. Protein composition of milk and their antiviral activity

There are in general two groups of proteins found in milk, casein and whey. The casein family accounts for approximately 80% of the protein mass and includes several types of casein, e.g. $\alpha s1$, $\alpha s2$, β and κ, which form micelle complexes in the water phase of milk. The whey proteins account for the remaining 20%, and include β-lactoglobulin (not present in human milk), α-lactalbumin, serum albumin, immunoglobulins, lactoferrin, transferring, and many minor proteins. Most of the whey proteins have been demonstrated to effectively prevent viral infection. For example, milk derived proteins including α-lactalbumin, β-lactoglobulin, apo-lactoferrin (iron free), and homo-lactoferrin (Fe^{3+} carrying), were able to inhibit rotavirus attachment to cellular receptors by binding to the viral particle [30]. Among these proteins, apo-lactoferrin was proven to be the most active. Studies also showed that immunoglobulins of raw milk from non-immunized cows and camels, as well as from a commercially available bovine macromolecular whey protein fraction, have specific antibodies against human rotavirus, which are capable of inhibiting replication of rotaviruses in tissue culture and protect mice from infection in a murine model of rotavirus infection [31-33]. Lactoperoxidase, a haem-containing glycoprotein of the mammalian peroxidase family, is an important enzyme in the whey fraction of milk. In combination with its physiological substrates hydrogen peroxide and thiocyanate, lactoperoxidase manifests a wide spectrum of virucidal activities against human immunodeficiency virus, herpes simplex virus 1, respiratory syncytial virus and echovirus [34, 35]. Oral administration of lactoperoxidase also attenuate pneumonia in influenza virus infected mice through suppression of infiltration of the inflammatory cells in the lungs [36]. Furthermore, the 46kD glycoprotein termed lactadherin, also known as milk fat globule-EGF factor 8 protein, inhibited rotavirus binding to cellular receptors (acetylneuraminic acid and/or integrin) on target and/or specifically interacting with viral structural glycoprotein VP4 of rotavirus, blocking host-pathogen interaction [37-40].

Lactoferrin, first isolated in 1960 from both human [41, 42] and bovine milk [43], has been demonstrated to exhibit antiviral activity against many viruses [8, 10-16, 18, 19, 40, 44-74] (Table 2). Most studies indicate that lactoferrin and its derived peptides are likely to interfere in the virus host cell interaction (Figure 1). For example it has been demonstrated that lactoferrin is able to bind both viral receptor and the viral surface protein VP1 on enterovirus (enterovirus 71 and echovirus 6), thus interfering with viral entry [51, 66, 67, 75, 76]. Similarly, both apo- and holo-lactoferrin has been demonstrated to interact both with canine herpes virus and surface receptors on the Madin-Darby canine kidney cells, thus inhibiting canine herpes virus infection [74]. With regard to the anti-herpes simplex virus 1 ability of lactoferrin, both bovine and human lactoferrin and lactoferricin have demonstrated the ability to block viral entry and also inhibit viral cell-to-cell spread in a dose dependent manner [55, 77-79], through interaction with negatively charged glycosaminoglycans like heparan sulphate on the cell surface [55, 80-83] and elements of the viral particle [55]. Differently from herpes simplex virus 1, Marchetti et. al. found that lactoferrin inhibited herpes simplex virus 2 plaque forming activity also in cells without glycosaminoglycans suggesting that lactoferrin might block one of the specific herpes simplex virus 2 entry receptors [10].

Many of the traditional entry blocking effects observed by lactoferrin involve electrostatic interaction with anionic heparan sulphate molecules on the host cell surface [82]. The ability to interact with anionic heparan sulphate is maybe not that surprising, when evaluating the three dimensional structural composition of lactoferrin, demonstrating a rather striking cationic patch on the N-terminal lobe of the molecule [84] (Figure 1). Similarly, other highly cationic peptides have also been demonstrated to effectively interfere with herpes simplex virus attachment and entry [80, 85].

Conversely, several other milk proteins i.e. β-lactoglobulin [86], α-lactalbumin [87] are described with anionic patches on their surfaces, while the casein homologues like αs2-casein [88] have both specific anionic and cationic patches on the surface (Figure 1). Thus, charge modification of milk proteins may increase their ability to interfere with virus host cell interactions. 3-hydroxyphthalic anhydride modification of human and bovine serum albumin, and bovine β-lactoglobulin, increased the proteins negative charges in addition to their ability to prevent interaction between human immunodeficiency virus 1 envelope glycoprotein gp120 and the CD4 host cell receptor, by direct interaction and blocking of the CD4 receptor [89, 90]. Similar effects have also been observed for 3-hydroxyphthalic anhydride modified α-lactalbumin and αs2-casein, as well as for maleylated- and succinylated-human serum albumin, indicating that human immunodeficiency virus inhibition was a general property of negatively charged polypeptides [9, 91, 92]. Among the inhibitory proteins, 3-hydroxyphthalic anhydride β-lactoglobulin also demonstrated a broad spectrum activity affecting herpes simplex virus 1 and 2 in addition to human cytomegalovirus by binding to the virus particles, inhibiting particularly the binding of monoclonal antibodies towards glycoprotein E and glycoprotein C [93]. Comparative results have been shown for anionic-modified human serum albumin and β-lactoglobulin which prevents influenza virus membrane fusion with the host cell membrane, a process mediated by the viral glycoprotein hemagglutinin [94]. Interestingly, this anti-influenza effect has not been observed for other milk proteins carrying negative charges, like succinylated bovine serum albumin, lactalbumin, lactoferrin, lysozyme and transferrin [94]. It is said that inhibition of viral fusion demonstrates a certain degree of specificity for negative charged proteins. However, addition of net negative charges to lactoferrin by acylation with either succinic- or acetic anhydride abolished its anti-poliovirus and anti-feline calicivirus activity, which may be attributed to the obliterate binding of acylated lactoferrin to the surfaces of susceptible cells [95]. Also, when negatively charged groups were added to lactoferrin by succinylation, the antiviral effect on human immunodeficiency virus 1 was increased, but the antiviral potency against human cytomegalovirus was mostly decreased [96], illustrating the proteins different modes of action. Similar results were also obtained by Florisa *et. al.* which demonstrated a stronger antiviral effect against human immunodeficiency virus by developing poly-anionic milk proteins, while stronger effects could be obtained against human cytomegalovirus by creating poly-cationic milk proteins [12].

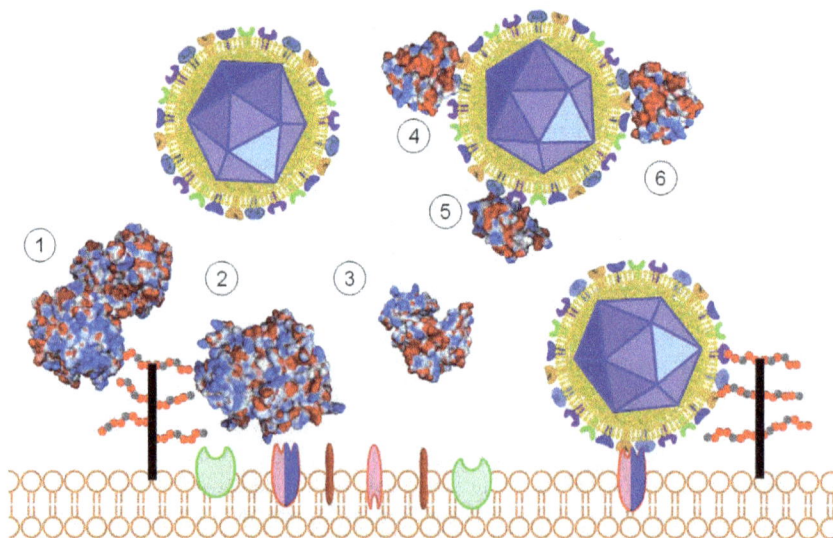

Figure 1. The traditional direct antiviral mechanisms of selected milk proteins. Several proteins are characterized to interact directly with cell surface heparan sulphate, like lactoferrin (1) and lactoperoxidase. Casein species like γ-casein (3) and αs2-casein (4) are despite high cationic character on their surface not described to interact with heparan sulphate, and the latter in stead been demonstated to interact with the virus pareticle. Anionic milk proteins like α-lactalbumine (5) and β-lactoglobuline (6) are also illustrated to interact directly with the virus particle, thus preventing host receptor interaction.

Source	Protein	% of whey protein	% of casein	Molecular size (kDa)	Nature	PDB code
Whey	β-lactoglobulin	50-55	NA	~18.4	Apolipoprotein	1DV9
	α-lactalbumin	20-25	NA	~14.1	Albumin	1A4V
	Immunoglobulins	10-15	NA	~150	Glycoprotein	---
	Lactoferrin	1-2	NA	~80	Glycoprotein	1BLF
	Lactoperoxidase	0.5	NA	~70	Glycoprotein	3GC1
	Serum albumin	5-10	NA	~66	Albumin	
	Glycomacropeptide	10-15	NA	---	Phosphoprotein	
Casein	αs1-casein	NA	40-50	~23	Phosphoprotein	
	αs2-casein	NA	10-15	~23	Phosphoprotein	1NA7
	β-casein	NA	30-35	~24	Phosphoprotein	
	κ-casein	NA	10-15	~19	Phosphoprotein	
	γ-casein	NA	5-10	~75-100	Phosphoprotein	2CHL

Note. The PDB extention codes are for crystal structure files for the respective milk proteins. The structures have been used when generating the graphic illustration on figure 1.

Table 2. Bioactive proteins from bovine milk.

The strong antiviral activity of poly-cationic compounds is generally explained by the compounds ability to interact with anionic heparan sulphate on the host cell surface, which works as a broad spectrum attachment receptor for several viruses [97, 98]. Thus, it is not surprising that methylated or ethylated α-lactalbumin and β-lactoglobulin demonstrate antiviral activity against the bacteriophage M13 through the inhibition of the phage DNA replication, as well as against herpes simplex virus 1 replication, with increasing activity proportional to the extent of esterification or increased basicity of the modified proteins [99, 100]. The net positive charge-modified human serum albumin had a similar antiviral effect as lactoferrin, against heparan sulphate adapted sindbis virus and semliki forest virus, by blocking the virus receptor on the cell surface, indicating that the antiviral activity of lactoferrin mainly is related to its net positive charge [73]. Methylated α-lactalbumin, β-lactoglobulin and lactoferrin also demonstrate enhanced antiviral activity against human influenza virus A subtype H3N2 and subtype H1N1 [101, 102], and lethal avian influenza A (H5N1) [103]. This effect is most likely linked to the disruption of the electrostatic interactions within hemagglutinin, by the esterified whey proteins, thus affecting the proteins stability and capacity to trigger envelope fusion with the host cell. Furthermore, methylation of β-lactoglobulin does also enhance the proteins antiviral activity against coxsackie virus and poliovirus type 1 in a dose dependent manner [104]. This illustrates that chemically modified whey proteins with added negative or positive charges can exert increased antiviral effect against a diverse group of viruses, through different antiviral mechanisms. The virucidal activity of the modified milk proteins, with additional negative charges, may attribute to a stronger interaction of these proteins with the viral envelope proteins. Esterification of whey proteins with methanol or ethanol would increase their cationic charge, thus increasing their affinity for negatively charged macromolecules such as host cell receptors and viral DNA or RNA, thus inhibiting viral attachment to cellular membranes or inhibiting viral replication and transcription, respectively. The structural differences between enveloped and non-enveloped viruses in addition to the unique protein composition in milk from different species preclude a generalized conclusion of the milk proteins potential. Thus, further studies should be carried out to identify the underlying molecular interactions involved, and the true therapeutic potential of these milk derived molecules.

3. Traditional antiviral mechanisms of milk-derived proteins

The life cycle of a virus comprises several phases such as binding to the host cell surface, entry or fusion, replication of the viral genome, viral protein synthesis, virus progeny assembly and release. All these steps may be targeted by antiviral agents or milk derived proteins.

Binding to structural virus proteins prevent virus host cell interaction. For the non-enveloped viruses, structural proteins on the surface of the virion protruding as spikes, such as glycoprotein VP4 of rotavirus [105] or fibers associated with each penton base of the capsid on for example adenovirus [106]. These proteins recognize host cell surface receptors, and are involved in facilitating the initial virus to host cell attachment. Enveloped viruses, meaning the viral capsid is coated with a lipid membrane known as the viral envelop, infect host cells via the interaction between envelop proteins and cellular receptors. The envelop proteins include E1 and E2 of hepatitis C virus, F protein of respiratory syncytial virus, hemagglutinin of influenza viruses, etc.

Many of the antiviral milk proteins can bind to structural proteins of the virion in order to prevent binding of the virus to the target cell and subsequently inhibit entry of the viral genome into the host cell. Human lactoferrin (apo- or Fe^{3+}), α-lactalbumin, β-lactoglobulin, human lactadherin, mucin, and immunoglobulin from milk could prevent rotavirus infection through the binding to structural viral protein VP4 [30, 37, 38, 40, 107, 108]. Also, the antiviral activity of lactoferrin against adenovirus has been attributed to the interaction of the milk protein with viral capsid proteins [60, 61, 69].

Furthermore, Ikeda *et. al.* has also demonstrated that lactoferrin effectively protect against hepatitis C virus infection in hepatocytes and lymphocytes by neutralizing the virus, while a basic N-terminal loop of lactoferrin named lactoferricin exhibited no antiviral properties in the same experiments [63]. Lactoferrin has also been demonstrated to inhibit the absorption and growth of respiratory syncytial virus in cell culture through direct interaction with the F(1) subunit of the viral F protein, which is the most important surface glycoprotein participating in viral penetration [44, 45]. Blocking of viral entry like this, leads to down-regulation of respiratory syncytial virus induced interleulin-8 secretion from the HEp-2 cells, which consequently leads to a dampening of the immune response as the low levels of interleukine-8 is inadequate to recruit neutrophils to phagocytose the viral antigen [46].

Hemagglutinin is an antigenic glycoprotein found on the surface of influenza viruses. The glycoprotein has two main functions; recognition of target cells through the binding of sialic acid-containing receptors and facilitating entry of the viral genome into the target cells by initiating fusion of host endosomal membrane with the viral membrane. Thus, targeting the hemagglutination activity of hemaglutinin could be a robust mechanism in fighting influenza virus infections. Influenza hemaglutinin has also successfully been targeted by both human and bovine lactoferrin (apo- and holo-), as well as κ-casein glycomacropeptide, reduce viral hemagglutination [30, 53]. Moreover, the addition of methylated β-lactoglobulin in the medium of Madin-Darby canine kidney cell lines infected with influenza virus H1N1 reduced hemagglutination in a concentration-dependent manner [101].

Interference with viral entry, through virus and/or cell surface interaction. Viruses recognize and conjugate to specific host cell receptors. These receptor molecules are mainly of protein nature, including glycoprotein, lipoprotein and glycolipid-protein. Hence, host cell specificity or preference is guided by the level of expression of these individual receptor molecules on the different cells. For example, the main goal of human immunodeficiency virus is to infect CD4+ T-lymphocytes and initiate replication of a large number of progeny virions. However, the initial infection with this virus is usually of epithelial dendritic cells, which then are used for transport to the lymph nodes. Human immunodeficiency virus attachment to for example emigrating dendritic cells, is mediated by the successive interactions of the viral envelop glycoprotein gp120 with CD4 (a glycoprotein known as cluster of differentiation 4) and a co-receptor, CXCR4 (C-X-C chemokine receptor type 4, also known as fusin or CD184) or CCR5 (C-C chemokine receptor type 5, also known as CD195) [109-111]. However, in cells like macrophages and skin dendritic cells that are lacking or weakly expressing CD4, many other cell surface molecules such as heparan sulphate proteoglycans [112, 113], mannose receptor [48, 114], or dendritic cells-specific intracellular adhesion molecule-3-grabbing non-integrin (DC-SIGN) [62, 109] can play a key role in the initial multistep interaction between the virus and host cell surface. Consequently, one might hypothesis that human immunodeficiency virus entry into the host cell might be efficiently inhibited via the interaction between antiviral milk proteins from bovine or human sources and some of the receptors described above. This has also been demonstrated to be true, for lactoferrin which effectively can bind heparan sulphate as well as mannose receptor like nucleolin, both which will inhibit virus attachment [18, 48]. Other studies have also indicated that a peptide fragment (hLF1-33) of human lactoferrin (residue 1-33) constituting the glycosaminoglycan recognizing site of the human lactoferrin, exhibit inhibitory effect on human immunodeficiency virus 1 attachment to epithelial cells, though its activity was lower as compared to the native protein [19]. Interestingly however, hLF1-33 had no inhibitory effect on transferring human immunodeficiency virus 1 from immature dendritic cells to CD4 T-lymphocytes, but enhancing virus transmission in contrast to human lactoferrin. This may suggest that the hLF1-33 exposed domain is not involved in human lactoferrin associated inhibition of human immunodeficiency virus 1 transfer to CD4 T-cells [19]. Moreover, bovine lactoferrin could bind strongly to DC-SIGN to prevent human immunodeficiency virus 1 capture and subsequent transmission on dendritic cells, and bovine lactoferrin was a much more efficient inhibitor than human lactoferrin on blocking not only dendritic cell mediated human immunodeficiency virus 1 transmission to - but also replication in CD4 T-cells [49].

Although it has been identified that a metallopeptidase, angiotensin-converting enzyme 2, is a functional receptor for severe acute respiratory syndrome coronavirus infection [115], other reports have demonstrated that DC-SIGN, L-SIGN (also called CD209L, specific for liver/lymph node) [116-119], and heparan sulphate [120] also are involved in the virus pathogenesis. Thus, there are reasons to believe that lactoferrin could prevent severe acute respiratory syndrome coronavirus spread in the host through the same mechanism as described for human immunodeficiency virus, by interacting with DC-SIGN or heparan sulphate receptors. Recently, Lang et. al. also described that lactoferrin curtailed the entry of severe acute respiratory syndrome coronavirus into HEK293E/ angiotensin-converting enzyme 2-Myc cells by binding to heparan sulphate [120].

Human papillomavirus can also use heparan sulphate on the target cell surface as a receptor. Thus, by incubating HaCaT cells and papillomavirus 16 virus like particles with human and bovine lactoferrin Drobni et. al. have confirmed that human papillomavirus entry can be inhibited by lactoferrin in a dose-dependent fashion [65]. Subsequently, they also demonstrated that bovine lactoferrin peptide (bLF17-33) region 17-33 was a more potent inhibitor of both human papillomavirus 5 and 16 pseudovirus infection than the native protein, while human lactoferricin (hLF1-49) region 1-49 from human lactoferrin, showed modest antiviral activity against the same viruses and bLF17-42 prevented only papillomavirus 5 pseudovirus infection. With regard to the viral attachment, only hLF1-49 and bLF17-42 exhibited antiviral effect [121].

In a classical pre-incubation study on Vero E6 cells, it was demonstrated that lactoferrin had enhanced antiviral activity against hantavirus infection when added prior to infection. However, this boost in activity could be removed if the cells were subsequently washed with phosphate buffered saline prior to infection [72, 122]. These results might be explained by the weak interaction between lactoferrin and other cellular molecules rather than heparan sulphate, as the interaction between lactoferrin and heparan sulphate should withstand phosphate buffered saline washing [80]. Further research should be developed to identify whether β3 integrin and/or β1 integrin molecules are binding to lactoferrin [123-125].

Similarly, using indirect immunofluorescence, McCann et. al. found that bovine lactoferrin could bind to Crandell-Reese feline kidney cells used for propagation of feline calicivirus, as well as Monkey Embryo kidney cells used with poliovirus, indicating that the interference of viral infection might be attributed to lactoferrin binding to the cellular receptor on the respective cells, though the related cell receptors for feline calicivirus and poliovirus have not yet been identified [50]. Contradicting this, it was demonstrated that lactoferricin decreased feline calicivirus but not poliovirus infection. Moreover, feline herpes virus-1 replication could be prevented by exposing cultured Crandell-Reese feline kidney cells to lactoferrin prior to or during viral adsorption, but not following viral adsorption, suggesting that the inhibitory effect on feline herpes virus 1 adsorption to the cell surface and/or viral penetration into the cell might be related to the interaction between lactoferrin and cellular receptor on the Crandell-Reese feline kidney cells [54].

Interference with certain viral enzymes required for virus replication. The process of viral replication will involve a myriad of enzymes, such as DNA- or RNA-polymerases, reverse transcriptase, integrase, etc. The necessity of viral enzymes for viral replication means that interference with any of them potentially could result in a selective antiviral mode of action. Ng *et. al.* has assayed the inhibitory effect of proteins from bovine milk on the crucial enzymes for the human immunodeficiency virus type 1 life cycle [126]. They demonstrated that lactoferrin strongly inhibited reverse transcriptase but only slightly inhibited the viral protease and integrase. In parallel, α-lactalbumin, β-lactoglobulin and casein were demonstrated to affect human immunodeficiency virus protease and integrase, while not affecting the reverse transcriptase [126].

4. Modulation of innate immune responses - A novel antiviral strategy

The immune system consists of the innate and the adaptive branch, which exerts its functions through recognition of foreign pathogen resulting in a series of responses to eliminate the infectious material. Both innate leukocytes (including macrophages, dendritic cells, and natural killer cells) and adaptive immune cells (B-cells and T-cells) are involved in host immune protection and bridging these two pathways is a variety of traditional signal molecules (cytokines and chemokines). Recently it has also been documented that natural occurring host defence peptides (and proteins) are involved in the orchestration of a well balanced and effective immune response [127-129]. Lactoferrin is one such host defence protein, and it has been demonstrated that lactoferrin can increase the cytotoxic functions of natural killer cells and lymphokine-actived killer cells especially in infants, which normally have low activity in these cell populations [130]. Lactoferrin can also enhance the mobility of polymorphonuclear leucocytes and increase the production of superoxide [131], activate macrophages and stimulate the release of both pro- and anti-inflammatory cytokines, i.e. interleukin-1,-6,-8,-18, interferon-γ and tumor necrosis factor-α [132]. The antiviral effect of lactoferrin on cytomegalovirus in a murine infection model has been demonstrated to be a result of augmentation of natural killer cell activity rather than of the cytolytic T-lymphocytes [14]. Similarly, human lactoferrin has also been proven to have an effect on natural killer cell cytotoxicity against haematopoietic and epithelial tumor cells [133].

Furthermore, the antiviral activity of lactoferrin against vesicular stomatitis virus has been related to its capacity of up-regulating the accumulation of interferon-β in peritoneal macrophages from mice [11]. Another experiment with interperitoneal administration of lactoferrin to CBA mice demonstrated enhanced production of tumor necrosis factor-α and interleukin-12. Similar results were also reported after *in vitro* stimulation of J774A.1 murine macrophages by lactoferrin [134]. Increased expression of interleukin-12, in addition to interferon-β and NOD2, were also observed in mice that were administered lactoferrin orally after being infected with influenza virus, thus suggesting that lactoferrin potentially can promote systemic host immunity [135]. As an important inductor of interferon-γ production in T-cells and natural killer cells, interleukin-12 exhibited a marked synergism with interferon-γ in activating monocytes and macrophages, promoting the differentiation of B-cells and T-cells, and increasing the induction of major histocompatibility complex I and II molecules by up-regulating expression of the interleukin-18 receptor on cells producing interferon-γ [136-138].

With regard to modulation of the adaptive immune system, lactoferrin could exert higher growth stimulatory activity on lymphocytes than transferrin [139], induce phenotypic changes of immature B- and T-cells from newborn or chromosome X-linked immunodeficient mice, as well as enable B-cells to present antigen to an antigen-specific T-helper type 2 cell line [140, 141]. Immature B-cells cultured with lactoferrin will also increase their ability to promote antigen-specific T-cell proliferation, indirectly indicating enhanced B-cell antigen presentation [140].

In summary, the effects of lactoferrin on the activation, maturation, migration and antigen presentation of the innate and adaptive immune cells, suggest that lactoferrin have the potential to associate the cellular functions and responses of the innate and adaptive immune cells, respectively. The modulating effects of lactoferrin on cytokine levels, especially of interleukin-12 and interleukin-18 illuminates the milk proteins role in connecting the innate and adaptive immune response.

5. Milk derived peptides as immune modulators

There is a great quantity of milk proteins and peptides other than lactoferrin that can lead to immune regulation, involving in both up- and down-regulation of the immune system. Peptides from casein [4, 6, 142], β-lactoglobulin [143, 144], and α-lactalbumin [7] also enhance and/or suppress immune cell function (Table 3).

Precursor protein	Fragment	Peptide sequence	Name	Function	Reference
αs1-casein	23-27	FFVAP	α-casokinin-5	ACE-inhibition	[222, 223]
	28-34	FPEVFGK	α-casokinin-7	ACE-inhibition	[222]
	23-34	FFVAP FPEVFGK		ACE-inhibition	[224]
	104-109	YKVPQL		ACE-inhibition	[225]
	158-162	YVPFP	αs1-casomorphin	Opioid agonist, immunomodulation	[159, 226]
	169-193	LGTQYTDAPSFSDIPNPIGSENSEK		ACE-inhibition	[227]
	194-199	TTMPLW	α-casokinin-6	ACE-inhibition, immunostimulatory activity	[5, 228]
	201-212	IGSENSEKTTMP		ACE-inhibition	[229]
αs2-casein	94-103	QKALNEINQF		ACE-inhibition	[166]
	163-176	TKKTKLTEEEKNRL		ACE-inhibition	[166]
β-casein	1-25	RELEELNVPGEIVES(P)LS(P)S(P)S(P)EESITR	casein phosphopeptide	Immunostimulatory activity	[3]
	54-59	VEPIPY		Immunostimulatory activity	[6, 149]
	60-66	YPFPGPI	β-casomorphin-7	ACE-inhibition, immunomodulation activity	[1, 230, 231]
	63-68	PGPIPN		Immunomodulation	[6, 232]
	60-70	YPFPGPIPN	β-casomorphin-11	Immunostimulatory activity, opioid and ACE-inhibitory activities	[233]

Precursor protein	Fragment	Peptide sequence	Name	Function	Reference
	73-89	NIPPLTQTPVVVPPFIQ		ACE-inhibition	[229]
	114-118	YPVEP	β-casochemotide-1	Promote innate host immune response	[142]
	124-133	MPFPKYPVEP		ACE-inhibition	[229]
	169-175	KVLPVPQ		ACE-inhibition	[225, 229]
	177-183	AVPYPQR	β-casokinin-7	ACE-inhibition	[223]
	191-193	LLY		Immunomodulation activity	[6, 234]
	193-202	YQQPVLGPVR	β-casokinin-10	ACE-inhibition, immunostimulatory activity	[1]
	210-221	EPVLGPVRGPFP		ACE-inhibition	[229]
κ-casein	108-110	IPP		ACE-inhibition	[235]
	106-116	MAIPPKKNQDK	casoplatelin	Antithrombotic activity	[236]
		YIPIQYVLSR	Casoxin C	Opioid agonist	[237]
		YPSY	Casoxin 4	Opioid agonist	[238]
α-lactalbumin	18-20	YGG		Immunomodulation	[1]
	50-53	YGLF	α-lactorphin	ACE-inhibition	[143, 144]
	99-108	LDDDLTDDI		ACE-inhibition	[239]
	104-108	LTDDI		ACE-inhibition	[239]
β-lactoglobulin	22-25	TMKG		ACE-inhibition	[239]
	32-40	AGTWYSLAM		ACE-inhibition	[239]
	94-100	IPAVFKI		ACE-inhibition	[239]
	106-111	NKVLVL		ACE-inhibition	[239]
	102-105	YLLF	β-lactorphin	ACE-inhibition	[143, 144]
	142-148	ALPMHIR		ACE-inhibition	[240]
Bovine serum albumin	399-404	YGFQNA	serorphin	Opioid	[241]
	208-216	ALKAWSVAR	albutensin A	ACE-inhibition	[242]
Bovine lactoferrin	17-41	FKCRRWQWRMKKLGAPSITCVRRAF	lactoferricin	Anti-herpes simplex virus activity, ACE-inhibition, immunomodulation activity	[2, 80]
	17-26	FKCRRWQWRW		immunomodulation activity	[21]
Human lactoferrin	1-49	GRRRRSVQWCAVSQPEATKCFQWQR NMRKVRGPPVSCIKRDSPIQCI	lactoferricin	Anti-herpes simplex virus activity	[80]
	1-32	GRRRRSVQWCAVSQPEATKCFQWQR NMRKVRGP	LF-33 (human)	Anti-human immunodeficiency virus activity	[18, 19]
	222-230; 264-268	ADRDQYELL; EDLIWK		Inhibit herpes simplex virus 1 infection	[243]
	268-284	KWNLLRQAQEKFGKDKS	Lactoferrampin	Immunomodulation activity	[148]
	318-323	YLGSGY	Lactoferroxin A	Opioid agonist	[244]
	536-540	RYYGY	Lactoferroxin B	Opioid agonist	[244]
	673-679	KYLGPQY	Lactoferroxin C	Opioid agonist	[244]

Note: ACE-inhibition, Angiotensin-converting enzyme-inhibition.

Table 3. Milk proteins-derived peptides with antiviral and immunemodulatory activity

Up-regulation of immune system. Bovine and human lactoferricin, from N-terminal end of bovine and human lactoferrin, respectively, are known for their ability to improve and modulate the function of host immune system [145-147]. Additional, deletion fragments of lactoferrin like the peptide sequence FKCRRWQWRW, corresponding to N-terminal fragment 17-26 of bovine lactoferrin, has demonstrated potent activation of polymorphonuclear leukocytes [21]. Another lactoferrin derived peptide termed lactoferrampin, containing residues 268-284 from the N-terminal domain of lactoferrin, is located in close proximity to the cationic lactoferricin sequence, in the three dimensional structure. Lactoferrampin has been shown to exhibit antimicrobial activity and can improving immune function and gut health in the present of lactoferricin. Dietary supplementation of piglets with an expressed fusion peptide composed of bovine lactoferricin linked to lactoferrampin demonstrated the ability to increased serum levels of IgA, IgG, and IgM, while decreasing the incidence of diarrhea in the piglets [148].

Immunomodulating casein peptides have been found to stimulate the proliferation of human lymphocytes and the phagocytic activities of macrophages [4]. Casein phosphopeptides from fermented milk products, such as plain yogurts and cheeses, has shown beneficial effects on the immune system including the mitogenic effect and IgA enhancing effect in mouse spleen cell cultures [3]. According to the results of other studies, human β-casein fraction 54-59 has demonstrated to enhance the phagocytic activity of macrophages both in mice and humans and increase resistance against certain bacteria in mice [6, 149].

Chemotactic factors in the tissue do also play an essential role in host defence against microbial infection by inducing leukocyte infiltration. A pentapeptide (β-casochemotide-1) with amino acid sequence (YPVEP) matching an actinase E digest peptide from bovine β-casein (corresponding to fraction 114-118), has been tested and demonstrated to both chemoattract and activate, human and mouse monocytes and macrophages by using a unique G-protein coupled receptors [142].

In addition, Colostrinin, also known as PRP, is a naturally occurring mixture of proline-rich polypeptides derived from colostrums and it can stimulate the immune response in animal and *in vitro* studies by causing differentiation of murine thymocytes into functionally active T-cells [150], as well as inhibit autoimmune disorders. Subsequent studies have shown that Colostrinin largely consists of the peptides derived from proteolytic processing of the milk proteins β-casein and β-casein homolog's [151]. Among the Colostrinin digestion peptides are an active nona-peptide fragment VESYVPLFP demonstrating a full spectrum of biological activities [150]. Furthermore, fermentation of milk by *Lactobacillus helveticus* has also proven to generate novel peptide fragments, i.e. three derived from β-casein and one peptide from α-lactalbumin, and the peptides have demonstrated the ability to stimulate the production of tumor necrosis factor-α and modulate macrophage activity [7].

Down-regulation of immune responses. It has been demonstrated that a modified whey protein concentrate, developed as a by-product from commercial manufacturing of cheese, not only suppress B- and T-lymphocyte proliferating responses to mitogens in a dose-dependent fashion, but also suppress alloantigen-induced lymphocyte proliferation during a mixed leukocyte reaction. Moreover, the modified whey protein concentrate could also have been demonstrated to suppress other indices of lymphocyte activation, e.g. cytokine secretion and the formation of activated (CD25+) T-cell blasts, showing that the mechanism of suppression may be related to an inhibition of the lymphocyte activation process. However, the interleukin-2 cytokine-mediated response was not affected by the presence of the modified whey protein concentrate in culture [152]. Similarly, intact κ-casein and κ-caseinoglycopeptide (fragment 106-169), which have been prepared from κ-casein digested with rennin, in addition to a commercial whey protein concentrate, all significantly inhibit the mitogen-induced proliferative response of mouse spleen lymphocytes and Peyer's patch cells [153-155]. As a result of this it has been proposed that κ-caseinoglycopeptide fragment 106-169 can inhibit the phytohaemagglutinin induced proliferation of mouse splenocytes via at least two different models; production of an inhibitory component that reacts with the anti-interleukin-1 antibody or through suppression of interleukin-2 receptor expression on CD4+ T-cells [156].

The opioid system plays a major role in immune modulation, both through classical opioid receptor, but also through other mechanisems. For example, opioid peptides have been demonstrated to inhibit phagocytosis [157], decrease natural killer cell number and activity and decrease cell-mediated hypersensitivity [158]. Also, αs1-casomorphin, an opioid agonist, can modulate antibody and cytokine secretion by multiple myeloma cells in a cell line-dependent and opioid receptor-independent manner, but it was shown to decrease the antibody secretion by normal B-lymphocytes and the proliferation rate of multiple myeloma cells through opioid receptor activation [159]. In other words, there might be two different opioid mechanisms, mediated by parallel signalling pathways, i.e. one early non-opioid receptor related effect modulating the constitutive secretion of immunoglobulin and cytokine, as well as a second long lasting receptor-mediated action of cell growth. Thus, opioids might be employed in controlling the humoral immunity.

Furthermore, the rennin-angiotensin-aldosterone system is not only a major regulator of blood pressure; it also plays a key role in autoimmunity. The angiotensin peptide (AII), is one component of the rennin-angiotensin-aldosterone system, and has direct activity on T-cell function, including activation, expression level of tissue-homing markers and production of tumor necrosis factor-α [160]. Inhibitors of angiotensin-converting enzyme will dampen the proteolytic process of the larger angiotensin peptides (AI) to the active AII. Thus, inhibition of T-cell angiotensin-converting enzyme blocks production of tumor necrosis factor-α, which modulates the proliferation of human immunodeficiency virus [161] and regulates the helper activity in B-cell activation [162]. This will also suppress the auto-reactive T-helper 1 and T-helper 17 cells and promotes antigen-specific CD4+FoxP3+ regulatory T-cells through inhibition of the canonical NF-κB1 transcription factor complex and activation of the alternative NF-κB2 pathway [163]. Moreover, angiotensin-converting enzyme inhibitors play a pivotal role in immune defence by decreasing the degradation of bradykinin and enkephalin [4, 164].

A variety of angiotensin-converting enzyme inhibitory peptides have been found in the hydrolysates of milk using different enzymes; the bovine αs1-casein (fragment) f24-47, f104-

109, f169-193, f194-199 and f201-212, αs2- casein f94-103 and f163-176, β-casein f60-66, f60-70, f169-175, f177-183 and f193-202, α-lactalbumin f18-20 and f50-53; β-lactoglobulin f102-105 and f142-148, bovine serum albumin f208-216, lactoferrin f17-41 (Table 3). Among all these angiotensin-converting enzyme inhibitors, it should be emphasised that peptides αs1-casein f194-199, β-casein f60-66 and f193-202 have shown to have both angiotensin-converting enzyme inhibitory activities and immune stimulatory effect.

Moreover, recombinant human αs1-casein expressed in *Escherichia coli* has been purified and digested with trypsin in an attempt to find peptides with angiotensin-I-converting enzyme inhibitory activity. Three novel angiotensin-converting enzyme inhibitory peptides, A-II, B-II and C, have been isolated and their amino acid sequences identified as YPER (residues 8-11), YYPQIMQY (residues 136-143) and NNVMLQW (residues 164-170), respectively [165]. Two other sequences QKALNEINQF and TKKTKLTEEEKNRL from bovine milk αs2-casein have even stronger inhibitory effects on the angiotensin-converting enzyme [166]. Regardless, no structure-function relationship study for milk-derived peptides in respect to their angiotensin-converting enzyme inhibitory effect has yet been described. However, it has been suggested that peptides with angiotensin-converting enzyme inhibitory function show some common features. First, the interaction between different inhibitory peptides and the angiotensin-converting enzyme is strongly influenced by the C-terminal tripeptide residues of the substrate, which interacts with the active sites of the enzyme [167]. The inhibitory potency of the peptides is further attributed by the hydrophobic (e.g. proline) as well as the positive charged (e.g. arginine and lysine) amino acids in the C-terminal end [168]. Additionally, the model of angiotensin-converting enzyme inhibition involves interaction, not only to the active site but also to the anionic inhibitor binding sites, which are different from the catalytic sites of the enzyme.

6. Synergy between milk proteins and conventional antiviral drugs

A combination of human lactoferrin with recombinant murine interferon-γ resulted in synergistic suppressive effects on disease progression in friend virus complex infected mice [56]. The experiment concluded that natural killer cell activity decreased by friend virus complex, and that the cellular activity returned to normal levels and survival rates increased upon treatment with lactoferrin and interferon-γ. Another study also supporting immune cell regulation by lactoferrin was performed by Spadaro *et. al.* In this study the anticancer activity of a recombinant form of human lactoferrin, talactoferrin-alfa (Agennix, Houston, TX) was evaluated. Talactoferrin-alfa was administered orally to BALB/c mice and the results showed an increase in the intestinal mucosal interferon-γ production, CD8+ T-cell cytotoxicity and the Peyer's patch cellularity which included expansion of CD8+ T lymphocytes and nature killer T-cells, whereas no such phenomena were showed in interferon-γ knockout mice [169]. Thus, the inhibition of friend virus complex infection and tumor growth by lactoferrin/talactoferrin-alfa seems to be mediated by an interferon-γ-dependent enhancement of CD8+ T-cells and natural killer T-cell activity, leading to diversified functions like antiviral defence, immune activation and cell growth regulation [170-173].

Also, combined pre-infection administration of lactoferrin with post-infection administration of Ribavirin on Vero E6 cells could completely inhibit focus formation during hantavirus infection (similar to the traditional plaque formation). This combination therapy also demonstrated significantly increased survival rates in an *in vivo* mice model, not particularly surprising as lactoferrin inhibits viral adsorption to cells and Ribavirin interferes with viral RNA synthesis [122]. Moreover, the antiviral synergy of lactoferrin/lactoferricin with Cidofovir, Ribavirin, Zidovudine, and Acyclovir as all been well documented against human cytomegalovirus, hepatitis C virus, human immunodeficiency virus 1 and herpes simplex virus, respectively [13, 174-176].

Although it is known that lactoferrin has been used to inhibit initial viral infection by interfering with viral attachment and/or entry, the mode of antiviral activity against lots of viruses needs to be clarified in the future, e.g. the infection with hepatitis G virus in human MT-2C T-cells was prevented by bovine lactoferrin with no clear mechanism [63]. With regard to friend virus complex infection, most researchers have confirmed that human lactoferrin has anti-friend virus complex activity in a mouse leukemia model [57, 64], but have no direct effect on friend virus complex infection *in vitro*, indicating a mechanism involving immune regulationg rather than direct viral affects. They discovered that human lactoferrin prolonged survival rates and decreased viral titres in the spleen of infected mice by administering human lactoferrin intraperitoneally in the early phase of friend virus complex infection. Probably, the anti-friend virus complex mechanism of human lactoferrin related to the regulation on the myelopoiesis [64], which should be verified thoroughly.

7. Commercial potential of milk derived proteins and peptides

Immune regulation. There is emerging evidence that the utility of many immune mediators originating from milk represents novel therapeutic approaches depending on their activity of immune stimulation, immune suppression and induction of immunological tolerance. Hence, milk-derived proteins and peptides with immune modulating activity are claimed to be a health enhancing nutritional dietary supplement in functional food and pharmaceutical preparations. For instance, Colostrinin from bovine colostrum have demonstrated possible efficacy against various illnesses including viral infections, and ailments characterized by an overactive immune system, such as allergies, autoimmune diseases, neurodegenerative diseases like Alzheimer's disease, etc. Capsules or chewable tablets containing Colostrinin are sold as an over the counter dietary supplement and are available in many countries in the world under names like Colostrinin, MemoryAid, Cognisure, Cognase, Cognate and Dyna (ReGen Therapetutics Limited, London, England) [151]. Moreover, whey proteins are used as common ingredients in various products including infant formulas, specialized enteral and clinical protein supplements, sports nutrition products, and specific weight management- and mood control products.

Additionally, synthetic peptides derivatives tailored on natural milk proteins or fragments there of, may be another powerful way for design of immune regulating pharmaceutical candidates. For example, synthetic peptides tailored from milk proteins have been shown to enhance proliferation of human peripheral blood lymphocytes. In particular, two fragments (YG and YGG), of bovine α-lactalbumin (fraction 18-19 and 18-20) can significantly stimulate the lymphocyte proliferation, while β-casomorphin-7 and β-casokinin-10, corresponding to fragments 60-66 and 193-202 of bovine β-casein, respectively, suppresses lymphocyte proliferation at low concentrations while enhanced proliferation at high concentrations [1] (Table 3).

Recent studies have shown that synthetic innate defence regulator peptides offer protection by enhancing innate immune defenses of the host while suppressing potentially harmful excessive inflammatory response triggered by the invading pathogen. For example, innate defence regulator peptide 1 was chemotactic for T-helper cells type 1 [23], monocytes [27] and neutrophil response [28], acting in a mitogen-activated protein kinase-dependent manner, while reducing pro-inflammatory cytokine responses. Another peptide, innate defence regulator 1002, induces chemokines in human peripheral blood mononuclear cells [24] which prevents the production of interleukin-1β-induced matrix metalloproteinase 3 and monocyte chemotactic protein-1 and selectively suppresses the inflammatory response [25].

With the aid of computational molecular modeling technologies, theoretical prediction of immune regulatory peptides has become available and practical. For example, RDP58, a novel d-amino acid decapeptide (r-(nle)$_3$-r-(nle)$_3$-gy-CONH$_2$), which was developed by computer-aided rational based design on human leukocyte antigen-derived peptides [177], has been discovered to suppress the T-helper 1 cytokine profile, decrease production of inflammatory cytokines including tumor necrosis factor-α, interferon-γ, interleukin-2 and interleukin-12 in both cell lines and animal models [26, 178, 179]. Several clinical trials on human including phase I safety in normal volunteers, phase II mild-moderate active ulcerative colitis, phase II moderate active ulcerative colitis and phase IICrohn's disease had been completed (Genzyme Corporation; Sanofi, Bridgewater, NJ). Moreover, quantitative structure-activity relationship analysis has been done for peptide design and optimization in developing novel antimicrobial drugs [180-183], and the numerical improvements of quantitative structure-activity relationship studies has been exemplified recently [184], though there are limitations to the predictive ability of the models [185, 186] this technology clearly accelerates lead peptide discovery.

Suppression of immunological functions by milk derived proteins is thought to be important in the ontogeny of the neonatal gastrointestinal immune system, specifically by ensuring a state of tolerance with respect to food proteins. Kulkarni and Karlsson has demonstrated the essential role of milk-derived immunosuppressive factors (i.e. growth factor-β) during early development, and that neonatal mice deficient in transforming growth factor-β remain viable only as long as they receive maternal milk containing this same growth factor [187].

Also, it is envisaged that most of the potential immunosuppressive activity of milk-derived peptides would be effective on chronic inflammatory diseases and organ transplant patients by decreasing allergy, autoimmunity, and organ rejection. For example, lactoferrin could enhance the production of anti-inflammatory factors, like interleukin-11, not only in a hepatitis mouse model, but also in human intestinal myofibroblasts [188]. Additionally, hydrolysis of caseins with L. casei GG-derived enzymes has generated molecules with suppressive effects on lymphocyte proliferation and benefited the intestinal bacteria in the down-regulation of hypersensitivity reactions to ingested proteins in patients with food allergy [189]. Furthermore, two synthesized analogs of the hexapeptide of human β-casein (fraction 54-59) with modification at the N-terminal region not only showed inhibition in alloantigen inducing lymphocyte proliferation and production of interferon-γ in a SRBC mice model, but also demonstrated increased production of interleukin-4 and improved the skin graft survival. Thus, these peptides might serve as good templates for development of safe and effective immunosuppressant drugs [17]. Similarly, two other synthetic β-casein peptides HLPLP and WSVPQPK, have demonstrated potent antioxidant activity and inhibitory activity of angiotensin-converting enzyme [190]. Accordingly, proteins and peptides from milk could potentially be used in production of immune stimulating- and immunosuppressant agents for both prophylaxis and treatment of infectious diseases and immune related illnesses.

Antiviral therapy. Lactoferrin might be a useful addition to conventional antiretroviral therapy, as traditional therapy supplemented with lactoferrin has demonstrated a more effective increase in CD4+ cell count than either treatments alone [191]. A potential oral vaccine resulting in expression of enterovirus 71 VP1 capsid protein in a transgenic animal system, under the control of an α-lactalbumin promoter and an α-casein leader sequence has demonstrated protection against enterovirus 71 [192]. Also, the combined treatment with human lactoferrin and recombinant murine interferon-γ on feline calicivirus infection might be of significance as a potential therapy for patients with leukemia and those infected with retroviruses [56].

Most of the proteins and peptides with antiviral potential has also demonstrated synergy with conventional antiviral drugs, reducing the dose of the antiviral drugs, and limiting the development of drug-resistant viruses on account of the selective targeting of the host rather than infectious pathogens. At the present time, many peptides with immune regulating effects have been approved for clinical use against virus infection, such as Zadaxin, IM862, SCV-07 and so on [193]. Similarly, two peptide inhibitors of interleukin-10 may be applied to increase anti-hepatitis C virus immune response by restoring the immune stimulatory capabilities of dendritic cells, which have been suppressed by high levels of interleukin-10 [194]. Moreover, candidacidal activities of a synthetic peptide from human lactoferrin fraction 1-11 and 21-31 have been investigated for killing of multidrug-resistant pathogens [195-199]. Present research results, such as phase I safety and tolerability trials of human lactoferrin by AM Pharma [22], indicate that human lactoferrin 1-11 acts by selectively stimulating the innate immune system [200]. Thus, human lactoferrin 1-11 is more likely to be an interesting candidate for further exploration in various clinical tests, such as coating for dental or bone implants, in biosensing applications or in radiopharmaceutical therapy [199].

Vaccine adjuvant. Vaccine adjuvants, such as an immune potentiator or immunomodulator, have been used for decades to improve the immune response to vaccine antigens. This involves presentation of the antigen to the immune system, regulation of both quantitative and qualitative aspects of the immune responses, targeting of specific cells, etc. Many adjuvants had been developed in the past, but were never accepted for routine vaccination because of safety concerns, such as acute toxicity and the possibility of delayed side effects. Thus, novel vaccine adjuvants without side effects should be proposed. Despite numerous publications on milk proteins and milk derived peptides with immune regulating activity, there are scarce reports of their adjuvant potential to vaccine. Lactoferrin could function as an effective adjuvant as it has been documented to enhance efficacy of the Bacillus Calmette-Guerin, the current vaccine for tuberculosis disease by promoting host protection and decreasing disease manifestation [201, 202]. Additionally, recombinant porcine lactoferrin significantly increased serum IgA, IgG and infectious bursal disease virus-specific antibody, as well as enhanced interferon-γ and interleukin-12 expressed in chicken T-lymphocytes, suggesting that porcine lactoferrin could enhance cell-mediated immunity and strengthen the ability of vaccinating against infectious bursal disease infection [203, 204].

An innate defence regulator peptide, HH2, has shown synergy with oligonucleotides containing CpG motifs, when used as an immunoadjuvant to enhance the immune response through stimulation of T-helper 1 and T-helper 2 responses in newborn piglets which were vaccinated with a pseudorabies attenuated virus vaccine [29]. Recently, Brown et. al. found that the combination of oligonucleotides contain CpG motifs and HH2 displayed robust adjuvant effects on induction of T-helper 1 cellular immune response in mice by formulating with a booster recombinant Chlamydia antigen subunit vaccine [205]. Another synthetic peptide, WKYMVm, originally identified as a peptide that stimulated the activity of monocytes, neutrophils and dendritic cells [206-210], has demonstrated to selectively enhance the vaccine-induced CD8+ T-cell responses in a dose-dependent manner, in terms of interferon-γ secretion and cytolytic activity when it was co-delivered with human immunodeficiency virus, hepatitis B virus and influenza virus vaccines [211]. It is indicated that WKYMVm may function as a novel adjuvant for DNA vaccine.

Cancer inhibition. More recently, a widely-read article focused on the amazing cases in which milk proteins and derived peptides were used in the treatment of different kinds of cancers. Whey protein is superior to other dietary proteins for suppression of tumour development in animal models usually for colon and mammary tumorigenesis [212]. Furthermore, lactoferrin and its peptides, for example, lactoferricin [212-214], both possess anticancer activity by inducing apoptosis; inhibit angiogenesis, modulating the carcinogen metabolizing enzymes, and so on. Casein and casein derived peptides have antimutagenic properties, and other whey protein components, such as β-lactoglobulin, α-lactalbumin and serum albumin, have also demonstrated anticancer potential [212, 215-217]. Moreover, a recombinant adenovirus containing the human lactoferrin cDNA has been constructed and its effects against tumor growth have been investigated in mice bearing EMT6 breast cancer. The results showed that recombinant delivery of human lactoferrin cDNA could induce apoptosis of the tumor cells by triggering the mitochondrial-dependent pathway and activation of caspase 3, suggesting that this recombinant cDNA delivery might be a promising drug strategy for breast cancer gene therapy [218].

In addition, the synthetic peptide, P60 (RDFQSFRKMWPFFAM) [219], has demonstrated potential of inhibiting the immunosuppressive activity of murine and human derived regulatory T-cells and enhances the effector T-cell stimulation *in vitro* by binding to regulatory T-cells specific forkhead or winged helix transcription factor 3. Thus, P60 can improve the immunogenicity of cancer and viral vaccines against CT26 tumor challenge and hepatitis C virus infection. Also, the *in vivo* antitumoral effects of LTX-302, a 9-mer peptide derived from bovine lactoferricin, have been examined by intratumoral injection. The results showed that LTX-302 induced tumor necrosis and infiltration of inflammatory cells followed by complete regression of the tumors, as it results in long term and specific cellular immunity against the A20 B-cell lymphoma that is CD4+ and CD8+ T-cells dependent [20].

8. Conclusion

Most of the milk proteins and peptides that have been identified with antiviral properties are broad spectrum components targeting general features and mechanisms involved in a viral infection cycle. Hence, many of these milk proteins do also demonstrate synergy with conventional antiviral drugs. Recently, the diverse immunomodulatory activities of milk proteins/peptides have illustrated these molecules interesting potential as antiviral therapeutics, though the precise mechansiems of immune regulation needs to be thoroughly described. Although the synthetic peptides usually are shorter than natural proteins, the antiviral immune regulating properties of many of these synthetic derivatives appear to be similar as for the entire proteins. Thus we would argue that milk proteins and peptides, have great potential to serve as templates for design of more potent antiviral drugs. With proper scientific effort these molecules may have great therapeutic potential as supplements for current antiviral and anticancer therapy, as novel vaccine adjuvants for both human and far animals, and as immunosuppressants for autoimmune diseases and allergy treatment.

Author details

Haiyan Sun
China Animal Husbandry Zhihe (Beijing) Biotech Co., Ltd, Beijing, China

Håvard Jenssen[*]
Roskilde University, Dept. of Science, Systems & Models, Roskilde, Denmark

9. References

[1] Kayser, H., and Meisel, H. (1996) Stimulation of human peripheral blood lymphocytes by bioactive peptides derived from bovine milk proteins, *FEBS Lett 383*, 18-20.

[2] Bellamy, W., Takase, M., Yamauchi, K., Wakabayashi, H., Kawase, K., and Tomita, M. (1992) Identification of the bactericidal domain of lactoferrin, *Biochim Biophys Acta 1121*, 130-136.

[*] Corresponding Author

[3] Kawahara, T., Aruga, K., and Otani, H. (2005) Characterization of casein phosphopeptides from fermented milk products, *J Nutr Sci Vitaminol (Tokyo) 51*, 377-381.

[4] Meisel, H. (1997) Biochemical properties of regulatory peptides derived from milk proteins, *Biopolymers 43*, 119-128.

[5] Fiat, A. M., Levy-Toledano, S., Caen, J. P., and Jolles, P. (1989) Biologically active peptides of casein and lactotransferrin implicated in platelet function, *J Dairy Res 56*, 351-355.

[6] Migliore-Samour, D., Floc'h, F., and Jolles, P. (1989) Biologically active casein peptides implicated in immunomodulation, *J Dairy Res 56*, 357-362.

[7] Tellez, A., Corredig, M., Brovko, L. Y., and Griffiths, M. W. (2010) Characterization of immune-active peptides obtained from milk fermented by Lactobacillus helveticus, *J Dairy Res 77*, 129-136.

[8] Berkhout, B., van Wamel, J. L., Beljaars, L., Meijer, D. K., Visser, S., and Floris, R. (2002) Characterization of the anti-HIV effects of native lactoferrin and other milk proteins and protein-derived peptides, *Antiviral Res 55*, 341-355.

[9] Berkhout, B., Derksen, G. C., Back, N. K., Klaver, B., de Kruif, C. G., and Visser, S. (1997) Structural and functional analysis of negatively charged milk proteins with anti-HIV activity, *AIDS Res Hum Retroviruses 13*, 1101-1107.

[10] Marchetti, M., Ammendolia, M. G., and Superti, F. (2009) Glycosaminoglycans are not indispensable for the anti-herpes simplex virus type 2 activity of lactoferrin, *Biochimie 91*, 155-159.

[11] Puddu, P., Carollo, M. G., Belardelli, F., Valenti, P., and Gessani, S. (2007) Role of endogenous interferon and LPS in the immunomodulatory effects of bovine lactoferrin in murine peritoneal macrophages, *J Leukoc Biol 82*, 347-353.

[12] Florisa, R., Recio, I., Berkhout, B., and Visser, S. (2003) Antibacterial and antiviral effects of milk proteins and derivatives thereof, *Curr Pharm Des 9*, 1257-1275.

[13] van der Strate, B. W., De Boer, F. M., Bakker, H. I., Meijer, D. K., Molema, G., and Harmsen, M. C. (2003) Synergy of bovine lactoferrin with the anti-cytomegalovirus drug cidofovir in vitro, *Antiviral Res 58*, 159-165.

[14] Shimizu, K., Matsuzawa, H., Okada, K., Tazume, S., Dosako, S., Kawasaki, Y., Hashimoto, K., and Koga, Y. (1996) Lactoferrin-mediated protection of the host from murine cytomegalovirus infection by a T-cell-dependent augmentation of natural killer cell activity, *Arch Virol 141*, 1875-1889.

[15] Beljaars, L., van der Strate, B. W., Bakker, H. I., Reker-Smit, C., van Loenen-Weemaes, A. M., Wiegmans, F. C., Harmsen, M. C., Molema, G., and Meijer, D. K. (2004) Inhibition of cytomegalovirus infection by lactoferrin in vitro and in vivo, *Antiviral Res 63*, 197-208.

[16] Andersen, J. H., Osbakk, S. A., Vorland, L. H., Traavik, T., and Gutteberg, T. J. (2001) Lactoferrin and cyclic lactoferricin inhibit the entry of human cytomegalovirus into human fibroblasts, *Antiviral Res 51*, 141-149.

[17] Puri, A., Bhattacharya, M., Tripathi, L. M., and Haq, W. (2009) Derivatives of human beta-casein fragments (54-59) exhibit highly potent immunosuppressant activity, *Int Immunopharmacol 9*, 1092-1096.

[18] Saidi, H., Eslahpazir, J., Carbonneil, C., Carthagena, L., Requena, M., Nassreddine, N., and Belec, L. (2006) Differential modulation of human lactoferrin activity against both R5 and X4-HIV-1 adsorption on epithelial cells and dendritic cells by natural antibodies, *J Immunol 177*, 5540-5549.

[19] Carthagena, L., Becquart, P., Hocini, H., Kazatchkine, M. D., Bouhlal, H., and Belec, L. (2011) Modulation of HIV Binding to Epithelial Cells and HIV Transfer from Immature Dendritic Cells to CD4 T Lymphocytes by Human Lactoferrin and its Major Exposed LF-33 Peptide, *Open Virol J 5*, 27-34.

[20] Berge, G., Eliassen, L. T., Camilio, K. A., Bartnes, K., Sveinbjornsson, B., and Rekdal, O. (2010) Therapeutic vaccination against a murine lymphoma by intratumoral injection of a cationic anticancer peptide, *Cancer Immunol Immunother 59*, 1285-1294.

[21] Ueta, E., Tanida, T., and Osaki, T. (2001) A novel bovine lactoferrin peptide, FKCRRWQWRM, suppresses Candida cell growth and activates neutrophils, *J Pept Res 57*, 240-249.

[22] Velden, W. J., van Iersel, T. M., Blijlevens, N. M., and Donnelly, J. P. (2009) Safety and tolerability of the antimicrobial peptide human lactoferrin 1-11 (hLF1-11), *BMC Med 7*, 44.

[23] Bowdish, D. M., Davidson, D. J., Scott, M. G., and Hancock, R. E. (2005) Immunomodulatory activities of small host defense peptides, *Antimicrob Agents Chemother 49*, 1727-1732.

[24] Nijnik, A., Madera, L., Ma, S., Waldbrook, M., Elliott, M. R., Easton, D. M., Mayer, M. L., Mullaly, S. C., Kindrachuk, J., Jenssen, H., and Hancock, R. E. (2010) Synthetic cationic peptide IDR-1002 provides protection against bacterial infections through chemokine induction and enhanced leukocyte recruitment, *J Immunol 184*, 2539-2550.

[25] Turner-Brannen, E., Choi, K. Y., Lippert, D. N., Cortens, J. P., Hancock, R. E., El-Gabalawy, H., and Mookherjee, N. (2011) Modulation of interleukin-1beta-induced inflammatory responses by a synthetic cationic innate defence regulator peptide, IDR-1002, in synovial fibroblasts, *Arthritis Res Ther 13*, R129.

[26] Iyer, S., Lahana, R., and Buelow, R. (2002) Rational design and development of RDP58, *Curr Pharm Des 8*, 2217-2229.

[27] Scott, M. G., Dullaghan, E., Mookherjee, N., Glavas, N., Waldbrook, M., Thompson, A., Wang, A., Lee, K., Doria, S., Hamill, P., Yu, J. J., Li, Y., Donini, O., Guarna, M. M., Finlay, B. B., North, J. R., and Hancock, R. E. (2007) An anti-infective peptide that selectively modulates the innate immune response, *Nat Biotechnol 25*, 465-472.

[28] Lee, H. Y., and Bae, Y. S. (2008) The anti-infective peptide, innate defense-regulator peptide, stimulates neutrophil chemotaxis via a formyl peptide receptor, *Biochem Biophys Res Commun 369*, 573-578.

[29] Cao, D., Li, H., Jiang, Z., Cheng, Q., Yang, Z., Xu, C., Cao, G., and Zhang, L. (2011) CpG oligodeoxynucleotide synergizes innate defense regulator peptide for enhancing the systemic and mucosal immune responses to pseudorabies attenuated virus vaccine in piglets in vivo, *Int Immunopharmacol 11*, 748-754.

[30] Superti, F., Ammendolia, M. G., Valenti, P., and Seganti, L. (1997) Antirotaviral activity of milk proteins: lactoferrin prevents rotavirus infection in the enterocyte-like cell line HT-29, *Med Microbiol Immunol 186*, 83-91.

[31] Yolken, R. H., Losonsky, G. A., Vonderfecht, S., Leister, F., and Wee, S. B. (1985) Antibody to human rotavirus in cow's milk, *N Engl J Med 312*, 605-610.

[32] el Agamy, E. I., Ruppanner, R., Ismail, A., Champagne, C. P., and Assaf, R. (1992) Antibacterial and antiviral activity of camel milk protective proteins, *J Dairy Res 59*, 169-175.

[33] Bojsen, A., Buesa, J., Montava, R., Kvistgaard, A. S., Kongsbak, M. B., Petersen, T. E., Heegaard, C. W., and Rasmussen, J. T. (2007) Inhibitory activities of bovine macromolecular whey proteins on rotavirus infections in vitro and in vivo, *J Dairy Sci 90*, 66-74.

[34] Pourtois, M., Binet, C., Van Tieghem, N., Courtois, P., Vandenabbeele, A., and Thiry, L. (1990) Inhibition of HIV infectivity by lactoperoxidase-produced hypothiocyanite, *J Biol Buccale 18*, 251-253.

[35] Mikola, H., Waris, M., and Tenovuo, J. (1995) Inhibition of herpes simplex virus type 1, respiratory syncytial virus and echovirus type 11 by peroxidase-generated hypothiocyanite, *Antiviral Res 26*, 161-171.

[36] Shin, K., Wakabayashi, H., Yamauchi, K., Teraguchi, S., Tamura, Y., Kurokawa, M., and Shiraki, K. (2005) Effects of orally administered bovine lactoferrin and lactoperoxidase on influenza virus infection in mice, *J Med Microbiol 54*, 717-723.

[37] Newburg, D. S., Peterson, J. A., Ruiz-Palacios, G. M., Matson, D. O., Morrow, A. L., Shults, J., Guerrero, M. L., Chaturvedi, P., Newburg, S. O., Scallan, C. D., Taylor, M. R., Ceriani, R. L., and Pickering, L. K. (1998) Role of human-milk lactadherin in protection against symptomatic rotavirus infection, *Lancet 351*, 1160-1164.

[38] Newburg, D. S. (1999) Human milk glycoconjugates that inhibit pathogens, *Curr Med Chem 6*, 117-127.

[39] Hewish, M. J., Takada, Y., and Coulson, B. S. (2000) Integrins alpha2beta1 and alpha4beta1 can mediate SA11 rotavirus attachment and entry into cells, *J Virol 74*, 228-236.

[40] Kvistgaard, A. S., Pallesen, L. T., Arias, C. F., Lopez, S., Petersen, T. E., Heegaard, C. W., and Rasmussen, J. T. (2004) Inhibitory effects of human and bovine milk constituents on rotavirus infections, *J Dairy Sci 87*, 4088-4096.

[41] Johansson, B. (1960) Isolation of an iron-containing red protein from human milk, *Acta Chem Scand 14*, 510-512.

[42] Montreuil, J., Tonnelat, J., and Mullet, S. (1960) Preparation and properties of lactosiderophilin (lactotransferrin) of human milk, *Biochim Biophys Acta 45*, 413-421.

[43] Groves, M. L. (1960) The isolation of a red protein from milk, *J. Am. Chem. Sco. 82*, 3345-3350.

[44] Grover, M., Giouzeppos, O., Schnagl, R. D., and May, J. T. (1997) Effect of human milk prostaglandins and lactoferrin on respiratory syncytial virus and rotavirus, *Acta Paediatr 86*, 315-316.

[45] Portelli, J., Gordon, A., and May, J. T. (1998) Effect of compounds with antibacterial activities in human milk on respiratory syncytial virus and cytomegalovirus in vitro, *J Med Microbiol 47*, 1015-1018.

[46] Sano, H., Nagai, K., Tsutsumi, H., and Kuroki, Y. (2003) Lactoferrin and surfactant protein A exhibit distinct binding specificity to F protein and differently modulate respiratory syncytial virus infection, *Eur J Immunol 33*, 2894-2902.

[47] Li, S., Zhou, H., Huang, G., and Liu, N. (2009) Inhibition of HBV infection by bovine lactoferrin and iron-, zinc-saturated lactoferrin, *Med Microbiol Immunol 198*, 19-25.

[48] Legrand, D., Vigie, K., Said, E. A., Elass, E., Masson, M., Slomianny, M. C., Carpentier, M., Briand, J. P., Mazurier, J., and Hovanessian, A. G. (2004) Surface nucleolin participates in both the binding and endocytosis of lactoferrin in target cells, *Eur J Biochem 271*, 303-317.

[49] Groot, F., Geijtenbeek, T. B., Sanders, R. W., Baldwin, C. E., Sanchez-Hernandez, M., Floris, R., van Kooyk, Y., de Jong, E. C., and Berkhout, B. (2005) Lactoferrin prevents dendritic cell-mediated human immunodeficiency virus type 1 transmission by blocking the DC-SIGN--gp120 interaction, *J Virol 79*, 3009-3015.

[50] McCann, K. B., Lee, A., Wan, J., Roginski, H., and Coventry, M. J. (2003) The effect of bovine lactoferrin and lactoferricin B on the ability of feline calicivirus (a norovirus surrogate) and poliovirus to infect cell cultures, *J Appl Microbiol 95*, 1026-1033.

[51] Lin, T. Y., Chu, C., and Chiu, C. H. (2002) Lactoferrin inhibits enterovirus 71 infection of human embryonal rhabdomyosarcoma cells in vitro, *J Infect Dis 186*, 1161-1164.

[52] Ikeda, M., Sugiyama, K., Tanaka, T., Tanaka, K., Sekihara, H., Shimotohno, K., and Kato, N. (1998) Lactoferrin markedly inhibits hepatitis C virus infection in cultured human hepatocytes, *Biochem Biophys Res Commun 245*, 549-553.

[53] Kawasaki, Y., Isoda, H., Shinmoto, H., Tanimoto, M., Dosako, S., Idota, T., and Nakajima, I. (1993) Inhibition by kappa-casein glycomacropeptide and lactoferrin of influenza virus hemagglutination, *Biosci Biotechnol Biochem 57*, 1214-1215.

[54] Beaumont, S. L., Maggs, D. J., and Clarke, H. E. (2003) Effects of bovine lactoferrin on in vitro replication of feline herpesvirus, *Vet Ophthalmol 6*, 245-250.

[55] Valimaa, H., Tenovuo, J., Waris, M., and Hukkanen, V. (2009) Human lactoferrin but not lysozyme neutralizes HSV-1 and inhibits HSV-1 replication and cell-to-cell spread, *Virol J 6*, 53.

[56] Lu, L., Shen, R. N., Zhou, S. Z., Srivastava, C., Harrington, M., Miyazawa, K., Wu, B., Lin, Z. H., Ruscetti, S., and Broxmeyer, H. E. (1991) Synergistic effect of human lactoferrin and recombinant murine interferon-gamma on disease progression in mice infected with the polycythemia-inducing strain of the Friend virus complex, *Int J Hematol 54*, 117-124.

[57] Chen, L. T., Lu, L., and Broxmeyer, H. E. (1987) Effects of purified iron-saturated human lactoferrin on spleen morphology in mice infected with Friend virus complex, *Am J Pathol 126*, 285-292.

[58] Marchetti, M., Superti, F., Ammendolia, M. G., Rossi, P., Valenti, P., and Seganti, L. (1999) Inhibition of poliovirus type 1 infection by iron-, manganese- and zinc-saturated lactoferrin, *Med Microbiol Immunol 187*, 199-204.

[59] Hara, K., Ikeda, M., Saito, S., Matsumoto, S., Numata, K., Kato, N., Tanaka, K., and Sekihara, H. (2002) Lactoferrin inhibits hepatitis B virus infection in cultured human hepatocytes, *Hepatol Res 24*, 228.

[60] Di Biase, A. M., Pietrantoni, A., Tinari, A., Siciliano, R., Valenti, P., Antonini, G., Seganti, L., and Superti, F. (2003) Heparin-interacting sites of bovine lactoferrin are involved in anti-adenovirus activity, *J Med Virol 69*, 495-502.

[61] Pietrantoni, A., Di Biase, A. M., Tinari, A., Marchetti, M., Valenti, P., Seganti, L., and Superti, F. (2003) Bovine lactoferrin inhibits adenovirus infection by interacting with viral structural polypeptides, *Antimicrob Agents Chemother 47*, 2688-2691.

[62] Berkhout, B., Floris, R., Recio, I., and Visser, S. (2004) The antiviral activity of the milk protein lactoferrin against the human immunodeficiency virus type 1, *Biometals 17*, 291-294.

[63] Ikeda, M., Nozaki, A., Sugiyama, K., Tanaka, T., Naganuma, A., Tanaka, K., Sekihara, H., Shimotohno, K., Saito, M., and Kato, N. (2000) Characterization of antiviral activity of lactoferrin against hepatitis C virus infection in human cultured cells, *Virus Res 66*, 51-63.

[64] Lu, L., Hangoc, G., Oliff, A., Chen, L. T., Shen, R. N., and Broxmeyer, H. E. (1987) Protective influence of lactoferrin on mice infected with the polycythemia-inducing strain of Friend virus complex, *Cancer Res 47*, 4184-4188.

[65] Drobni, P., Naslund, J., and Evander, M. (2004) Lactoferrin inhibits human papillomavirus binding and uptake in vitro, *Antiviral Res 64*, 63-68.

[66] Tinari, A., Pietrantoni, A., Ammendolia, M. G., Valenti, P., and Superti, F. (2005) Inhibitory activity of bovine lactoferrin against echovirus induced programmed cell death in vitro, *Int J Antimicrob Agents 25*, 433-438.

[67] Weng, T. Y., Chen, L. C., Shyu, H. W., Chen, S. H., Wang, J. R., Yu, C. K., Lei, H. Y., and Yeh, T. M. (2005) Lactoferrin inhibits enterovirus 71 infection by binding to VP1 protein and host cells, *Antiviral Res 67*, 31-37.

[68] Swart, P. J., Kuipers, M. E., Smit, C., Pauwels, R., deBethune, M. P., de Clercq, E., Meijer, D. K., and Huisman, J. G. (1996) Antiviral effects of milk proteins: acylation results in polyanionic compounds with potent activity against human immunodeficiency virus types 1 and 2 in vitro, *AIDS Res Hum Retroviruses 12*, 769-775.

[69] Arnold, D., Di Biase, A. M., Marchetti, M., Pietrantoni, A., Valenti, P., Seganti, L., and Superti, F. (2002) Antiadenovirus activity of milk proteins: lactoferrin prevents viral infection, *Antiviral Res 53*, 153-158.

[70] Puddu, P., Borghi, P., Gessani, S., Valenti, P., Belardelli, F., and Seganti, L. (1998) Antiviral effect of bovine lactoferrin saturated with metal ions on early steps of human immunodeficiency virus type 1 infection, *Int J Biochem Cell Biol 30*, 1055-1062.

[71] Yi, M., Kaneko, S., Yu, D. Y., and Murakami, S. (1997) Hepatitis C virus envelope proteins bind lactoferrin, *J Virol 71*, 5997-6002.

[72] Murphy, M. E., Kariwa, H., Mizutani, T., Yoshimatsu, K., Arikawa, J., and Takashima, I. (2000) In vitro antiviral activity of lactoferrin and ribavirin upon hantavirus, *Arch Virol 145*, 1571-1582.

[73] Waarts, B. L., Aneke, O. J., Smit, J. M., Kimata, K., Bittman, R., Meijer, D. K., and Wilschut, J. (2005) Antiviral activity of human lactoferrin: inhibition of alphavirus interaction with heparan sulfate, *Virology 333*, 284-292.

[74] Tanaka, T., Nakatani, S., Xuan, X., Kumura, H., Igarashi, I., and Shimazaki, K. (2003) Antiviral activity of lactoferrin against canine herpesvirus, *Antiviral Res 60*, 193-199.

[75] Pietrantoni, A., Ammendolia, M. G., Tinari, A., Siciliano, R., Valenti, P., and Superti, F. (2006) Bovine lactoferrin peptidic fragments involved in inhibition of Echovirus 6 in vitro infection, *Antiviral Res 69*, 98-106.

[76] Ammendolia, M. G., Pietrantoni, A., Tinari, A., Valenti, P., and Superti, F. (2007) Bovine lactoferrin inhibits echovirus endocytic pathway by interacting with viral structural polypeptides, *Antiviral Res 73*, 151-160.

[77] Ammendolia, M. G., Marchetti, M., and Superti, F. (2007) Bovine lactoferrin prevents the entry and intercellular spread of herpes simplex virus type 1 in Green Monkey Kidney cells, *Antiviral Res 76*, 252-262.

[78] Jenssen, H., Sandvik, K., Andersen, J. H., Hancock, R. E., and Gutteberg, T. J. (2008) Inhibition of HSV cell-to-cell spread by lactoferrin and lactoferricin, *Antiviral Res 79*, 192-198.

[79] Marr, A. K., Jenssen, H., Moniri, M. R., Hancock, R. E., and Pante, N. (2009) Bovine lactoferrin and lactoferricin interfere with intracellular trafficking of Herpes simplex virus-1, *Biochimie 91*, 160-164.

[80] Jenssen, H., Andersen, J. H., Uhlin-Hansen, L., Gutteberg, T. J., and Rekdal, O. (2004) Anti-HSV activity of lactoferricin analogues is only partly related to their affinity for heparan sulfate, *Antiviral Res 61*, 101-109.

[81] Marchetti, M., Longhi, C., Conte, M. P., Pisani, S., Valenti, P., and Seganti, L. (1996) Lactoferrin inhibits herpes simplex virus type 1 adsorption to Vero cells, *Antiviral Res 29*, 221-231.

[82] Andersen, J. H., Jenssen, H., Sandvik, K., and Gutteberg, T. J. (2004) Anti-HSV activity of lactoferrin and lactoferricin is dependent on the presence of heparan sulphate at the cell surface, *J Med Virol 74*, 262-271.

[83] Marchetti, M., Trybala, E., Superti, F., Johansson, M., and Bergstrom, T. (2004) Inhibition of herpes simplex virus infection by lactoferrin is dependent on interference with the virus binding to glycosaminoglycans, *Virology 318*, 405-413.

[84] Moore, S. A., Anderson, B. F., Groom, C. R., Haridas, M., and Baker, E. N. (1997) Three-dimensional structure of diferric bovine lactoferrin at 2.8 A resolution, *J Mol Biol 274*, 222-236.

[85] Jenssen, H., Andersen, J. H., Mantzilas, D., and Gutteberg, T. J. (2004) A wide range of medium-sized, highly cationic, alpha-helical peptides show antiviral activity against herpes simplex virus, *Antiviral Res 64*, 119-126.

[86] Uhrinova, S., Smith, M. H., Jameson, G. B., Uhrin, D., Sawyer, L., and Barlow, P. N. (2000) Structural changes accompanying pH-induced dissociation of the beta-lactoglobulin dimer, *Biochemistry 39*, 3565-3574.

[87] Chandra, N., Brew, K., and Acharya, K. R. (1998) Structural evidence for the presence of a secondary calcium binding site in human alpha-lactalbumin, *Biochemistry 37*, 4767-4772.

[88] Pechkova, E., Zanotti, G., and Nicolini, C. (2003) Three-dimensional atomic structure of a catalytic subunit mutant of human protein kinase CK2, *Acta Crystallogr D Biol Crystallogr 59*, 2133-2139.

[89] Neurath, A. R., Debnath, A. K., Strick, N., Li, Y. Y., Lin, K., and Jiang, S. (1995) Blocking of CD4 cell receptors for the human immunodeficiency virus type 1 (HIV-1) by

chemically modified bovine milk proteins: potential for AIDS prophylaxis, *J Mol Recognit 8*, 304-316.

[90] Neurath, A. R., Jiang, S., Strick, N., Lin, K., Li, Y. Y., and Debnath, A. K. (1996) Bovine beta-lactoglobulin modified by 3-hydroxyphthalic anhydride blocks the CD4 cell receptor for HIV, *Nat Med 2*, 230-234.

[91] Jansen, R. W., Molema, G., Pauwels, R., Schols, D., De Clercq, E., and Meijer, D. K. (1991) Potent in vitro anti-human immunodeficiency virus-1 activity of modified human serum albumins, *Mol Pharmacol 39*, 818-823.

[92] Takami, M., Sone, T., Mizumoto, K., Kino, K., and Tsunoo, H. (1992) Maleylated human serum albumin inhibits HIV-1 infection in vitro, *Biochim Biophys Acta 1180*, 180-186.

[93] Neurath, A. R., Strick, N., and Li, Y. Y. (1998) 3-Hydroxyphthaloyl beta-lactoglobulin. III. Antiviral activity against herpesviruses, *Antivir Chem Chemother 9*, 177-184.

[94] Schoen, P., Corver, J., Meijer, D. K., Wilschut, J., and Swart, P. J. (1997) Inhibition of influenza virus fusion by polyanionic proteins, *Biochem Pharmacol 53*, 995-1003.

[95] Pan, Y., Wan, J., Roginski, H., Lee, A., Shiell, B., Michalski, W. P., and Coventry, M. J. (2007) Comparison of the effects of acylation and amidation on the antimicrobial and antiviral properties of lactoferrin, *Lett Appl Microbiol 44*, 229-234.

[96] Harmsen, M. C., Swart, P. J., de Bethune, M. P., Pauwels, R., De Clercq, E., The, T. H., and Meijer, D. K. (1995) Antiviral effects of plasma and milk proteins: lactoferrin shows potent activity against both human immunodeficiency virus and human cytomegalovirus replication in vitro, *J Infect Dis 172*, 380-388.

[97] Mettenleiter, T. C. (2002) Brief overview on cellular virus receptors, *Virus Res 82*, 3-8.

[98] Spillmann, D. (2001) Heparan sulfate: anchor for viral intruders?, *Biochimie 83*, 811-817.

[99] Sitohy, M., Chobert, J. M., Karwowska, U., Gozdzicka-Jozefiak, A., and Haertle, T. (2006) Inhibition of bacteriophage m13 replication with esterified milk proteins, *J Agric Food Chem 54*, 3800-3806.

[100] Sitohy, M., Billaudel, S., Haertle, T., and Chobert, J. M. (2007) Antiviral activity of esterified alpha-lactalbumin and beta-lactoglobulin against herpes simplex virus type 1. Comparison with the effect of acyclovir and L-polylysines, *J Agric Food Chem 55*, 10214-10220.

[101] Sitohy, M., Besse, B., Billaudel, S., Haertle, T., and Chobert, J. M. (2010) Antiviral action of methylated beta-lactoglobuline on the human influenza virus A subtype H3N2., *Probiotics Antimicrobial Prot. 2*, 104-111.

[102] Sitohy, M., Scanu, M., Besse, B., Mollat, C., Billaudel, S., Haertle, T., and Chobert, J. M. (2010) Influenza virus A subtype H1N1 is inhibited by methylated beta-lactoglobulin, *J Dairy Res 77*, 411-418.

[103] Taha, S. H., Mehrez, M. A., Sitohy, M. Z., Abou Dawood, A. G., Abd-El Hamid, M. M., and Kilany, W. H. (2010) Effectiveness of esterified whey proteins fractions against Egyptian Lethal Avian Influenza A (H5N1), *Virol J 7*, 330.

[104] Sitohy, M., Dalgalarrondo, M., Nowoczin, M., Besse, B., Billaudel, S., Haertle, T., and Chobert, J. M. (2008) The effect of bovine whey proteins on the ability of poliovirus and Coxsackie virus to infect Vero cell cultures., *Int Dairy J 18*, 658-668.

[105] Arias, C. F., Isa, P., Guerrero, C. A., Mendez, E., Zarate, S., Lopez, T., Espinosa, R., Romero, P., and Lopez, S. (2002) Molecular biology of rotavirus cell entry, *Arch Med Res* 33, 356-361.

[106] Bilbao, R., Srinivasan, S., Reay, D., Goldberg, L., Hughes, T., Roelvink, P. W., Einfeld, D. A., Wickham, T. J., and Clemens, P. R. (2003) Binding of adenoviral fiber knob to the coxsackievirus-adenovirus receptor is crucial for transduction of fetal muscle, *Hum Gene Ther 14*, 645-649.

[107] Yolken, R., Kinney, J., Wilde, J., Willoughby, R., and Eiden, J. (1990) Immunoglobulins and other modalities for the prevention and treatment of enteric viral infections, *J Clin Immunol 10*, 80S-86S; discussion 86S-87S.

[108] Yolken, R. H., Peterson, J. A., Vonderfecht, S. L., Fouts, E. T., Midthun, K., and Newburg, D. S. (1992) Human milk mucin inhibits rotavirus replication and prevents experimental gastroenteritis, *J Clin Invest 90*, 1984-1991.

[109] Turville, S. G., Cameron, P. U., Handley, A., Lin, G., Pohlmann, S., Doms, R. W., and Cunningham, A. L. (2002) Diversity of receptors binding HIV on dendritic cell subsets, *Nat Immunol 3*, 975-983.

[110] Turville, S. G., Vermeire, K., Balzarini, J., and Schols, D. (2005) Sugar-binding proteins potently inhibit dendritic cell human immunodeficiency virus type 1 (HIV-1) infection and dendritic-cell-directed HIV-1 transfer, *J Virol 79*, 13519-13527.

[111] Turville, S., Wilkinson, J., Cameron, P., Dable, J., and Cunningham, A. L. (2003) The role of dendritic cell C-type lectin receptors in HIV pathogenesis, *J Leukoc Biol 74*, 710-718.

[112] Saphire, A. C., Bobardt, M. D., Zhang, Z., David, G., and Gallay, P. A. (2001) Syndecans serve as attachment receptors for human immunodeficiency virus type 1 on macrophages, *J Virol 75*, 9187-9200.

[113] Vives, R. R., Imberty, A., Sattentau, Q. J., and Lortat-Jacob, H. (2005) Heparan sulfate targets the HIV-1 envelope glycoprotein gp120 coreceptor binding site, *J Biol Chem 280*, 21353-21357.

[114] Nguyen, D. G., and Hildreth, J. E. (2003) Involvement of macrophage mannose receptor in the binding and transmission of HIV by macrophages, *Eur J Immunol 33*, 483-493.

[115] Li, W., Moore, M. J., Vasilieva, N., Sui, J., Wong, S. K., Berne, M. A., Somasundaran, M., Sullivan, J. L., Luzuriaga, K., Greenough, T. C., Choe, H., and Farzan, M. (2003) Angiotensin-converting enzyme 2 is a functional receptor for the SARS coronavirus, *Nature 426*, 450-454.

[116] Chen, J., and Subbarao, K. (2007) The Immunobiology of SARS*, *Annu Rev Immunol 25*, 443-472.

[117] Han, D. P., Lohani, M., and Cho, M. W. (2007) Specific asparagine-linked glycosylation sites are critical for DC-SIGN- and L-SIGN-mediated severe acute respiratory syndrome coronavirus entry, *J Virol 81*, 12029-12039.

[118] Jeffers, S. A., Tusell, S. M., Gillim-Ross, L., Hemmila, E. M., Achenbach, J. E., Babcock, G. J., Thomas, W. D., Jr., Thackray, L. B., Young, M. D., Mason, R. J., Ambrosino, D. M., Wentworth, D. E., Demartini, J. C., and Holmes, K. V. (2004) CD209L (L-SIGN) is a

receptor for severe acute respiratory syndrome coronavirus, *Proc Natl Acad Sci U S A* *101*, 15748-15753.

[119] Yang, Z. Y., Huang, Y., Ganesh, L., Leung, K., Kong, W. P., Schwartz, O., Subbarao, K., and Nabel, G. J. (2004) pH-dependent entry of severe acute respiratory syndrome coronavirus is mediated by the spike glycoprotein and enhanced by dendritic cell transfer through DC-SIGN, *J Virol 78*, 5642-5650.

[120] Lang, J., Yang, N., Deng, J., Liu, K., Yang, P., Zhang, G., and Jiang, C. (2011) Inhibition of SARS pseudovirus cell entry by lactoferrin binding to heparan sulfate proteoglycans, *PLoS One 6*, e23710.

[121] Mistry, N., Drobni, P., Naslund, J., Sunkari, V. G., Jenssen, H., and Evander, M. (2007) The anti-papillomavirus activity of human and bovine lactoferricin, *Antiviral Res 75*, 258-265.

[122] Murphy, M. E., Kariwa, H., Mizutani, T., Tanabe, H., Yoshimatsu, K., Arikawa, J., and Takashima, I. (2001) Characterization of in vitro and in vivo antiviral activity of lactoferrin and ribavirin upon hantavirus, *J Vet Med Sci 63*, 637-645.

[123] Jin, M., Park, J., Lee, S., Park, B., Shin, J., Song, K. J., Ahn, T. I., Hwang, S. Y., Ahn, B. Y., and Ahn, K. (2002) Hantaan virus enters cells by clathrin-dependent receptor-mediated endocytosis, *Virology 294*, 60-69.

[124] Song, J. W., Song, K. J., Baek, L. J., Frost, B., Poncz, M., and Park, K. (2005) In vivo characterization of the integrin beta3 as a receptor for Hantaan virus cellular entry, *Exp Mol Med 37*, 121-127.

[125] Hall, P. R., Leitao, A., Ye, C., Kilpatrick, K., Hjelle, B., Oprea, T. I., and Larson, R. S. (2010) Small molecule inhibitors of hantavirus infection, *Bioorg Med Chem Lett 20*, 7085-7091.

[126] Ng, T. B., Lam, T. L., Au, T. K., Ye, X. Y., and Wan, C. C. (2001) Inhibition of human immunodeficiency virus type 1 reverse transcriptase, protease and integrase by bovine milk proteins, *Life Sci 69*, 2217-2223.

[127] Hancock, R. E., and Sahl, H. G. (2006) Antimicrobial and host-defense peptides as new anti-infective therapeutic strategies, *Nat Biotechnol 24*, 1551-1557.

[128] Oppenheim, J. J., Tewary, P., de la Rosa, G., and Yang, D. (2007) Alarmins initiate host defense, *Adv Exp Med Biol 601*, 185-194.

[129] Oppenheim, J. J., and Yang, D. (2005) Alarmins: chemotactic activators of immune responses, *Curr Opin Immunol 17*, 359-365.

[130] Shau, H., Kim, A., and Golub, S. H. (1992) Modulation of natural killer and lymphokine-activated killer cell cytotoxicity by lactoferrin, *J Leukoc Biol 51*, 343-349.

[131] Gahr, M., Speer, C. P., Damerau, B., and Sawatzki, G. (1991) Influence of lactoferrin on the function of human polymorphonuclear leukocytes and monocytes, *J Leukoc Biol 49*, 427-433.

[132] Artym, J. (2006) Antitumor and chemopreventive activity of lactoferrin, *Postepy Hig Med Dosw 60*, 352-369.

[133] Damiens, E., Mazurier, J., el Yazidi, I., Masson, M., Duthille, I., Spik, G., and Boilly-Marer, Y. (1998) Effects of human lactoferrin on NK cell cytotoxicity against haematopoietic and epithelial tumour cells, *Biochim Biophys Acta 1402*, 277-287.

[134] Actor, J. K., Hwang, S. A., Olsen, M., Zimecki, M., Hunter, R. L., Jr., and Kruzel, M. L. (2002) Lactoferrin immunomodulation of DTH response in mice, *Int Immunopharmacol 2*, 475-486.

[135] Yamauchi, K., Wakabayashi, H., Shin, K., and Takase, M. (2006) Bovine lactoferrin: benefits and mechanism of action against infections, *Biochem Cell Biol 84*, 291-296.

[136] Farrar, M. A., and Schreiber, R. D. (1993) The molecular cell biology of interferon-gamma and its receptor, *Annu Rev Immunol 11*, 571-611.

[137] Young, H. A., and Hardy, K. J. (1995) Role of interferon-gamma in immune cell regulation, *J Leukoc Biol 58*, 373-381.

[138] Okamura, H., Kashiwamura, S., Tsutsui, H., Yoshimoto, T., and Nakanishi, K. (1998) Regulation of interferon-gamma production by IL-12 and IL-18, *Curr Opin Immunol 10*, 259-264.

[139] Hashizume, S., Kuroda, K., and Murakami, H. (1983) Identification of lactoferrin as an essential growth factor for human lymphocytic cell lines in serum-free medium, *Biochim Biophys Acta 763*, 377-382.

[140] Zimecki, M., Mazurier, J., Spik, G., and Kapp, J. A. (1995) Human lactoferrin induces phenotypic and functional changes in murine splenic B cells, *Immunology 86*, 122-127.

[141] Adamik, B., and Wlaszczyk, A. (1996) Lactoferrin - its role in defense against infection and immunotropic properties, *Postepy Hig Med Dosw 50*, 33-41.

[142] Kitazawa, H., Yonezawa, K., Tohno, M., Shimosato, T., Kawai, Y., Saito, T., and Wang, J. M. (2007) Enzymatic digestion of the milk protein beta-casein releases potent chemotactic peptide(s) for monocytes and macrophages, *Int Immunopharmacol 7*, 1150-1159.

[143] Antila, P., Paakkari, I., Järvinen, A., Mattila, M. J., Laukkanen, M., Pihlanto-Leppälä, A., Mäntsälä, P., and Hellman, P. (1991) Opioid peptides derived from in vitro proteolysis of bovine whey proteins, *Int Dairy J 1*, 251-229.

[144] Mullally, M. M., Meisel, H., and FitzGerald, R. J. (1996) Synthetic peptides corresponding to alpha-lactalbumin and beta-lactoglobulin sequences with angiotensin-I-converting enzyme inhibitory activity, *Biol Chem Hoppe Seyler 377*, 259-260.

[145] Wakabayashi, H., Takase, M., and Tomita, M. (2003) Lactoferricin derived from milk protein lactoferrin, *Curr Pharm Des 9*, 1277-1287.

[146] Gifford, J. L., Hunter, H. N., and Vogel, H. J. (2005) Lactoferricin: a lactoferrin-derived peptide with antimicrobial, antiviral, antitumor and immunological properties, *Cell Mol Life Sci 62*, 2588-2598.

[147] Tomita, M., Wakabayashi, H., Yamauchi, K., Teraguchi, S., and Hayasawa, H. (2002) Bovine lactoferrin and lactoferricin derived from milk: production and applications, *Biochem Cell Biol 80*, 109-112.

[148] Tang, Z., Yin, Y., Zhang, Y., Huang, R., Sun, Z., Li, T., Chu, W., Kong, X., Li, L., Geng, M., and Tu, Q. (2009) Effects of dietary supplementation with an expressed fusion peptide bovine lactoferricin-lactoferrampin on performance, immune function and intestinal mucosal morphology in piglets weaned at age 21 d, *Br J Nutr 101*, 998-1005.

[149] Parker, F., Migliore-Samour, D., Floc'h, F., Zerial, A., Werner, G. H., Jolles, J., Casaretto, M., Zahn, H., and Jolles, P. (1984) Immunostimulating hexapeptide from

human casein: amino acid sequence, synthesis and biological properties, *Eur J Biochem 145*, 677-682.

[150] Janusz, M., and Lisowski, J. (1993) Proline-rich polypeptide (PRP)--an immunomodulatory peptide from ovine colostrum, *Arch Immunol Ther Exp (Warsz) 41*, 275-279.

[151] Rattray, M. (2005) Technology evaluation: colostrinin, ReGen, *Curr Opin Mol Ther 7*, 78-84.

[152] Cross, M. L., and Gill, H. S. (1999) Modulation of immune function by a modified bovine whey protein concentrate, *Immunol Cell Biol 77*, 345-350.

[153] Otani, H., Monnai, M., Kawasaki, Y., Kawakami, H., and Tanimoto, M. (1995) Inhibition of mitogen-induced proliferative responses of lymphocytes by bovine kappa-caseinoglycopeptides having different carbohydrate chains, *J Dairy Res 62*, 349-357.

[154] Otani, H., and Monnai, M. (1995) Induction of an interleukin-1 receptor antagonist-like component produced from mouse spleen cells by bovine kappa-caseinoglycopeptide, *Biosci Biotechnol Biochem 59*, 1166-1168.

[155] Otani, H., and Hata, I. (1995) Inhibition of proliferative responses of mouse spleen lymphocytes and rabbit Peyer's patch cells by bovine milk caseins and their digests, *J Dairy Res 62*, 339-348.

[156] Otani, H., Horimoto, Y., and Monnai, M. (1996) Suppression of interleukin-2 receptor expression on mouse CD4+ T cells by bovine kappa-caseinoglycopeptide, *Biosci Biotechnol Biochem 60*, 1017-1019.

[157] Casellas, A. M., Guardiola, H., and Renaud, F. L. (1991) Inhibition by opioids of phagocytosis in peritoneal macrophages, *Neuropeptides 18*, 35-40.

[158] Castilla-Cortazar, I., Castilla, A., and Gurpegui, M. (1998) Opioid peptides and immunodysfunction in patients with major depression and anxiety disorders, *J Physiol Biochem 54*, 203-215.

[159] Vassou, D., Bakogeorgou, E., Kampa, M., Dimitriou, H., Hatzoglou, A., and Castanas, E. (2008) Opioids modulate constitutive B-lymphocyte secretion, *Int Immunopharmacol 8*, 634-644.

[160] Hoch, N. E., Guzik, T. J., Chen, W., Deans, T., Maalouf, S. A., Gratze, P., Weyand, C., and Harrison, D. G. (2009) Regulation of T-cell function by endogenously produced angiotensin II, *Am J Physiol Regul Integr Comp Physiol 296*, R208-216.

[161] Munoz-Fernandez, M. A., Navarro, J., Garcia, A., Punzon, C., Fernandez-Cruz, E., and Fresno, M. (1997) Replication of human immunodeficiency virus-1 in primary human T cells is dependent on the autocrine secretion of tumor necrosis factor through the control of nuclear factor-kappa B activation, *J Allergy Clin Immunol 100*, 838-845.

[162] Higuchi, M., Nagasawa, K., Horiuchi, T., Oike, M., Ito, Y., Yasukawa, M., and Niho, Y. (1997) Membrane tumor necrosis factor-alpha (TNF-alpha) expressed on HTLV-I-infected T cells mediates a costimulatory signal for B cell activation--characterization of membrane TNF-alpha, *Clin Immunol Immunopathol 82*, 133-140.

[163] Platten, M., Youssef, S., Hur, E. M., Ho, P. P., Han, M. H., Lanz, T. V., Phillips, L. K., Goldstein, M. J., Bhat, R., Raine, C. S., Sobel, R. A., and Steinman, L. (2009) Blocking angiotensin-converting enzyme induces potent regulatory T cells and modulates TH1- and TH17-mediated autoimmunity, *Proc Natl Acad Sci U S A 106*, 14948-14953.

[164] Schlimme, E., and Meisel, H. (1995) Bioactive peptides derived from milk proteins. Structural, physiological and analytical aspects, *Nahrung 39*, 1-20.

[165] Kim, Y. K., Yoon, S., Yu, D. Y., Lonnerdal, B., and Chung, B. H. (1999) Novel angiotensin-I-converting enzyme inhibitory peptides derived from recombinant human alpha s1-casein expressed in Escherichia coli, *J Dairy Res 66*, 431-439.

[166] Srinivas, S., and Prakash, V. (2010) Bioactive peptides from bovine milk alpha-casein: Isolation, characterization and multifunctional properties., *Int. J. Pept. Res. Ther. 16*, 7-15.

[167] Ondetti, M. A., and Cushman, D. W. (1982) Enzymes of the renin-angiotensin system and their inhibitors, *Annu Rev Biochem 51*, 283-308.

[168] Ondetti, M. A., Rubin, B., and Cushman, D. W. (1977) Design of specific inhibitors of angiotensin-converting enzyme: new class of orally active antihypertensive agents, *Science 196*, 441-444.

[169] Spadaro, M., Curcio, C., Varadhachary, A., Cavallo, F., Engelmayer, J., Blezinger, P., Pericle, F., and Forni, G. (2007) Requirement for IFN-gamma, CD8+ T lymphocytes, and NKT cells in talactoferrin-induced inhibition of neu+ tumors, *Cancer Res 67*, 6425-6432.

[170] Muller, U., Steinhoff, U., Reis, L. F., Hemmi, S., Pavlovic, J., Zinkernagel, R. M., and Aguet, M. (1994) Functional role of type I and type II interferons in antiviral defense, *Science 264*, 1918-1921.

[171] Chen, X., Xue, Q., Zhu, R., Fu, X., Yang, L., Sun, L., and Liu, W. (2009) Comparison of antiviral activities of porcine interferon type I and type II, *Sheng Wu Gong Cheng Xue Bao 25*, 806-812.

[172] Modestou, M. A., Manzel, L. J., El-Mahdy, S., and Look, D. C. (2010) Inhibition of IFN-gamma-dependent antiviral airway epithelial defense by cigarette smoke, *Respir Res 11*, 64.

[173] Schroder, K., Hertzog, P. J., Ravasi, T., and Hume, D. A. (2004) Interferon-gamma: an overview of signals, mechanisms and functions, *J Leukoc Biol 75*, 163-189.

[174] Ishibashi, Y., Takeda, K., Tsukidate, N., Miyazaki, H., Ohira, K., Dosaka-Akita, H., and Nishimura, M. (2005) Randomized placebo-controlled trial of interferon alpha-2b plus ribavirin with and without lactoferrin for chronic hepatitis C, *Hepatol Res 32*, 218-223.

[175] Viani, R. M., Gutteberg, T. J., Lathey, J. L., and Spector, S. A. (1999) Lactoferrin inhibits HIV-1 replication in vitro and exhibits synergy when combined with zidovudine, *AIDS 13*, 1273-1274.

[176] Andersen, J. H., Jenssen, H., and Gutteberg, T. J. (2003) Lactoferrin and lactoferricin inhibit Herpes simplex 1 and 2 infection and exhibit synergy when combined with acyclovir, *Antiviral Res 58*, 209-215.

[177] Grassy, G., Calas, B., Yasri, A., Lahana, R., Woo, J., Iyer, S., Kaczorek, M., Floc'h, R., and Buelow, R. (1998) Computer-assisted rational design of immunosuppressive compounds, *Nat Biotechnol 16*, 748-752.

[178] Travis, S., Yap, L. M., Hawkey, C., Warren, B., Lazarov, M., Fong, T., and Tesi, R. J. (2005) RDP58 is a novel and potentially effective oral therapy for ulcerative colitis, *Inflamm Bowel Dis 11*, 713-719.

[179] Liu, W., Deyoung, B. R., Chen, X., Evanoff, D. P., and Luo, Y. (2008) RDP58 inhibits T cell-mediated bladder inflammation in an autoimmune cystitis model, *J Autoimmun 30*, 257-265.

[180] Hilpert, K., Elliott, M. R., Volkmer-Engert, R., Henklein, P., Donini, O., Zhou, Q., Winkler, D. F., and Hancock, R. E. (2006) Sequence requirements and an optimization strategy for short antimicrobial peptides, *Chem Biol 13*, 1101-1107.

[181] Jenssen, H., Fjell, C. D., Cherkasov, A., and Hancock, R. E. (2008) QSAR modeling and computer-aided design of antimicrobial peptides, *J Pept Sci 14*, 110-114.

[182] Fjell, C. D., Jenssen, H., Hilpert, K., Cheung, W. A., Pante, N., Hancock, R. E., and Cherkasov, A. (2009) Identification of novel antibacterial peptides by chemoinformatics and machine learning, *J Med Chem 52*, 2006-2015.

[183] Prado-Prado, F. J., Borges, F., Uriarte, E., Perez-Montoto, L. G., and Gonzalez-Diaz, H. (2009) Multi-target spectral moment: QSAR for antiviral drugs vs. different viral species, *Anal Chim Acta 651*, 159-164.

[184] Linusson, A., Elofsson, M., Andersson, I. E., and Dahlgren, M. K. (2010) Statistical molecular design of balanced compound libraries for QSAR modeling, *Curr Med Chem 17*, 2001-2016.

[185] Fjell, C. D., Hancock, R. E., and Jenssen, H. (2010) Computer-aided design of antimicrobial peptides, *Curr Pharmacutical Anal 6*, 66-75.

[186] Fjell, C. D., Jenssen, H., Cheung, W. A., Hancock, R. E., and Cherkasov, A. (2011) Optimization of antibacterial peptides by genetic algorithms and cheminformatics, *Chem Biol Drug Des 77*, 48-56.

[187] Kulkarni, A. B., and Karlsson, S. (1993) Transforming growth factor-beta 1 knockout mice. A mutation in one cytokine gene causes a dramatic inflammatory disease, *Am J Pathol 143*, 3-9.

[188] Kuhara, T., Yamauchi, K., and Iwatsuki, K. (2012) Bovine lactoferrin induces interleukin-11 production in a hepatitis mouse model and human intestinal myofibroblasts, *Eur J Nutr 51*, 343-351.

[189] Sutas, Y., Soppi, E., Korhonen, H., Syvaoja, E. L., Saxelin, M., Rokka, T., and Isolauri, E. (1996) Suppression of lymphocyte proliferation in vitro by bovine caseins hydrolyzed with Lactobacillus casei GG-derived enzymes, *J Allergy Clin Immunol 98*, 216-224.

[190] Hernández-Ledesma, B., Quirós, A., Amigo, L., and Recio, I. (2007) Identification of bioactive peptides after digestion of human milk and infant formula with peptis and pancreatin., *Int Dairy J 17*, 42-49.

[191] Zuccotti, G. V., Salvini, F., Riva, E., and Agostoni, C. (2006) Oral lactoferrin in HIV-1 vertically infected children: an observational follow-up of plasma viral load and immune parameters, *J Int Med Res 34*, 88-94.

[192] Chen, H. L., Huang, J. Y., Chu, T. W., Tsai, T. C., Hung, C. M., Lin, C. C., Liu, F. C., Wang, L. C., Chen, Y. J., Lin, M. F., and Chen, C. M. (2008) Expression of VP1 protein in the milk of transgenic mice: a potential oral vaccine protects against enterovirus 71 infection, *Vaccine 26*, 2882-2889.

[193] Hamill, P., Brown, K., Jenssen, H., and Hancock, R. E. (2008) Novel anti-infectives: is host defence the answer?, *Curr Opin Biotechnol 19*, 628-636.

[194] Diaz-Valdes, N., Manterola, L., Belsue, V., Riezu-Boj, J. I., Larrea, E., Echeverria, I., Llopiz, D., Lopez-Sagaseta, J., Lerat, H., Pawlotsky, J. M., Prieto, J., Lasarte, J. J., Borras-Cuesta, F., and Sarobe, P. (2011) Improved dendritic cell-based immunization against hepatitis C virus using peptide inhibitors of interleukin 10, *Hepatology 53*, 23-31.

[195] Lupetti, A., Paulusma-Annema, A., Welling, M. M., Senesi, S., van Dissel, J. T., and Nibbering, P. H. (2000) Candidacidal activities of human lactoferrin peptides derived from the N terminus, *Antimicrob Agents Chemother 44*, 3257-3263.

[196] Nibbering, P. H., Ravensbergen, E., Welling, M. M., van Berkel, L. A., van Berkel, P. H., Pauwels, E. K., and Nuijens, J. H. (2001) Human lactoferrin and peptides derived from its N terminus are highly effective against infections with antibiotic-resistant bacteria, *Infect Immun 69*, 1469-1476.

[197] Lupetti, A., Paulusma-Annema, A., Welling, M. M., Dogterom-Ballering, H., Brouwer, C. P., Senesi, S., Van Dissel, J. T., and Nibbering, P. H. (2003) Synergistic activity of the N-terminal peptide of human lactoferrin and fluconazole against Candida species, *Antimicrob Agents Chemother 47*, 262-267.

[198] Lupetti, A., van Dissel, J. T., Brouwer, C. P., and Nibbering, P. H. (2008) Human antimicrobial peptides' antifungal activity against Aspergillus fumigatus, *Eur J Clin Microbiol Infect Dis 27*, 1125-1129.

[199] Brouwer, C. P., Rahman, M., and Welling, M. M. (2011) Discovery and development of a synthetic peptide derived from lactoferrin for clinical use, *Peptides 32*, 1953-1963.

[200] Kruse, T., and Kristensen, H. H. (2008) Using antimicrobial host defense peptides as anti-infective and immunomodulatory agents, *Expert Rev Anti Infect Ther 6*, 887-895.

[201] Hwang, S. A., Welsh, K. J., Boyd, S., Kruzel, M. L., and Actor, J. K. (2011) Comparing efficacy of BCG/lactoferrin primary vaccination versus booster regimen, *Tuberculosis (Edinb) 91 Suppl 1*, S90-95.

[202] Hwang, S. A., Arora, R., Kruzel, M. L., and Actor, J. K. (2009) Lactoferrin enhances efficacy of the BCG vaccine: comparison between two inbred mice strains (C57BL/6 and BALB/c), *Tuberculosis (Edinb) 89 Suppl 1*, S49-54.

[203] Hung, C. M., Wu, S. C., Yen, C. C., Lin, M. F., Lai, Y. W., Tung, Y. T., Chen, H. L., and Chen, C. M. (2010) Porcine lactoferrin as feedstuff additive elevates avian immunity and potentiates vaccination, *Biometals 23*, 579-587.

[204] Hung, C. M., Yeh, C. C., Chen, H. L., Lai, C. W., Kuo, M. F., Yeh, M. H., Lin, W., Tu, M. Y., Cheng, H. C., and Chen, C. M. (2010) Porcine lactoferrin administration enhances peripheral lymphocyte proliferation and assists infectious bursal disease vaccination in native chickens, *Vaccine 28*, 2895-2902.

[205] Brown, T. H., David, J., Acosta-Ramirez, E., Moore, J. M., Lee, S., Zhong, G., Hancock, R. E., Xing, Z., Halperin, S. A., and Wang, J. (2012) Comparison of immune responses and protective efficacy of intranasal prime-boost immunization regimens using adenovirus-based and CpG/HH2 adjuvanted-subunit vaccines against genital Chlamydia muridarum infection, *Vaccine 30*, 350-360.

[206] Bae, Y. S., Ju, S. A., Kim, J. Y., Seo, J. K., Baek, S. H., Kwak, J. Y., Kim, B. S., Suh, P. G., and Ryu, S. H. (1999) Trp-Lys-Tyr-Met-Val-D-Met stimulates superoxide generation and killing of Staphylococcus aureus via phospholipase D activation in human monocytes, *J Leukoc Biol 65*, 241-248.

[207] Bae, Y. S., Kim, Y., Kim, J. H., Lee, T. G., Suh, P. G., and Ryu, S. H. (2000) Independent functioning of cytosolic phospholipase A2 and phospholipase D1 in Trp-Lys-Tyr-Met-Val-D-Met-induced superoxide generation in human monocytes, *J Immunol 164*, 4089-4096.

[208] Bae, Y. S., Kim, Y., Kim, J. H., Suh, P. G., and Ryu, S. H. (1999) Trp-Lys-Tyr-Met-Val-D-Met is a chemoattractant for human phagocytic cells, *J Leukoc Biol 66*, 915-922.

[209] Baek, S. H., Seo, J. K., Chae, C. B., Suh, P. G., and Ryu, S. H. (1996) Identification of the peptides that stimulate the phosphoinositide hydrolysis in lymphocyte cell lines from peptide libraries, *J Biol Chem 271*, 8170-8175.

[210] Seo, J. K., Choi, S. Y., Kim, Y., Baek, S. H., Kim, K. T., Chae, C. B., Lambeth, J. D., Suh, P. G., and Ryu, S. H. (1997) A peptide with unique receptor specificity: stimulation of phosphoinositide hydrolysis and induction of superoxide generation in human neutrophils, *J Immunol 158*, 1895-1901.

[211] Lee, C. G., Choi, S. Y., Park, S. H., Park, K. S., Ryu, S. H., and Sung, Y. C. (2005) The synthetic peptide Trp-Lys-Tyr-Met-Val-D-Met as a novel adjuvant for DNA vaccine, *Vaccine 23*, 4703-4710.

[212] Parodi, P. W. (2007) A role for milk proteins and their peptides in cancer prevention, *Curr Pharm Des 13*, 813-828.

[213] Mader, J. S., Salsman, J., Conrad, D. M., and Hoskin, D. W. (2005) Bovine lactoferricin selectively induces apoptosis in human leukemia and carcinoma cell lines, *Mol Cancer Ther 4*, 612-624.

[214] Richardson, A., de Antueno, R., Duncan, R., and Hoskin, D. W. (2009) Intracellular delivery of bovine lactoferricin's antimicrobial core (RRWQWR) kills T-leukemia cells, *Biochem Biophys Res Commun 388*, 736-741.

[215] Pettersson, J., Mossberg, A. K., and Svanborg, C. (2006) alpha-Lactalbumin species variation, HAMLET formation, and tumor cell death, *Biochem Biophys Res Commun 345*, 260-270.

[216] Perego, S., Cosentino, S., Fiorilli, A., Tettamanti, G., and Ferraretto, A. (2011) Casein phosphopeptides modulate proliferation and apoptosis in HT-29 cell line through their interaction with voltage-operated L-type calcium channels, *J Nutr Biochem*.

[217] Raja, R. B., and Arunachalam, K. D. (2011) Anti-genotoxic potential of casein phosphopeptides (CPPs): a class of fermented milk peptides against low background radiation and prevention of cancer in radiation workers, *Toxicol Ind Health 27*, 867-872.

[218] Wang, J., Li, Q., Ou, Y., Han, Z., Li, K., Wang, P., and Zhou, S. (2011) Inhibition of tumor growth by recombinant adenovirus containing human lactoferrin through inducing tumor cell apoptosis in mice bearing EMT6 breast cancer, *Arch Pharm Res 34*, 987-995.

[219] Casares, N., Rudilla, F., Arribillaga, L., Llopiz, D., Riezu-Boj, J. I., Lozano, T., Lopez-Sagaseta, J., Guembe, L., Sarobe, P., Prieto, J., Borras-Cuesta, F., and Lasarte, J. J. (2010) A peptide inhibitor of FOXP3 impairs regulatory T cell activity and improves vaccine efficacy in mice, *J Immunol 185*, 5150-5159.

[220] Redwan el, R. M., and Tabll, A. (2007) Camel lactoferrin markedly inhibits hepatitis C virus genotype 4 infection of human peripheral blood leukocytes, *J Immunoassay Immunochem 28*, 267-277.

[221] Chobert, J. M., Sitohy, M., Billaudel, S., Dalgalarrondo, M., and Haertle, T. (2007) Anticytomegaloviral activity of esterified milk proteins and L-polylysines, *J Mol Microbiol Biotechnol 13*, 255-258.

[222] Maruyama, S., Mitachi, H., Tanaka, H., Tomizuka, N., and Suzuki, H. (1987) Studies of the active site and antihypertensive activity of angiotensin I-converting enzyme inhibitors derived from casein., *Agric. Biol. Chem 51*, 1581-1586.

[223] Maruyama, S., Nakagomi, K., Tomizuka, N., and Suzuki, H. (1985) Angiotensin I-converting enzyme inhibitor derived from and enzymatic hydrolysate of casein. II. Isolation and bradykinin-potentiating activity on the uterus of the ileum of rats. , *Agric. Biol. Chem 49*, 1405-1409.

[224] Maruyama, S., and Suzuki, H. (1982) A peptide inhibitor of angiotensin I-converting enzyme in the tryptic hydrolysate of casein., *Agric. Biol. Chem 46*, 1393-1394.

[225] Maeno, M., Yamamoto, N., and Takano, T. (1996) Identification of an antihypertensive peptide from casein hydrolysate produced by a proteinase from Lactobacillus helveticus CP790, *J Dairy Sci 79*, 1316-1321.

[226] Kampa, M., Loukas, S., Hatzoglou, A., Martin, P., Martin, P. M., and Castanas, E. (1996) Identification of a novel opioid peptide (Tyr-Val-Pro-Phe-Pro) derived from human alpha S1 casein (alpha S1-casomorphin, and alpha S1-casomorphin amide), *Biochem J 319 (Pt 3)*, 903-908.

[227] Minervini, F., Algaron, F., Rizzello, C. G., Fox, P. F., Monnet, V., and Gobbetti, M. (2003) Angiotensin I-converting-enzyme-inhibitory and antibacterial peptides from Lactobacillus helveticus PR4 proteinase-hydrolyzed caseins of milk from six species, *Appl Environ Microbiol 69*, 5297-5305.

[228] Maruyama, S., Mitachi, H., Awaya, J., Kurono, M., Tomizuka, N., and Suzuki, H. (1987) Angiotnsin I-converting enzyme inhibitor activity of the C-terminal hexapeptide of αs1-casein., *Agric. Biol. Chem 51*, 2557-2561.

[229] Hayes, M., Stanton, C., Slattery, H., O'Sullivan, O., Hill, C., Fitzgerald, G. F., and Ross, R. P. (2007) Casein fermentate of Lactobacillus animalis DPC6134 contains a range of novel propeptide angiotensin-converting enzyme inhibitors, *Appl Environ Microbiol 73*, 4658-4667.

[230] Brantl, V., Teschemacher, H., Henschen, A., and Lottspeich, F. (1979) Novel opioid peptides derived from casein (beta-casomorphins). I. Isolation from bovine casein peptone, *Hoppe Seylers Z Physiol Chem 360*, 1211-1216.

[231] Brantl, V., Teschemacher, H., Blasig, J., Henschen, A., and Lottspeich, F. (1981) Opioid activities of beta-casomorphins, *Life Sci 28*, 1903-1909.

[232] Migliore-Samour, D., and Jolles, P. (1988) Casein, a prohormone with an immunomodulating role for the newborn?, *Experientia 44*, 188-193.

[233] Meisel, H., and Frister, H. (1989) Chemical characterization of bioactive peptides from in vivo digests of casein, *J Dairy Res 56*, 343-349.

[234] Berthou, J., Migliore-Samour, D., Lifchitz, A., Delettre, J., Floc'h, F., and Jolles, P. (1987) Immunostimulating properties and three-dimensional structure of two tripeptides from human and cow caseins, *FEBS Lett 218*, 55-58.

[235] Nakamura, Y., Yamamoto, N., Sakai, K., Okubo, A., Yamazaki, S., and Takano, T. (1995) Purification and characterization of angiotensin I-converting enzyme inhibitors from sour milk, *J Dairy Sci 78*, 777-783.

[236] Jollés, P., and Caen, J. P. (1991) Parallels between milk clotting and blood clotting: opportunities for milk-derived products., *Trends Food Sci Technol.*

[237] Takahashi, M., Moriguchi, S., Suganuma, H., Shiota, A., Tani, F., Usui, H., Kurahashi, K., Sasaki, R., and Yoshikawa, M. (1997) Identification of casoxin C, an ileum-contracting peptide derived from bovine kappa-casein, as an agonist for C3a receptors, *Peptides 18*, 329-336.
[238] Patten, G. S., Head, R. J., and Abeywardena, M. Y. (2011) Effects of casoxin 4 on morphine inhibition of small animal intestinal contractility and gut transit in the mouse, *Clin Exp Gastroenterol 4*, 23-31.
[239] Pihlanto-Leppala, A., Koskinen, P., Piilola, K., Tupasela, T., and Korhonen, H. (2000) Angiotensin I-converting enzyme inhibitory properties of whey protein digests: concentration and characterization of active peptides, *J Dairy Res 67*, 53-64.
[240] Mullally, M. M., Meisel, H., and FitzGerald, R. J. (1997) Identification of a novel angiotensin-I-converting enzyme inhibitory peptide corresponding to a tryptic fragment of bovine beta-lactoglobulin, *FEBS Lett 402*, 99-101.
[241] Tani, F., Shiota, A., Chiba, H., and Yoshikawa, M. (1994) Serorphin, an opiod peptide derived from bovine serum albumin, in β-casomorphins and related peptides, In *Recent developments*, (Brantl, V., and Teschemacher, H., Eds.), pp 49-53, VCH-Verlag, Weinheim.
[242] Yamauchi, K. (1992) Biologically functional proteins of milk and peptides derived from milk proteins., *Bulletin of the IDF 272*, 51-58.
[243] Siciliano, R., Rega, B., Marchetti, M., Seganti, L., Antonini, G., and Valenti, P. (1999) Bovine lactoferrin peptidic fragments involved in inhibition of herpes simplex virus type 1 infection, *Biochem Biophys Res Commun 264*, 19-23.
[244] Tani, F., Iio, K., Chiba, H., and Yoshikawa, M. (1990) Isolation and characterization of opioid antagonist peptides derived from human lactoferrin, *Agric Biol Chem 54*, 1803-1810.

Chapter 2

Bioactive Casein Phosphopeptides in Dairy Products as Nutraceuticals for Functional Foods

Gabriella Pinto, Simonetta Caira, Marina Cuollo,
Sergio Lilla, Lina Chianese and Francesco Addeo

Additional information is available at the end of the chapter

1. Introduction

Nutraceuticals, a term combining the words "nutrition" and "pharmaceutical", is *a food or food product that provides medical or health benefits including the prevention and treatment of diseases*. A functional food essentially provides a health benefit beyond the basic nutrition, whereas nutraceutical is used to describe an isolated or concentrated molecular extract of bioactive compounds. Milk is a unique food providing a variety of essential nutrients necessary to properly fuel the body. Inactive food proteins can release encrypted bioactive peptides *in vivo* or *in vitro* by digestive enzymatic hydrolysis. Bioseparation protocols offer unique possibilities for a number of application areas, *e.g.*, hydrolyzate-based nutraceutical ingredients for functional foods, dietary supplements and medical foods.

Many ingredients are included in the wide range of nutraceuticals, such as essential amino acids, conjugated fatty acids, vitamins, minerals and polyphenols. They have already been patented and incorporated in functional foods and nutritional beverages. Such components are believed to improve overall health and well-being, reducing the risk of specific diseases or minimizing the effects of other health concerns. However, milk is devoid of flavonoids, the most common group of vegetable polyphenolic compounds, which act as antioxidants and free radical scavengers. In contrast, the ingestion of soy and green tea extract may reduce the risk of developing prostate cancer and may protect against various other types of cancer [1-2]. An interesting patented invention has made available an extended-release form of polyphenols and riboflavin (vitamin B2) coated with methylcellulose [3]. Coating slows down the release of polyphenols in the nutraceutical preparation [3]. Possible applications of coating technology could be extended to all of the bioactive peptides susceptible to digestive enzymes. For example, glutathione can be maintained in human blood at normal levels by supplying it as dry-filled capsules [4]. Nutrients and bioactive compounds may be microencapsulated by using mixtures of proteins or peptides and oils. Encapsulation of ω-3 *fatty acids* (FA) enhances the stability and bioavailability of bioactive food ingredients [5]. By these means, new transparent bioavailable beverages containing ω-3 rich oils, phospholipids and minerals in an oxidatively stable food system were created [6]. Iron or calcium casein phosphopeptides (CPP) were embedded in the chitosan lactate fiber as a protective agent against oil oxidation [6]. A recent patent relates to a nutraceutical composition consisting of a sweetener admixture for food or drink comprising calcium lactate, calcium acetate, vitamin D3 and sucralose (for fortified zero calorie formulation) or sugar (white/brown; for fortified sugar preparation) [7]. Milk proteins playing a physiological role include proteins such as β-lactoglobulin, α-lactalbumin, immunoglobulins, lactoferrin, heat-stable proteose peptones, serum albumin and various acid soluble phosphoglycoproteins. Casein (CN), representing 80% of total milk proteins, consists of four α_{s1}-, α_{s2}-, β-, and κ-CN families in the approximate ratio 38:11:38:13. Research performed in recent years has shown that caseins and whey proteins are rich in encrypted biologically active peptides such as exorphins (casomorphins), CPP and immunopeptides [8]. The peptides are released by enzymes in the form of mature bioactive components or the precursors thereof [9]. They are 3- to 20-residue long peptides released during *in vivo* gastrointestinal digestion. Historically, the opioid peptides were discovered as the result of a systematic search for exogenous substances, namely (i) first discovered in 1979, opioid agonist peptides derived from milk proteins were characterized in 1986; (ii) in 1982, angiotensin-converting enzyme (ACE)-inhibitory peptides were found to be antihypertensive peptides; (iii) then, fibrinogen-like sequences with antithrombotic activity were found; (iv) phagocytic activity and lymphocyte proliferation of numerous immunomodulating peptides were observed; (v) CPP facilitating the absorption of minerals, especially calcium, magnesium and iron were found; and (vi) antimicrobial peptides were discovered [10]. There are many milk peptides that possess multifunctional activities, *i.e.*, they can play two or more hormone-like roles. Bioactive peptides grouped according to their function in human well-being are shown in Figure 1.

Nutraceutical products comprising short bioactive peptides showing *in vitro* or *in vivo* antimicrobial, ACE-inhibitory activity and/or antihypertensive and/or antioxidant activity are being considered for possible use by the pharmaceutical industry. The CN hydrolyzates could serve as food preservatives to reinforce the body's natural defenses or as pharmaceutical products for facilitating the control of blood and/or bacterial infections [12]. Much research has been devoted to increasing mineral transport by phosphorylated groups of peptides [13]. CPP in commercial hydrolysed casein (Tatua Cooperative Dairy Co. Ltd, New Zealand and Arla Foods Ingredients and Sweden) seem to help the absorption of chelated calcium, iron, copper, zinc and manganese in the intestine (Table 1). Thus, CPP-bound amorphous calcium phosphate (ACP) displayed anticariogenic effect when added to dentifrices or oral care products by localizing calcium and phosphate ions at the tooth surface. Similarly, it has been claimed that a chewing gum or other confectionery product containing a combination of CPP-ACP and sodium bicarbonate as active ingredients can provide dental health benefits [14]. In experiments on humans, synthetic CPP-ACP nanocomplexes incorporated in mouth rinses and sugar-free chewing gums have been proven to be potential anticariogenic agents [15] (Table 1).

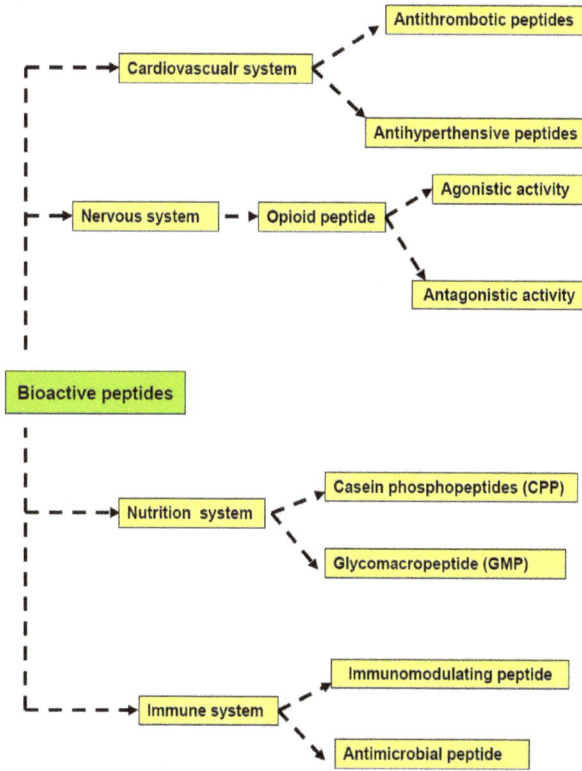

Figure 1. Main bioactivity of peptides formed by the enzymatic digestion of milk proteins (source: [11]).

Brand name	Product type	Claimed functional bioactive peptides	Health/function claims	Manufacturer
HCP102/HCP 105	Hydrolysate ingredient	CPP	Helps mineral absorption	Tatua (New Zealand)
Capolac	Hydrolysate ingredient	CPP	Helps mineral absorption	Arla Foods Ingredients (Sweden)
Recaldent	Chewing gum	ACP-CPP	Anticariogenic	Cadbury Adams (USA)
Recaldent	Toothpaste	ACP-CPP	Anticariogenic	GC Tooth Mousse (USA)

ACP= amorphous calcium phosphate; CPP= Casein phosphopeptides

Table 1. Commercial dairy products and ingredients with health or function claims based on CPP content (source: modified from [16]).

2. Bioactive peptides released *in vitro* by the hydrolysis of milk proteins

Peptides with various bioactivities can be produced according to two different methods: i) *in vitro* fermentation of milk inoculated with starter cultures and ii) *in vitro* digestion of milk proteins by one or more proteolytic enzymes.

(i) The proteolytic system of lactococci is able to degrade milk proteins using cell-wall-bound proteinases by releasing di-, tri-, and/or oligo-peptides and amino acids supporting the growth of bacteria. In addition, lactococcal peptidases released into the curd/cheese consequently to autolysis can further degrade the internalized peptides to amino acids [17]. The higher the exopeptidase activity in cheeses, the greater the age of the cheese [18]. Consistently, yogurt and other cultured dairy drinks have some of the highest counts of cells that actually survived and thus possessed the lowest number of peptides derived from the aminopeptidase activity. In a comprehensive review of literature, no enzyme with carboxypeptidase (CPase) activity has been reported for either lactococci or other LAB [19-20]. The bacterial peptidases have different and partly overlapping specificities (Figure 2).

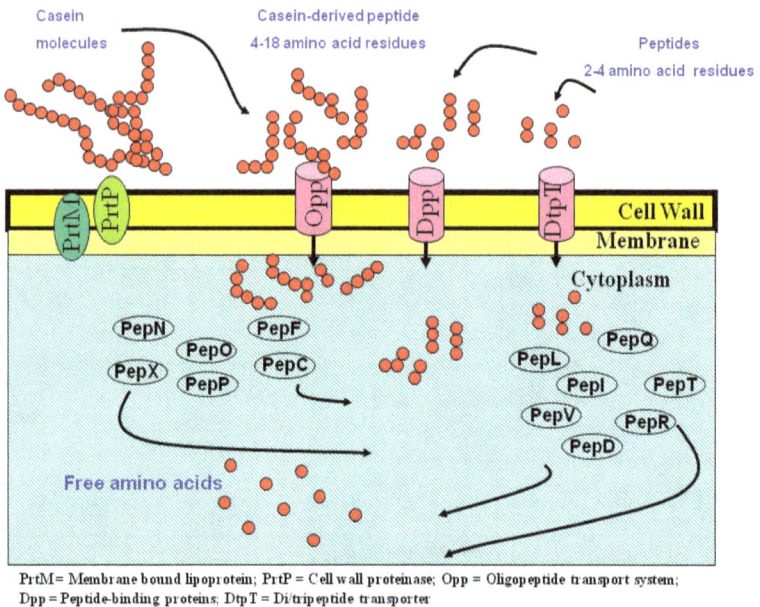

PrtM= Membrane bound lipoprotein; PrtP = Cell wall proteinase; Opp = Oligopeptide transport system; Dpp = Peptide-binding proteins; DtpT = Di/tripeptide transporter

Figure 2. A simplified model presenting proteolysis, transport, peptidolysis and regulation of the proteolytic system of *Lactococcus lactis* on casein breakdown [20-22]. Intracellular peptidases PepO and PepF are endopeptidases, PepN/PepC/PepP are general aminopeptidases, PepX is X-prolyldipeptidyl aminopeptidase, PepT is tripeptidase, PepQ is prolidase, PepR is prolinase, PepI is proline iminopeptidase, and PepD and PepV are dipeptidases D and V. The role of PepN, PepC, PepI, PepP and PepA and PepO, PepF, PepX, PepQ, PepR, PepV, PepT and peptidolytic cycles are depicted schematically (various alternative routes of breakdown are possible for most peptides).

Although the intracellular endopeptidases PepN and PepC are unable to hydrolyze casein molecules, the X-prolyl dipeptidyl aminopeptidases (PepX) are active on oligopeptides hydrolyzing the internal bonds of casein-derived peptides. Taken together, these enzymes are able to remove the N-terminal residues from peptides, with the specificity primarily depending on the nature of the N-terminal amino acid [20-21]. Di- and tripeptides generated by endopeptidases, general aminopeptidases and PepX are next subjected to additional cleavage by the tripeptidase, PepT, and dipeptidases, PepV and PepD (Figure 2). Other peptidases with more specific substrate specificities include PepA, which liberates N-terminal acidic residues from 3- to 9-residue long peptides; PepP, which prefers tripeptides carrying proline in the middle position; PepR and PepI, which act on dipeptides containing proline in the penultimate position; PepQ, which cleaves dipeptides carrying proline in the second position; and PepS, which shows preference for peptides containing two to five residues with Arg or aromatic amino acid residues in the N-terminal position [20-21,23] (Figure 2). This proteolytic system is able to support LAB growth to high cell densities (10^9–10^{10} cfu/mL) in milk containing only small amounts of hydrolytic products that are transportable into the cells for assimilation. *Lb. helveticus* and *Lb. delbrueckii* ssp. *bulgaricus* possess cell wall proteinase activity stronger than that of lactococci. The number of intracellular proteins released by *St. thermophilus* is greater than that by *Lb. helveticus* [24]. Proteins released in cheese from a starter-based thermophilic LAB, such as *Lb. helveticus, Lb. delbruecki* subsp. *lactis* and *Streptococcus salivarius* subsp. *thermophilus* and *Propionibacterium freudenreichii*, have been identified using 2D-PAGE and mass spectrometry (MS) analysis [24]. Similarly, bioactive peptides have been determined using High Performance Liquid Chromatography (HPLC) and offline Matrix-Assisted Laser Desorption/Ionization Mass Spectrometry-Time-Of-Flight (MALDI-MS-TOF) [25]. These peptides were all generated from CN and released upon proteolysis depending on the bacterial strain [26]. In this manner, proteinases play a primary role, as they are able to generate specific bioactive peptides. Recombinant human α_{s1}-casein digested by trypsin gave rise to several ACE-inhibitory peptides and calcium-binding CPP. These peptides did not form in cheese whey, although they can be formed from CN during fermentation using various commercial dairy starters [27]. The combination of the LAB bacteria and proteolytic enzymes could serve to increase the range of bioactive peptides.

(ii) *In vitro hydrolysis* of CN by pepsin and trypsin could produce many bioactive peptides. Pepsin, an endopeptidase with broad specificity, preferentially cleaves hydrophobic, preferably aromatic, residues. Trypsin specifically hydrolyses peptide bonds just after a lysine or an arginine residue of β-casein A^1 and A^2 variant (11 and 4, respectively); κ-casein A and B (9 and 5); α_{s2}-CN A (24 and 6); or α_{s1}-CN B (14 and 6). In this manner, tryptic hydrolysates of CN contain uneven peptides of up to 8159 Da and also free amino *acids* [28]. We have demonstrated that CN hydrolysate by pepsin (P) and trypsin (T) in succession does not contain peptides with molecular mass greater than 2431 Da. β-CN resulted extensively hydrolyzed into a high number of oligopeptides by using proteases with well defined cleavage specificity. Controlled partial hydrolysis by proteinases could lead to the formation of partially degraded proteins critical for obtaining new functional products. The choice of digestion enzymes needs to be evaluated carefully because influences the hydrolysate final composition. A high number of peptides with antimicrobial, anti-hypertensive and opiod-like activity has been identified (Table 2), some of which exactly matched those described in the literature for potential bioactivity (Table 2). The potent opioid β-CM7 peptide retained part of its original opioid activity-like food hormone when progressively shortened. The synthetic β-casomorphin derivatives have been shown to be highly specific and potent β-type opioid receptor ligands [29].

We also performed sequential milk protein (powder sample) digestions using various endoproteases facilitating a consistent partial hydrolysis. The high degree of specificity in terms of cleaving peptide bonds exhibited by a cocktail of enzymes (P, T and P432 from Biocatalysts, U.K) yielded a limited number of CPP. After fractionation on an HA column,

Identity of bioactive peptides	Bioactivity	References	PT peptides
β-CN (f60-70); (f59-61); (f59-64); (f60-68)	Opiod	[29-30]	(f58-68); (f59-63); (f59-62)
β-CN (f74-76)	ACE-Inhibitory	[31-33]	(f69-80); (f71-80); (f75-80)
β-CN (f80-90); (f84-86)			(f80-88); (f81-89); (f81-88); (f81-92); (f81-93); (f81-94)
β-CN (f108-113)			(f108-103)
β-CN (f140-143)			(f139-142)
β-CN (f177-183)			(f177-183); (f177-184); (f179-182)
β-CN (f169-174)			(f169-176); (f172-176)
β-CN (f193-198)			(f193-199)
β-CN (f193-202)			(f193-202)
β-CN (f1-25)4P	Mineral carrier, immunomodulatory, cytomodulatory	[34-38]	(f12-17)1P; (f12-25)3P; (f12-25)4P; (f15-25)2P; (f15-25)3P; (f19-25)1P; (f19-25)2P
β-CN (f29-41)1P	Mineral carrier	[13]	(f33-42)1P; (f33-43)1P; (f33-44)1P
β-CN (f84-86)	ACE-Inhibitory	[31-33]	(f80-87); (f81-88); (f81-89); (f81-92); (f81-93); (f81-94)
α_{s1}-CN (f23-34)			(f24-32)
α_{s1}-CN (f25-27)			(f25-31)
α_{s1}-CN (f90-96); (f90-95); (f91-95)	Opiod (agonist)	[39-41]	(f92-95); (f91-95); (f90-95)
α_{s1}-CN (f142-147)	ACE-Inhibitory	[31-33]	(f142-145); (f143-146); (f143-150); (f144-149)
α_{s1}-CN (f157-164)			(f155-164)
α_{s1}-CN (f194-199)			(f194-199); (f197-199); (f193-199)
α_{s1}-CN	Mineral carrier	[13]	(f110-119)1P
α_{s1}-CN			(f41-55)1P; (f68-79)2P
α_{s2}-CN (f174-179)	ACE-Inhibitory	[31-33]	(f174-179)
α_{s1}-CN (f189-193)			(f189-193)
α_{s2}-CN	Mineral carrier	[13]	(f138-146)1P
α_{s2}-CN			(f124-137)2P; (f126-136)2P; (f126-137)2P
κ-CN (f33-38)	Opiod (antagonist)		(f58-65)1P

PT= Pepsin and Trypsin action; P= Phosphate group.

Table 2. Identity of bioactive peptides found in the PT digest of milk protein powder sample. The bioactivity of the peptide from which they derive and the references are reported.

both non-CPP and CPP were identified as shown in Table 3. Different oxidation rates of Met residues in the same protein resulted in formation of the peptides with different molecular mass. In most cases, a single Met-containing peptide and its oxidized counterpart were identified. However, in many cases, when proteins contained several consecutive endoprotease sensitive bonds, different long peptides containing the same Met-oxidation site(s) were identified. The data in Table 3 suggest that the cocktail enzymes containing amino- and CPases, in addition to P and T, progressively reduce the size of peptides without altering the degree of phosphorylation. Evaluation of protein/peptide quality can take advantage of the tandem MS for the detection of native, partly oxidized and partly dephosphorylated peptides.

By this means, phosphorylated peptides (2 and 3P/mole) of a precursor of lactophorin (LP) (28 kDa milk glycoprotein), proteose-peptone component 3, and glycosylation-dependent adhesion molecule 1 were detected. In addition, three low molecular weight non-CPP derived from LP were detected.

Native milk proteins used as the substrate for digestion by enzymes did not form CPP. This suggests that denatured LP and other whey proteins could have the tendency to form low molecular-mass peptide aggregates characterized by a poor solubility. For this reason, the use of milk protein and/or any milk substrate powder must be discouraged to eliminate phosphates and salts from the substrate. The enzymatic hydrolysis of the casein implies the use of endoproteases. However, the protein hydrolysate with alcalase is used in infant formulae, dietetic foods, nutraceuticals, ice creams, dressings, fermented products, yogurts, and personal care products. CPP released by alcalase are truncated with respect to those released by trypsin. The identified peptides can be categorized into two groups, one containing multiphosphorylated peptides and the other tri-, di- and mono CPP. Each group contained a number of variously long peptides due to the broad specificity of alcalase cleaving peptide bonds mainly on the carboxyl side of Glu, Met, Leu, Tyr, Lys, and Gln. The exoproteases responsible for the hydrolysis are inactivated by heating for ~10 min to ~85 °C. The *in vitro* sequential use of pepsin, pancreatic proteases and extracts of human intestinal brush border membranes, mimicking the respective gastric, duodenal and jejuneal *in vivo* digestion of CN, exhibited significant bioactive effects. A limited number of CN and whey protein peptides survived the *in vitro* simulated gastro-intestinal digestion. The anionic character seems to confer a marked resistance to multi-phosphorylated CPP hydrolysis by endoprotease. Ten out of 19 CPP contained SerP available for binding minerals, and four of these peptides, α_{s1}-CN (f57-90)5P, α_{s1}-CN (f56-90)5P, α_{s1}-CN (f55-76)5P, β-CN (f1-52)5P, were reported for the first time in the CN digests [42]. Only β-CN (f1-25)4P, 3P and 2P survived the simulated gastrointestinal digest of CN [43].

The ingress of foreign material in general, such as CPP, across the mucosal brush-border into the enterocyte is conditioned by the efficient dephosphorylation of peptides by alkaline phosphatase. This aspect deserves more in-depth investigation.

Parent protein	Molecular mass (Da)		Start		End	Peptide sequence	Peptide modifications
	Expected	Calculated					
α_{s1}-CN	1222.5	1222.5	110	-	119	(L)EIVPNSAEER(L)	1P
	1517.8	1517.6	68	-	79	(S)SEEIVPNSVEQK(H)	2P
	1525.7	1525.5	41	-	53	(L)SKDIGSESTEDQA(M)	2P
	1586.6	1586.5	43	-	55	(K)DIGSESTEDQAME(D)	2P; Oxidation (Met)
	1672.7	1672.6	41	-	54	(L)SKDIGSESTEDQAM(E)	2P; Oxidation (Met)
	1785.8	1785.6	41	-	55	(L)SKDIGSESTEDQAME(D)	2P
	1801.7	1801.6	41	-	55	(L)SKDIGSESTEDQAME(D)	2P; Oxidation (Met)
	1963.9	1963.8	39	-	55	(N)ELSKDIGSESTEDQAME(D)	1P; Oxidation (Met)
	1989.9	1989.7	37	-	52	(K)VNELSKDIGSESTEDQ(A)	3P
	2060.7	2060.7	37	-	53	(K)VNELSKDIGSESTEDQA(M)	3P
α_{s2}-CN	900.4	900.3	58	-	65	(S)SEESAEVA(T)	P
	937.4	937.3	141	-	147	(D)MESTEVF(T)	1P; Oxidation (Met)
	1067.4	1067.3	57	-	65	(S)SSEESAEVA(T)	2P
	1089.5	1089.4	138	-	146	(K)TVDMESTEV(F)	1P
	1105.4	1105.4	138	-	146	(K)TVDMESTEV(F)	1P; Oxidation (Met)
	1252.6	1252.5	138	-	147	(K)TVDMESTEVF(T)	1P; Oxidation (Met)
	1410.6	1410.5	126	-	136	(R)EQLSTSEENSK(K)	2P
	1538.7	1538.6	126	-	137	(R)EQLSTSEENSKK(T)	2P
	1623.7	1623.6	1	-	13	KNTMEHVSSSEES(I)	2P
	1639.7	1639.5	1	-	13	KNTMEHVSSSEES(I)	2P; Oxidation (Met)
	1680.8	1680.6	124	-	136	(L)NREQLSTSEENSK(K)	2P
	1719.7	1719.5	1	-	13	KNTMEHVSSSEES(I)	3P; Oxidation (Met)
	1808.9	1808.7	124	-	137	(L)NREQLSTSEENSKK(T)	2P
β-CN	639.3	639.3	12	-	16	(E)IVESL(S)	1P
	900.5	900.4	19	-	25	(S)SEESITR(I)	1P
	1067.4	1067.4	18	-	25	(S)SSEESITR(I)	2P
	1354.6	1354.5	15	-	25	(E)SLSSSEESITR(I)	2P
	1434.6	1434.5	15	-	25	(E)SLSSSEESITR(I)	3P
	1447.7	1447.5	33	-	43	(K)FQSEEQQQTED(E)	1P
	1576.7	1576.6	33	-	44	(K)FQSEEQQQTEDE(L)	1P
	1689.8	1689.6	33	-	45	(K)FQSEEQQQTEDEL(Q)	1P
	1775.8	1775.7	12	-	25	(E)IVESLSSSEESITR(I)	3P
	1855.6	1855.6	12	-	25	(E)IVESLSSSEESITR(I)	4P
κ-CN	968.4	968.4	145	-	152	(A)TLEDSPEV(I)	1P
	1734.7	1734.7	147	-	161	(L)EDSPEVIESPPEINT(V)	1P
Lactophorin	1226.5	1226.51	34	-	43	(L)SKEPSISRED(L)	1P
	1306.5	1306.5	34	-	43	(L)SKEPSISRED(L)	2P
	1419.6	1419.56	34	-	44	(L)SKEPSISREDL(I)	2P
	1499.7	1499.5	34	-	44	(L)SKEPSISREDL(I)	3P

Phosphoserine residues are coloured red.

Table 3. Native and partly Met-oxidized CPP isolated from a three enzyme (Pepsin, trypsin and P432) milk protein hydrolyzate.

3. *In vivo* digestion of casein, formation of CPP and their physiological importance

Among the biologically active peptides, CPP characterized by SerP and/or ThrP residues account for ~30% of monoesters of hydroxyl amino acids. They mainly occur in the Ser/Thr-Xaa-SerP/Glu/Asp sequence consensus, where Xaa is any amino acid residue but Pro. The three-phosphorylated motif -SerP-SerP-SerP-Glu-Glu- occurs in α_{s1}-CN (f66-70), α_{s2}-CN (f8-12), α_{s2}-CN (f56-60), and β-CN (f17-21). According to the current CN nomenclature, bovine α_{s1}-CN, α_{s2}-CN, β-CN, and κ-CN possess 8-9, 11-13, 4-5, and 1-2 phosphate (P) residues, respectively, and the P number could change according to the casein variant [44]. For example, β-CN D has one SerP residue less than the A counterpart due to the substitution $Lys^{18} \rightarrow Ser^{18}$.

The *in vivo* digestion of milk proteins takes place mainly in the stomach under the action of pepsins, gastric digestive proteinases that are able to digest ~20% proteins. Afterwards, the pepsin digests pass to the duodenum where peptides are further hydrolyzed by pancreatic enzymes. The digestion is completed by membrane proteases and a variety of peptidases embedded in the brush border of the small intestine and released by the intestinal microflora. These peptidases release an amino acid residue or a dipeptide from the N- and C-terminal side of oligopeptides [45-47]. Phosphatase(s) located in the brush border of the apical membrane of enterocytes, act(s) in removing phosphate groups, thus promoting partial or full peptide dephosphorylation of peptides in different body districts. The phosphorylated sequence is responsible, at the intestinal pH, for binding Ca^{++}, Zn^{++}, and Mg^{++} and for the *in vivo* resistance of the complex to gastrointestinal proteases [48]. Fe complexed to β-CN (f15-25)4P was scarcely hydrolyzed throughout the digestion, suggesting that the coordination of iron ions to CPP inhibits the action of both phosphatase and peptidases [49]. Brush border enzyme alkaline phosphatase activity could improve the absorption of Fe complexed CPP by releasing Fe from peptides. Moreover, Fe complexed to β-CN CPP was absorbed more than Fe complexed to α_{s1}-CN CPP [50]. The differences in protein composition between cow and breast milk could explain some of the differences in the Fe bioavailability of the latter [50]. Iron deficiency, a major worldwide nutritional problem, can be reduced by CPP. Fe complexed CPP prevents the formation of poorly absorbed high molecular weight ferric hydroxides. Zinc absorption can also be enhanced by the formation of Zn complexed to CCP, in particular to β-CN (f1-25)4P [51]. Some portion of the mineral complexed CPP formed in the small intestine was resistant to the digestive and enteric bacteria enzymes and found in the feces of rats fed a casein-based diet [52]. Although literature data regarding intestinal CPP absorption are conflicting, the peptides seem to interact directly with the plasma membrane. One possible mode of CPP action on the transmembrane flux of calcium is that CPP might insert themselves into the plasma membrane and form their own calcium selective channels or act as calcium-carrier peptides rapidly internalized via endocytosis or other processes and eventually provide ionized calcium in the cytosol [53]. Cellular uptake studies of fluorine-18 labeled CPP in human colorectal adenocarcinoma cell line (HT-29) and human head and neck squamous cell carcinoma line (FaDu) cells at 37 °C and 48 °C showed a poor cell penetration because of the poor transport of the phosphopeptides through the cell membrane [54]. The results from *in vivo* studies are still too controversial, as there are many factors affecting Ca availability, such as the various co-present dietary compounds in the intestinal lumen [55]. Despite the vigor of the saturable active transport process by the duodenum, most of the absorption of ingested calcium occurs in the ileum (88% of calcium), jejunum (4%) and duodenum (8%) [56]. An important factor determining the contribution of the ileum to overall calcium absorption is the relatively long transit time of calcium in the segments of the small intestine, accounting for approximately 102 min in the ileum and 6 min in the duodenum [57]. The higher absorption of calcium occurred when inorganic P was added to the Ca-CPP preparation. CPP exhibit a potent ability to form soluble complexes with Ca^{2+} and other trace elements, preventing the formation of Ca-phosphate precipitate in the intestine. CPP could limit the inhibitory effect of phosphate on Ca availability and increase Ca transport across the distal small intestine [55]. All components of the diet reaching the ileum make calcium soluble or keep it in solution within the ileum. Several molecules, particularly CPP, stimulate the passive diffusion of minerals. CPP have been for the first time detected in human ileostomy fluid, confirming their ability to survive gastrointestinal passage into the human distal ileum [58]. CPP released during milk digestion appeared to be stable for up to 8 h in ileostomy contents [58]. The *in vivo* formation of bovine CPP was demonstrated in the small intestinal fluid of minipigs after ingestion of a diet containing casein [59] and in the stomach and duodenum after ingestion of milk or yogurt [60]. The *in vivo* survival of CPP to the prolonged intestinal passage in the distal small intestine is a prerequisite for their function as bioactive substances [58]. CPP are protected from degradation in the gut by the milk matrix, provided that they are ingested as milk constituents and not as isolated CPP. Whole casein or individual casein fractions are used as raw materials to obtain CPP as dietary supplements. Ca could be bound to either SerP or Glu residues [61], suggesting that CPP may enhance the solubility of calcium in the intestinal lumen, thereby increasing the mineral availability for absorption in the small intestine [62,13]. Chemically synthesized CPP, *i.e.*, β-CN (f1-25)4P and αs1-CN (f59-79)5P, carrying the characteristic cluster Ser(P)-Ser(P)-Ser(P)-Glu-Glu, increase the intracellular calcium uptake by the human cultured HT-29 tumor cells [63], Caco-2 cells [64] and osteoblasts [65]. A more pronounced effect has been observed for β-CN-derived peptides than for αs1-CN-counterparts. It has been suggested that CPP promote calcium binding, which would depend on the structural conformation conferred by the two phosphorylated 'acidic motif' and the N-terminal sequence of β-CN [63].

Dental caries are initiated via the demineralization of tooth hard tissue by organic acids directly from the diet or produced from fermentable carbohydrate by dental plaque cariogenic bacteria. CPP can help to replace the minerals that were previously lost consequently to caries [66-67]. Hence, there is a great interest in developing CPP as nutraceutical ingredients for the formulation of functional foods.

4. CPP enrichment by different techniques

CPP preferably comprise components released by four different casein families, each having a molecular weight greater than 500 Da. Multiply and singly, tryptic CPP can be simultaneously detected using MALDI-TOF, and the location of phosphate groups by a combination of tandem mass spectrometry and computer-assisted database search programs, such as SEQUEST (Trademark, University of Washington, Seattle Wash) [68-69].

Nano-electrospray MS/MS has been used for phosphopeptide sequencing and exact determination of phosphorylation sites [70]. However, mass spectrometric analysis of proteolytic digests of proteins rarely provides full coverage of the phosphorylated sequence, with parts of the sequence often going undetected. In addition, protein phosphorylation is often sub-stoichiometric, such that ionization of CPP present in lower abundance in complex hydrolysates is ordinarily suppressed by strongly ionizable non-phosphorylated peptides. The MALDI-MS desorption/ionization efficiency for phosphopeptides was reported to be an order of magnitude lower than that recorded for the non-phosphorylated counterpart, and ionization became more difficult as the number of phosphate groups increased [71]. Direct analysis of phosphopeptides utilizes two orthogonal MS scanning techniques, both based on the production of phosphopeptide-specific marker ions at m/z 63 and/or 79 in the negative ion mode. These scanning methods combined with the liquid chromatography (LC)-electrospray mass spectrometry (ESI) and nano-electrospray MS/MS allow the selective detection and identification of phosphopeptides even in complex proteolytic digests. Thus, even when the signal of the phosphopeptide is indistinguishable from the background, as in the conventional MS scan, low-abundant and low-stoichiometric phosphorylated peptides can be selectively determined in the presence of a large excess of non-phosphorylated peptides. This strategy is particularly well suited to phosphoproteins that are phosphorylated to varying degrees of stoichiometry at multiple sites [72]. However, the identification and characterization of phosphoproteins would be greatly improved using selective enrichment of CPP prior to MS analysis. An ancient technique for phosphoprotein enrichment consisted of the precipitation of phosphopeptides as insoluble barium salts and recovery by centrifugation, as according to the Manson & Annan method [73]. High-throughput phosphoproteome technologies currently rely on combining pre-separation of proteins, most commonly by high-resolution two-dimensional polyacrylamide gel, in-gel tryptic cleavage of proteins, and subsequent MALDI-TOF or ESI-MS/MS mass spectrometry analysis of peptides [74]. The high resolving power of 2-DE with the sensitive MS requires extensive manual manipulation of samples. Alternative methods are based on chemical derivatization. For example, for β-elimination, a strong base such as NaOH or Ba(OH)$_2$ is used to cleave the phosphoester bonds of phosphoserine and phosphothreonine and form dehydroalanine or dehydroaminobutyric acid, respectively, each able to react with different nucleophiles, such as ethanedithiol (EDT) or dithiothreitol (DTT). This procedure provides a considerably simpler method to enrich CPP. By using cross-linking reagents with affinity tags, such as biotin, interfering non-cross-linked peptides are eliminated, and CPP are highly enriched [75]. Although the chemical derivatization methods are highly selective, they are not widely applied in phosphoproteome studies due to sample loss by the multiple reaction steps and unavoidable side reactions [76]. Immobilized metal-ion affinity chromatography (IMAC, with Fe^{3+}, Ga^{3+}, Ni^{2+} and Zr^{4+} metal ions) and metal oxide affinity chromatography (MOAC, with TiO$_2$, ZrO$_2$, Al$_2$O$_3$ and Nb$_2$O$_5$) have been widely used for the quantitative binding of CPP on resin or adsorbent. Iminodiacetic acid (IDA, a tridentate metal-chelator) or nitrilotriacetic acid (NTA, a quadradentate metal chelator) are often used as IMAC functional matrices reacting with multivalent metal ions to form chelated ions with positive charges useful for the purification of phosphopeptides. Usually Fe^{3+}, Ga^{3+} and Al^{3+} are bound to a chelating support prior to fractionating the complex mixture of peptides before MS analysis [77-78]. Ga^{3+} showed selectivity for CPP higher than Fe^{3+} and Al^{3+} [77,79]. Phosphopeptides bound to IMAC resin are successively recovered in the column effluent by increasing either pH or the phosphate concentration in the buffer [80]. The negatively charged CPP selectively interact with TiO_2 microspheres via bidentate binding at the dioxide surface [81-83]. TiO_2-MOAC showed higher specificity than immobilized gallium (Ga^{3+}), immobilized iron (Fe^{3+}), or zirconium dioxide (ZrO_2) affinity chromatography for phosphopeptide enrichment. The main problem associated with the chelating resins is the metal-ion leaching, which leads to CPP loss during the enrichment procedure. The selectivity of these methods was somewhat compromised by the detection of several acidic non-CPP that were also retained by the TiO_2 column [84]. To overcome this drawback, the carboxyl groups are methyl esterified which eliminates the non-specific adsorption of acidic peptides on IMAC [85]. Considerable efforts have been expended to remove acidic non-CPP by washing the resin with 2,5-dihydroxybenzoic acid (DHB) [86] or phthalic acid [87]. It has been found that aliphatic hydroxyl acid modified metal oxide works more efficiently and more specifically than aromatic modifiers such as DHB and phthalic acid in titania and zirconia MOC [88]. However, all affinity techniques developed for the current enrichment strategies of CPP gave reproducible but incomplete results due to *poor binding* of low concentrations of CPP and the insufficient *recovery* of multiple phosphorylated peptides [89]. Recently, a specific hydroxyapatite (HA)-based enrichment procedure has been developed for complex mixtures of phosphoprotein/CPP [90]. Salt such as calcium phosphate, also occurring in bone and tooth tissue in the HA form, with the formula $[Ca_{10}(PO_4)_6(OH)_2]$, has been previously used to enrich bone proteins [91]. The phosphate groups of phosphoproteins interact with crystalline lattice Ca^{2+} [92] more strongly than do the carboxyl groups [93]. Moreover, increasing protein phosphorylation leads to tighter binding of the proteins/CPP to HA [92]. One might conclude that the affinity of the multi-phosphorylated proteins/peptides for HA is significantly higher than that of the same components with lower phosphorylation. Essentially, the HA-based protocol immobilizes on HA microgranules proteins/peptides through their phosphate groups, while the non phosphorylated components are washed out using various buffers. In a previous article, CPP immobilized on HA were progressively eluted, increasing phosphate in the elution buffer, and then identified by off-line MALDI-MS [94]. This procedure was accelerated, and loss during elution was minimized by spotting HA-CN/CPP microgranules onto a MALDI target and analyzing the peptides directly by MALDI-TOF [90]. This method was useful for measuring the phosphorylation level of phosphoproteins/CPP quickly, with less than 2 h elapsing from the fractionation of the protein/CPP to the readout of the MALDI spectra (excluding the trypsinolysis step).

The more important advantages of the procedure are the possibility of (1) detecting phosphorylated proteins/peptides even in complex mixtures, (2) determining phosphorylated sites and those dephosphorylated by phosphatase, (3) attaining information regarding weakly and heavily phosphorylated peptides and (4) adding the HA-CPP complex directly to food, which is enabled by the use of an edible resin such HA [90]. Moreover, use of available commercial CPP preparations by the food industry is difficult for three primary reasons: i) the matrix bound to CPP is often not edible and such products can be hazardous; ii) the preparation of CPP is a long and a laborious procedure that requires cumbersome and expensive manipulation; and iii) CPP have an unpleasant taste even in modest amounts, which disadvantageously limits their direct utilization as a human food ingredient. A novel HA-based method for food grade CPP preparation has been performed on tryptic digests of casein. HA captured all CPP free of non-CPP [90]. There were approximately 32 HA bound CPP, and all non-CPP peptides were eluted [90]. HA-based enrichment procedure has been successfully applied to phosphopeptide recovery from complex biological fluids such as human serum thus providing a great source of potential biomarkers of disease. Four primary phosphopeptides derived from fibrinogen were enriched from human serum (Figure 3a-b, Table 4). A similar set of phosphorylated peptides was previously obtained using a modified IMAC strategy coupled to iterative mass spectrometry-based scanning techniques [95], using the titanium ion-immobilized mesoporous silica particles and MALDI-TOF [96] and cerium ion-chelated magnetic silica microspheres [97].

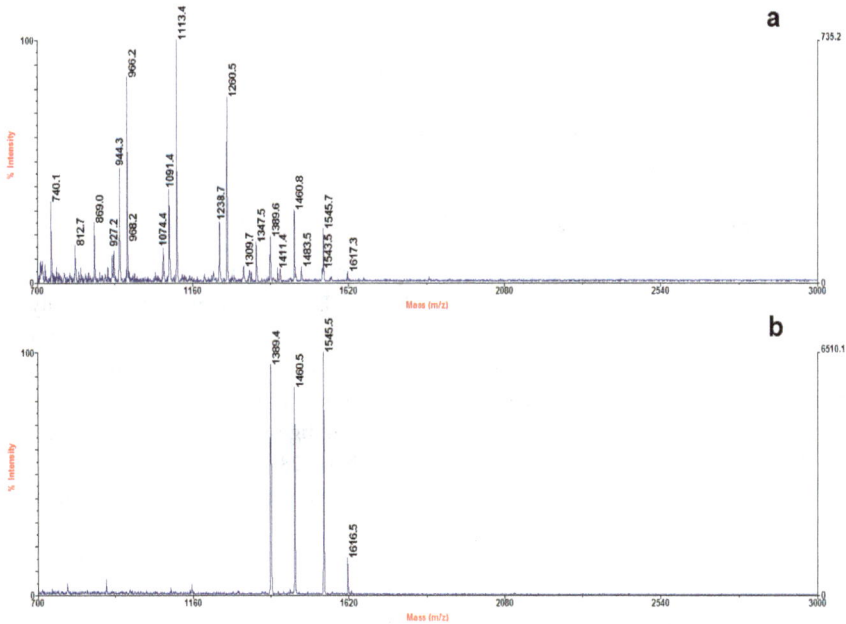

Figure 3. MALDI-MS-TOF spectra for the human serum before (a) and after (b) enrichment by HA (insert is the zoomed between 700 and 3000 Da).

Fibrinopeptide A (FPA) (f1-16)1P, (SerP^3), a 16-residue long peptide (1615 Da) (Table 4), is the segment anchored on the thrombin surface [98]. The other three phosphopeptides, (f1-

15)1P, (f2-15)1P and (f2-16)1P (Table 4), are hydrolytic products of FPA. The serum level of fibrinogen and its hydrolytic products may reflect the expression and activation of enzymes including kinase, phosphatase, and protease [99]. An altered ratio of FPA (f2-15) and FPA (f1-16) is detected in patients affected by hepatocellular carcinoma; the D^2[pS]GEGDFLAEGGGV[15] peptide is upregulated, and the A^1D[pS]GEGDFLAEGGGVR[16] peptide is down-regulated greatly. The other two peptides, A^1D[pS]GEGDFLAEGGGV[15] and D^2[pS]GEGDFLAEGGGVR[16], varied only slightly between the two groups [96]. The proportions of fibrinogen and their phosphorylation products offer new opportunities for basic research in exploring new frontiers in bio-marker discovery.

Molecular Mass (Da)		Fibrinogen α-chain sequence	Phosphorylation sites
Theoretical	Measured MH$^+$		
1388.5	1389.4	D^2SGEGDFLAEGGGV^{15}	1
1459.5	1460.5	A^1DSGEGDFLAEGGGV^{15}	1
1544.6	1545.5	D^2SGEGDFLAEGGGVR^{16}	1
1615.6	1616.5	A^1DSGEGDFLAEGGGVR^{16}	1

Phosphoserine residues are coloured red.

Table 4. Identification of phosphorylated fibrinogen fragments from human serum immobilized on HA.

4.1. CPP in commercial milk as specific indicators of heated milks

Because of the lower value, the addition of UHT and milk powder to raw or pasteurized milk is prohibited (EU Directives 92/46 CEE and 94/71 CEE) for cheese milk. The intensity of heat treatment was found to correlate with the furosine content. Glycated proteins and peptides formed during the initial stages of the Maillard reaction are indirectly evaluated through the furosine content [100-101]. The Amadori compound formed upon the reaction of lysine residue with a lactose molecule will prevent the digestive enzymes from reaching the binding sites. Native and lactosylated forms of β-CN (f1-28)4P, (f1-27)4P and $α_{s2}$-CN (f1-24)4P, although typical of UHT milk and milk powder, are missing in raw, pasteurized milk (71.7 °C for 15 s) and intensely pasteurized milk. The lactosylated peptides that varied with heat treatment characterize UHT milk added in amounts not lower than 10% to raw and pasteurized milk [102]. Milk delactosed with microbial β-galactosidase did not suppress the Maillard reaction; indeed, the furosine concentration increased to 35-400 mg/100 g of protein [103]. As expected, a lactose-reduced UHT milk had β-CN (f1-28)4P glycated mainly by its monosaccharides (Figure 4a).

Therefore, the nonenzymatically glycated CPP derived from the reaction of one molecule of glucose or galactose with a lysine residue (m/z 3641) can be considered to be the signature peptide of lactose-reduced milk (Figure 4).

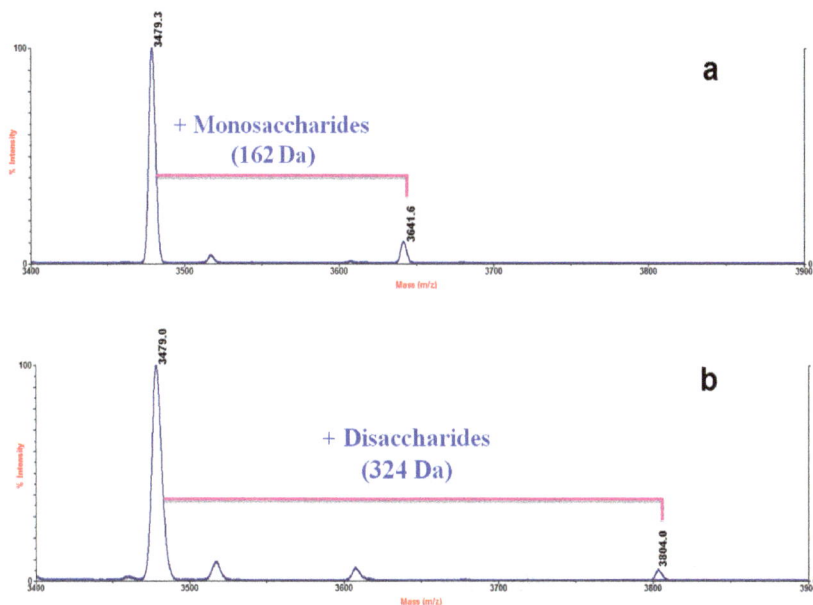

Figure 4. Enlarged view of the MALDI spectrum of β-CN (f1-28)4P (MH$^+$ = 3479 Da) signature glycosylated CPP in lactose-reduced UHT milk (a) and lactosylated CPP in UHT milk (b). The mass differences corresponded to lactose and glucose/galactose residues.

4.2. CPP in yogurt

Yogurt is a fermented milk defined as the "food produced by culturing one or more of the optional dairy ingredients (cream, milk, partially skimmed milk, and skim milk) with a characteristic bacteria culture that contains the lactic acid-producing bacteria, *Lactobacillus delbrueckii* subsp. *bulgaricus* and *Streptococcus thermophilus*". A heat treatment of 90 °C for 10 min is considered optimal to obtain a good quality yogurt [104], and the addition of milk powder increases the content of furosine to more than 300 mg/100 g protein [105]. In yogurt, enzymes could give rise to the liberation of a particularly high number of bioactive peptides, among them CPP, which could be partly due to LAB proteolytic activity. Comparison of CPP in raw, pasteurized and intensely heated milks has previously shown that there is a plethora of milk peptides among which a few were glycosylated CPP [102]. Yogurt is prepared from intensely heated milk instead of low-pasteurized drinking milk. The CPP of two preparations were enriched on HA and analyzed by MALDI-TOF (Table 5). The proportion of CPP with molecular masses between 2.5 and 4 kDa was significantly higher in yogurt than in pasteurized milk as shown in Table 5. Only four fragments of CPP-derived peptides produced during yogurt preparation occur in pasteurized milk (Table 5).

Molecular Mass (Da)		CPP sequence	Yogurt	Pasteurized milk
Measured MH⁺	Theoretical		Relative Intensity	
		α_{s1}-CN derived CPP		
1327.5	1326.4	α_{s1}-CN (f44-54)2P	0.3	n.d.
5009.1	5008.7	α_{s1}-CN (f39-79)6P	0.6	n.d.
5089.3	5088.7	α_{s1}-CN (f39-79)7P	1.2	n.d.
6959.0	6958.7	α_{s1}-CN (f35-90)8P	0.6	n.d.
7087.1	7086.9	α_{s1}-CN (f34-90)8P	0.2	n.d.
		α_{s2}-CN derived CPP		
1493.6	1492.7	α_{s2}-CN (f139-150)1P	1.2	n.d.
1617.5	1616.5	α_{s2}-CN (f1-12)3P	2.0	n.d.
1931.7	1930.8	α_{s2}-CN (f1-15)3P	8.6	n.d.
2007.9	2006.7	α_{s2}-CN (f7-21)4P	0.4	n.d.
2356.2	2355.1	α_{s2}-CN (f1-18)4P	20.8	n.d.
2666.7	2665.4	α_{s2}-CN (f51-72)4P	2.3	n.d.
2748.6	2747.5	α_{s2}-CN (f1-21)4P*	3.0	26.9
2876.8	2875.7	α_{s2}-CN (f1-22)4P	1.8	n.d.
3005.2	3004.8	α_{s2}-CN (f2-24)4P	0.4	n.d.
3134.3	3133.0	α_{s2}-CN (f1-24)4P*	24.2	46.0
3382.0	3381.1	α_{s2}-CN (f49-76)4P	8.4	n.d.
3461.9	3461.1	α_{s2}-CN (f49-76)5P	1.4	n.d.
3666.0	3665.9	α_{s2}-CN (f16-45)2P	0.9	n.d.
4166.4	4165.3	α_{s2}-CN (f115-149)3P*	0.6	n.d.
4294.7	4293.5	α_{s2}-CN (f115-150)3P*	0.6	n.d.
		β-CN derived CPP		
960.5	959.9	β-CN (f30-36)1P	0.8	n.d.
1462.5	1461.4	β-CN (f17-27)3P	1.8	n.d.
1511.6	1510.4	β-CN (f17-28)2P	0.6	n.d.
1515.5	1514.4	β-CN (f15-25)4P	1.3	n.d.
1591.4	1590.4	β-CN (f17-28)3P	19.5	n.d.
1628.9	1628.3	β-CN (f15-26)4P	0.7	n.d.
1645.4	1644.3	β-CN (f14-25)4P	1.1	n.d.
1743.8	1742.5	β-CN (f15-27)4P	16.0	n.d.
1791.4	1790.6	β-CN (f15-28)3P	7.9	n.d.
1871.6	1870.6	β-CN (f15-28)4P	100.0	n.d.
1999.8	1998.7	β-CN (f14-28)4P	0.4	n.d.
2240.4	2239.0	β-CN (f8-25)4P	1.6	n.d.
2709.4	2708.5	β-CN (f7-28)4P	4.1	n.d.
2967.4	2966.7	β-CN (f1-24)4P	0.2	n.d.
3080.5	3079.9	β-CN (f4-28)4P	0.6	n.d.
3123.2	3122.9	β-CN (f1-25)4P*	n.d.	5.8
3351.6	3350.2	β-CN (f1-27)4P	3.2	14.8
3479.9	3478.4	β-CN (f1-28)4P*	70.1	100.0
3607.8	3606.6	β-CN (f1-29)4P*	13.2	6.0
3803.8	3802.2	β-CN (f1-28)4P + 1 lactose	0.7	n.d.
3849.3	3848.8	β-CN (f1-31)4P	0.1	n.d.
3978.2	3977.0	β-CN (f1-32)4P	n.d.	0.5
		κ-CN derived CPP		
6229.8	6228.8	κ-CN (f106-163)1P	0.3	n.d.
6788.5	6787.5	κ-CN (f106-169)1P	0.1	n.d.

n.d. not detected; * CPP detected *in vitro* because it was liberated by plasmin in raw milk and enriched on HA.

Table 5. List of HA-enriched CPP identified in commercial samples of yogurt and pasteurized milk. Relative intensity of each peak is reported.

For example, β-CN (f1-28)4P, a peptide representing 100% intensity (assumed as base peak) of the signals in MALDI spectra, was reduced by approximately 30%; this loss was associated with the transformation of pasteurized milk into yogurt. One can deduce that the original peptide undergoes degradation even considering the higher number of formed CN peptides. In yogurt, β-CN (f1-28)4P, the most common CPP in pasteurized milk, was hydrolyzed into the peptide β-CN (f15-28)4P, which thus becomes the most abundant CPP.

β-CN (f1-29)4P, β-CN (f1-28)4P, β-CN (f1-25)4P, β-CN (f1-24)4P, α_{s2}-CN (f1-24)4P, α_{s2}-CN (f1-21)4P, resulting from the CN hydrolysis by plasmin and enriched on HA, were also found as C-terminally shortened peptides. During fermentation and storage, α_{s2}-CN (f115-150)3P and α_{s2}-CN (f115-149)3P derived by plasmin action did not react further to produce shorter peptides, most likely because of the absence of proteolytic enzymes. In the yogurt fraction recovered by centrifugation, only five multi-phosphorylated α_{s1}-CN and two low-phosphorylated κ-CN, κ-CN (f106-163)1P and κ-CN (f106-169)1P, were identified. Few CPP were less phosphorylated than the native peptides due to the presence of milk phosphatase, which was denatured in all pasteurized cheese-milks. The presence of lactosylated β-CN (f1-28)4P CPP was indicative of yogurt made with high-heat treated or milk fortified with milk protein powder [102]. Proteolysis of milk proteins in model yogurt systems has shown a similar set of primary CPP (Table 5). Therefore, the question is raised how CPP, derived from the enzymatic hydrolysis of yogurt CN are digested and absorbed in adult humans. For this reason, it is important to know the gastrointestinal resistance of CPP if used as a functional ingredient for fruit beverages. In the various stages of human digestion, a large quantity of CPP is produced in the stomach by partial hydrolysis of CN through pepsin action and in the small intestine by trypsin; these peptides are successively refined by endoproteases/exopeptidases. Although analysis of the intestinal contents of milk and yogurt ingestion has revealed the presence of CPP [60], their further resistance to gastrointestinal enzymes is poorly documented. Fragment β-CN (f1-24)4P has been previously identified in the lumen contents of rats after 60 min of digestion as a β-CN (f1-25)4P derived peptide [106]. Moreover, after yogurt ingestion, β-CN (f1-32)4P CPP was released in the human stomach [60] and β-CN (f1-31)4P was found in a yogurt sample. The fragment β-CN (f1-28)4P constitutes a clear example of the multi-functionality of milk-derived peptides because some regions in the primary structure of caseins contain overlapping peptide sequences that exert different biological effects, in this case both mineral binding and immunostimulatory action [107]. Even with the difference in the peptide pattern, it is evident that CPP binding iron (or other metal ions) remains soluble in the digestive tract, where they escape further enzyme digestion [106]. The authors have studied in depth the simulated digestion of CPP from peptide precursors. These studies greatly benefit from the knowledge of enzyme specificities and degradation mechanisms. CPase and chymotryptic activity of pancreatin exhibits broad specificity, cleaving bonds on the carboxyl side of several amino acid residues of CPP. The latter, which are in the mass range 960-7087 Da (Table 5), are good candidates for intestinal absorption and for playing a possible physiological role in mineral bioavailability. However, there are conflicting results on the lack of α_{s1}-CN (f43-52)2P and α_{s2}-CN (f1-19)4P identified by other authors after CN hydrolysis with pancreatin, an enzyme used during the intestinal step of simulated physiological digestion [108]. Generally, the physiological effects of CPP may not always be extended to precursor peptides, although they are structurally similar. There are not enough data concerning the effects of CPP addition to probiotic acid fermented milks. However, probiotic bacteria such as *Lactobacillus acidophilus* and *Bifidobacterium* spp., selected because of their beneficial action, which they may manifest on the health of the consumer, grow slowly in milk because of the lack of the proteolytic activity [109]. For this reason, and also to reduce the fermentation time, probiotic yogurt is manufactured by yogurt bacteria (*Streptococcus thermophilus* and *Lactobacillus delbrueckii* ssp. *bulgaricus*) with the addition of probiotic culture. In parallel, non-digestible food ingredients, *i.e.*, "prebiotics", resist digestion in the small intestine and reach the colon, where they act as a growth factor for *Bifidobacterium* species and are metabolized into short chain fatty acids by a limited number of the microorganisms also comprising the colonic microflora. Prebiotics are principally oligosaccharides (fructo-oligosaccharides, inulins, isomalto-oligosaccharides, lactitol, lactosucrose, lactulose, pyrodextrins, soy oligosaccharides, transgalacto-oligosaccharides, and xylo-oligosaccharides) that stimulate bifidobacteria growth. In probiotic yogurt containing inulin as a prebiotic, the number of bioactive peptides increased, which means that elevation in the proteolytic activity has a synergistic effect with probiotic counts of yogurt cultures. Therefore, the most proteolytic strains of *St. thermophilus* and *Lb. delbrueckii* subsp. *bulgaricus*, spp. enhance the growth of *Lb. acidophilus* and *Bifidobacterium*. In our studies, the overall opiate activity of the bio-yogurt preparation containing *Lb. acidophilus* and *Bifidobacterium* spp. and inulin as a prebiotic was approximately twice that typical of traditional yogurt. In addition, the above yogurt preparation contained a variety of opioid agonistic and antagonistic, immunomodulation, anti-thrombotic, ACE-inhibitor, and anti-microbial activities. Traditional and probiotic yogurt both possess a characteristic soluble fraction composed by peptides exhibiting biological activity, amongst others. This fraction was found to include CPP, β-casomorphins and antithrombotic peptide precursors that did not differ greatly from one another. In a study comparing the proteolytic, amino-, di-, tri- and endopeptidase activity of nine strains of *St. thermophilus*, six strains of *Lb. delbrueckii*, fourteen strains of *Lb. acidophilus* and thirteen strains of *Bifidobacterium* spp., aminopeptidase activity was detected for all bacterial strains – traditional yogurt strains and probiotic bifidobacteria - both at the extracellular and intracellular levels. High dipeptidase activity was demonstrated by all bacterial strains for *Lb. delbrueckii* ssp. *bulgaricus*, *Lb. acidophilus*, and *Bifidobacterium* spp., whereas *St. thermophilus* had greater dipeptidase activity at the extracellular level.

4.3. CPP in a few cheese varieties

Whole milk contains a variety of endogenous plasmin-mediated CN peptides. In addition, CPP were released following cell lysis and release of intracellular LAB enzymes. This phenomenon was observed especially at the end of ripening in long-ripened cheeses, such as Comté [110], Grana Padano [111], Parmigiano-Reggiano and semi-hard Herrgard cheese [112]. In Grana Padano cheese, 45 CPP were identified, of which 24 originated from β-casein, 16 from α_{s1}-casein and 5 from α_{s2}-casein. These CPP formally derive from three parent peptides, namely β-CN (f7-28)4P, α_{s1}-CN (f61-79)4P and α_{s2}-CN (f7-21)4P [111]. By comparing CPP of Grana Padano and Herrgard cheese, it was clear that CPP were all progressively shortened and dephosphorylated during ripening. CPP were very resistant to enzymatic degradation, especially when SerP residue was at the N-terminal end. The number of CPP identified according to the different procedures was comparable for Grana Padano [111] and Herrgard cheese [112]. In both the cheeses, CPP were progressively shortened and dephosphorylated during ripening, with both cheeses constituting heterogeneous mixtures of peptides phosphorylated at various sites sharing N- and C-terminally truncated CPP. However, some peptides proved very resistant to enzymatic degradation, especially when SerP residue was present at the N-terminal end of CPP [111]. The only SerP residue located at the N-terminus of CPP was subjected to dephosphorylation, exposing the dephosphorylated residue to aminopeptidase action. A heterogeneous CPP pattern also differentiated the cheese samples within a given form because of the phosphatase gradient amongst peripheral and central parts of the Grana Padano cheese form ($3 \cdot 10^5$ vs. $3 \cdot 10^2$). This is due to heat sensitivity in the temperature range 57 and 62 °C and the acid pH at which these enzymes are denatured. These data explain the discrepancy in the amount of serine, which varied by as much as 50% of SerP from the periphery towards the center of the cheese form [113]. In contrast, the CPP fraction of Herrgard cheese was more uniform, with the two cheese varieties sharing active plasmin and amino-peptidases from lactic acid bacteria. Because milk pasteurization denatures alkaline phosphatase while it activates plasmin [114], proteolysis in the above cheese is plasmin-dependent. It is therefore likely that CPP of pasteurized milk cheeses are intrinsically more stable than raw milk cheeses [111]. CPP in artisanal PDO ovine Fiore Sardo cheese have been previously reported [115]. Patterns of CPP similar to that observed for bovine cheese indicated that mechanisms of formation and degradation of CPP were similar regardless of the milk species and cheese variety. The dephosphorylation mechanism in Fiore Sardo was different from that found in Grana Padano cheese, most likely because of the use of different rennet types. In PDO Fiore Sardo cheese, no apparent difference in susceptibility to dephosphorylation was found amongst the differently located SerP peptide residues. This resulted in the simultaneous occurrence of partly dephosphorylated peptides, either internally or externally. CPP enrichment by HA, for example of pH 4.6 soluble fractions of hard Parmigiano Reggiano (PR) (30-mo-old), semi-hard, *pasta filata* Provolone del Monaco (PM) (6-mo-old), semi-cooked Asiago d'Allevo (AA) (3-mo-old) and mold-ripened cheese Gorgonzola (GR) (2-mo-old) cheese, has allowed the identification of CPP in high number (Figure 5) which may explain the broad-specificity of the cheese enzymes involved in CN proteolysis. Some CPP were derived from the Lys-X or Arg-X cleavage by plasmin primarily located in the N-terminal region of caseins, such as β-CN (f1-28)4P (Lys[28]-Lys[29]) or β-CN (f1-29)4P (Lys[29]-Ile[30]), α_{s1}-CN (f61-79)5P (Lys[79]-His[80]) and α_{s2}-CN (f1-24)4P (Lys[24]-Asn[25]). The native plasmin-derived CPP were then further hydrolyzed by cheese aminopeptidases and CPase into shorter peptides.

It is likely that ingested cheese carries a concentrated pH 4.6 soluble CPP fraction and a variable number of CPP according to the cheese variety. Above all, the presence and integrity of plasmin-mediated products of CN is a function of the milk, whether raw or pasteurized. Pasteurization reduces the milk plasmin activity only by ~15 percent, whereas plasmin activity increases during milk storage. UHT does not inactivate the plasmin in milk,

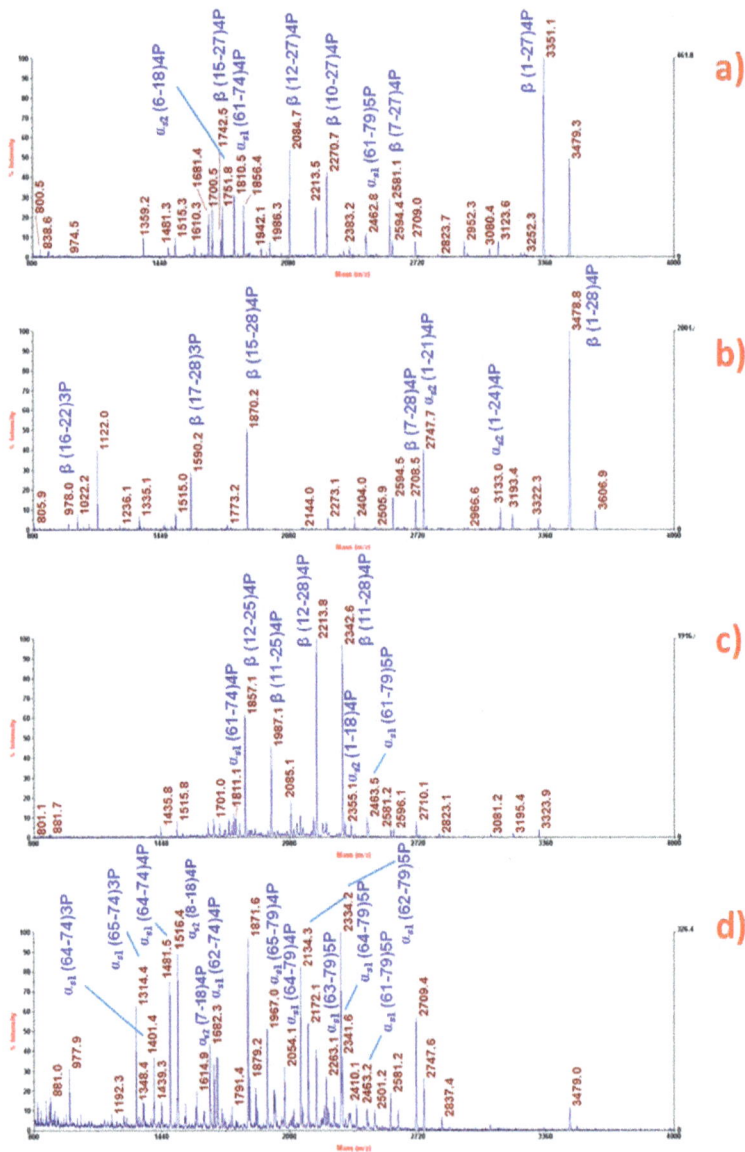

Figure 5. The MALDI spectra of CPP isolated by the addition of HA to pH 4.6 soluble fractions of Gorgonzola (a), Asiago (b), Provolone del Monaco (c), and Parmigiano Reggiano (d) cheeses. The inset magnifies the m/z values in the lower molecular mass peptide range 0.8-4 kDa.

and proteolytic activity will continue to damage milk. Heat treatments modify the peptide profile by increasing the content of larger peptides. The CN breakdown occurring during the ripening of PR cheese proceeded more slowly in PM cheese.. This means that eating PR cheese increases the quota of the co-ingested mineral bound CPP. In contrast, GR cheese show a different CPP level such as that of β-CN (f1-27)4P (3350.2 Da), resulting the most abundant CPP, when compared with hard cheeses (Figure 5a and 6); β-CN (f1-27)4P was further hydrolyzed into the shorter β-CN (f7-27)4P (2580.4 Da), β-CN (f10-27)4P (2270.0 Da), β- CN (f12-27)4P (2083.8 Da) and β-CN (f15-27)4 (1742.4 Da). The most abundant CPP in all three cheeses derived from the peptide β-CN (f1-28)4P, but long-ripened PR cheese was dissimilar from the other cheeses in its content of α_{s1}-CN (f62-79)5P (2332.9 Da) (Figure 5d and 6).

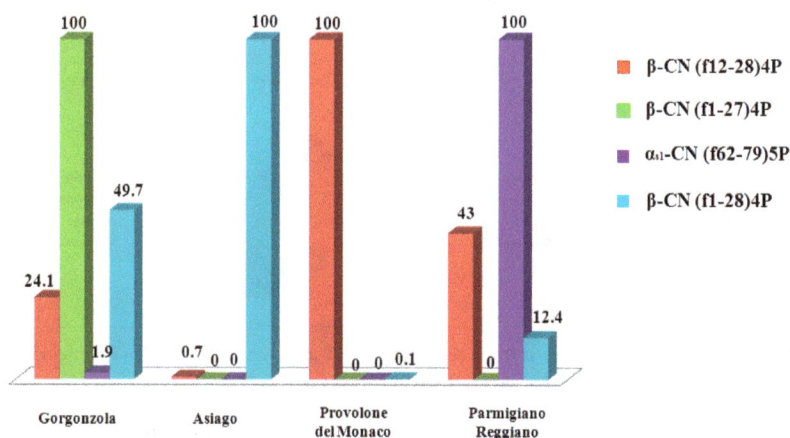

Figure 6. Histogram representation of CPP at 100% relative intensity and their performance in four cheeses.

β-CN (f1-28)4P (3478.4 Da) in AA cheese was the most abundant signal of the MALDI spectra and it was partially hydrolyzed into the shorter peptides β-CN (f7-28)4P (2708.5 Da), β-CN (f15-28)4P (1869.7 Da) and β-CN (f17-28)3P (1589.6 Da) (Figure 5b and 6). Considering exclusively the CPP molecular mass in the 3-3.5 kDa range of AA and GR cheese, the intensity of a high number of peptides transformed into a number of progressively lower molecular weight CPP, with accompanying liberation of peptides (Figure 5). The most common group of CPP occurred in the mass range of 1.7-2.9, reaching the maximum intensity for β-CN (f12-28)4P (2212.0 Da) in PM and α_{s1}-CN (f62-79)5P (2332.9 Da) in PR cheese (Figure 5c-d and 6). The presence of β-CN (f16-22)3P (977.7 Da) was discovered in both AA and PR cheeses and was not detected in the PM and GR cheeses (Figure 5). Our results show that longer plasmin-mediated peptides degraded into shorter CPP. These peptides became more evident when the chymosin retained in the cheese was largely inactivated by cooking the curd at high temperatures (~55 °C). The hydrolysis of CN by chymosin was covered by that of plasmin, which became the principal proteolytic enzyme in the cheese. This phenomenon is particularly evident in PM raw milk cheese for which the plasmin-mediated β-CN (f1-28)4P peptide, representing ~0.1% of the CPP, was almost completely hydrolyzed into the shorter peptides β-CN (f11-28)4P (2341.1 Da), β-CN (f12-28)4P (2212.0 Da), β-CN (f11-25)4P (1985.7 Da) and β-CN (f12-25)4P (1856.6 Da) (Figure 5c). When comparing the PR and PM cheese, the former had a high extent of β-CN (f1-28)4P as judged by the higher levels of the peptide. This demonstrates that CPP of PR cheese are progressively transformed into a number of lower-molecular-weight peptides. In contrast, the quasi-total absence in the PM cheese of β-CN (f1-28)4P and relatively few of the various sizes β-CN-derived CPP (Figure 6) could be the effect of the enzyme decline from the optimum level of activity to zero enzyme activity.

α_{s1}-CN CPP originated for the greater parts from the internal regions of the amino acid sequence, namely α_{s1}-CN (f61-79)5P and α_{s1}-CN (f33-60)3P.

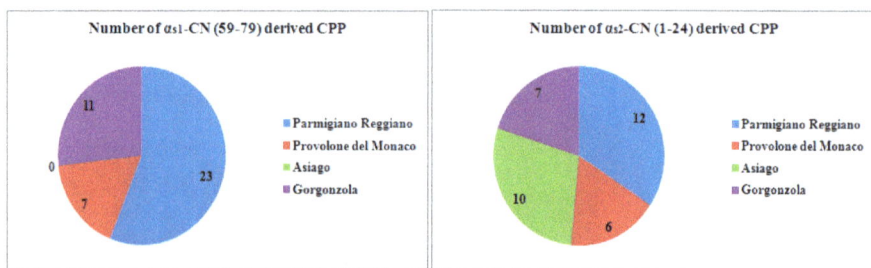

Figure 7. The number of CPP derived from α_{s1}-CN (f59-79)5P and α_{s2}-CN (f1-24)4P in Parmigiano Reggiano, Provolone del Monaco, Asiago and Gorgonzola cheeses.

In PR cheese, 23 casein-derived CPP were found to derive from the internal region of α_{s1}-CN, *i.e.*, α_{s1}-CN (f59-79)5P, whereas they were not detected in AA (Figure 7).

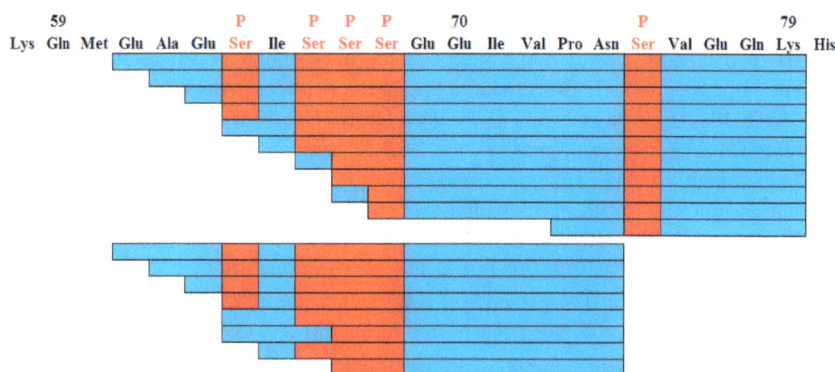

Figure 8. Amino acid sequence of the α_{s1}-CN (f59-79) 4P peptide and the CPP identified in PR cheese. Phosphoserine residues are indicated by red boxes.

The profile of the CPP depicts the mechanisms of both the proteolysis and dephosphorylation in a long ripened cheese. Peptide α_{s1}-CN (f61-79)5P, most likely arising from the parent peptide α_{s1}-CN (f1-79)7P through cleavage at Met^{60}-Glu^{61}, was dephosphorylated and concurrently hydrolyzed into shorter peptides. Alkaline and/or acid phosphatases acting on SerP residue dephosphorylated CPP. N-terminal Ser was then exposed to aminopeptidase and released as a free amino acid. Bacterial CPase or exopeptidase such as cathepsin D/chymosin release α_{s1}-CN (f61-74)4P and other derived CPP through cleavage at Asn^{74}-$SerP^{75}$ (Figure 8). In PR aged 30 months, α_{s1}-CN (f61-79)5P (2462.1 Da), α_{s1}-CN (f62-79)5P (2332.9 Da), α_{s1}-CN (f63-79)5P (2261.9 Da), α_{s1}-CN (f64-79)5P (2132.8 Da), α_{s1}-CN (f64-79)4P (2052.8 Da), α_{s1}-CN (f65-79)4P (1965.7 Da), α_{s1}-CN (f62-74)4P (1681.3 Da), α_{s1}-CN (f64-74)4P (1481.2 Da), α_{s1}-CN (f64-74)3P (1401.2 Da) and α_{s1}-CN (f65-74)3P (1314.1 Da) were the dominant CPP (Figure 5d and 8). Indeed, only 7 CPP for PM cheese and 11 CPP for GR were derived from α_{s1}-CN (f59-79)5P, namely α_{s1}-CN (f61-79)5P (2462.1 Da) and α_{s1}-CN (f61-74)4P (1810.5 Da) (Figure 5a and c and 7). Considering the α_{s2}-CN peptide, the α_{s2}-CN (f1-24)4P CPP were similar in number but significantly different in the case of the four cheese varieties (Figure 6). The dominant α_{s2}-CN-derived CPP were α_{s2}-CN (f1-24)4P (3132.9 Da) and α_{s2}-CN (f1-21)4P (2747.6 Da) in AA cheese (Figure 5b). The most abundant α_{s2}-CN-derived CPP were α_{s2}-CN (f1-18)4P (2355.1 Da) for PM and α_{s2}-CN (f6-18)4P (1751.4 Da), a shortened form of the primary CPP α_{s2}-CN (f1-18)4P, for GR cheese (Figure 5a and c). α_{s2}-CN (f7-18)4P (1614.3 Da) and α_{s2}-CN (f8-18)4P (1515.1 Da) characteristically accumulated in PR cheese (Figure 5d). Similar and discrete phosphorylated CPP derived species for α_{s2}-CN (f1-24) (3P and 4P) and β-CN (f1-28) (3P and 4P) occurred in all cheeses. For the other casein fractions, primary CPP are fully phosphorylated, such as α_{s1}-CN (f61-79)5P, whereas the derived peptides show a level of phosphorylation less than native form as observed in PM and GR. A higher dephosphorylation level characterized the CPP profile of PR cheese (Figure 8). The different profile of CPP could derive from the different length of ripening and from the cheese variety.

CPP in ovine cheeses

Cheeses that contain CPP are also manufactured from ovine milk. Proteolytic enzymes in Pecorino cheese originate from chymosin, pepsin and other clotting preparations such as paste rennet. These enzymatic activities are complemented by those secreted by the vegetative spores of *Penicillium roqueforti* during the maturation of blue-veined cheeses. The CPP patterns of Pecorino and Roquefort cheese have been characterized and main components identified. The sequence alignment of the CPP released throughout the hydrolysis of the β-CN (f1-28)4P in Pecorino and Roquefort cheeses are compared in Figure 9.

In PR cheese of different ages, the released CPP were progressively degraded at C- and N-terminal ends. CPases work from the C-terminal end and aminopeptidases from the N-terminal end, both removing the terminal amino acid residues incrementally. LAB does not produce CPases; thus, the ability to liberate the carboxyterminal amino acid and peptides is typical of the mold. The N-terminal amino acid seemed to be released faster than the C-terminal residues because of the lower activity of CPases. There are negligible differences in the CPP level of different cheese lots primarily because of the action of the enzymes from *P. roqueforti* after its sporulation in blue-veined cheese. More long-chain CPP β-CN, such as β-CN (f1-28)5P-4P, β-CN (f1-27)5P-4P and β-CN (f1-24)5P-4P, were detected in Pecorino cheese, whereas β-CN (f7-28)5P-4P and β-CN (f7-27)5P-4P resulted from the longer CPP in Roquefort cheese (Figure 9). In these cheese varieties, both αs1- and β-CN have been described as completely hydrolyzed at the end of ripening. This contradicts other findings indicating ~50% CN hydrolysis. Plasmin, NSLAB, and Lactobacilli contaminating flora proteinases are mainly responsible for extensive proteolysis in Parmigiano-Reggiano cheese, which is ripened for ~24 months at ~18-20 °C [116]. Here, chymosin is denatured by the high cooking temperature used during the manufacture of cheese. Molds develop at approximately 2 to 5 weeks of ripening, concurrently degrading CN into peptides of various sizes [117]. A similar mechanism for β-CN-derived CPP was found in Grana Padano cheese. Ser was proteolytically cleaved by aminopeptidases, and SerP hindered cleavage by the latter and continued its action after dephosphorylatation of SerP.

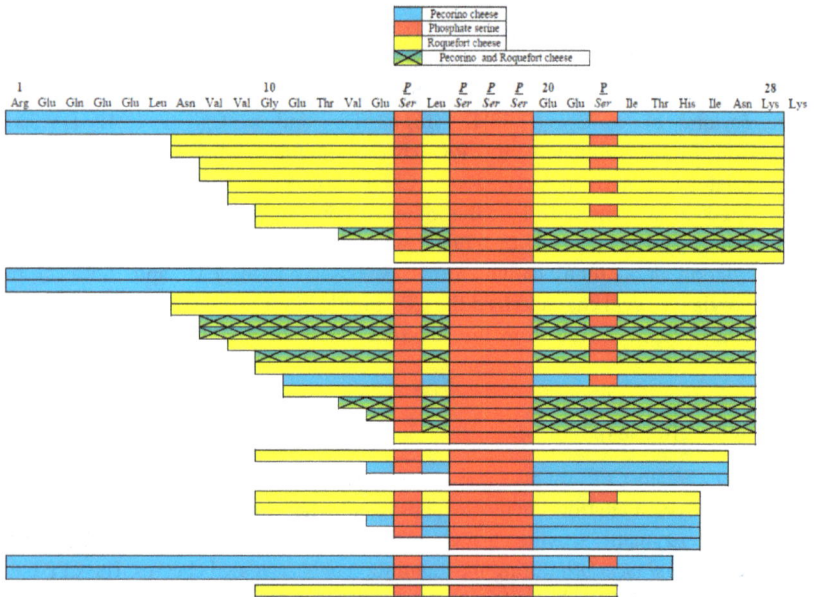

Figure 9. Amino acid sequence of the β-CN (f1-28)4P peptide and the CPP identified in Pecorino (blue line) and Roquefort (yellow line) cheeses. CPP common to the two cheeses are indicated by crosses, and phosphoserine residues are indicated by red boxes.

5. Anticariogenic effect of CPP in yogurt and cheese

Due to proteolytic activity, a great number of CPP are formed in raw milk cheese. In contrast, the enzymatic digestion of proteins to peptides can be reduced by milk pasteurization. Notwithstanding this, yogurt remains a consistent source of bioavailable CPP even if milk is heated to high temperatures (90 °C, 30 min) to create an inoculation medium in which the bacteria can grow and produce lactic acid. LAB provide plenty of CPP in the form of soluble complexes with Ca^{2+} that are effective in avoiding the life-threatening calcium phosphate precipitation, enhancing the intestinal absorption of minerals and retention in the human body [34,118]. The mineral-binding power of CPP also depends on the number of binding sites and their relative accessibility [119]. Dairy products such as cheese and yogurt are both rich in multi-phosphorylated peptides capable of interacting with colloidal calcium phosphate to manifest anticariogenic properties in human and animal [120-121,13]. The mechanism of anticariogenicity might be due to a direct chemical effect of casein and calcium phosphate components [122]. Tooth enamel is a polymeric substance consisting of crystalline calcium phosphate, embedded in a protein matrix. Thus, CPP can significantly enhance localization of ACP at the tooth surface, inhibiting enamel demineralization and promoting remineralization. In the development of teeth and bone, CPP act as hydroxyapatite nucleator and control the growth of the crystals, resulting in unique crystal morphology. A new calcium phosphate remineralization technology has been recently developed based on the complex CPP-ACP [Recaldent™ CASRN691364-49-5] [66]. This preparation is claimed to stabilize calcium and phosphate ions in high concentrations by binding ACP to pellicle and plaque of the tooth surface. Moreover, CPP-ACP inhibited the adhesion of *Streptococcus mutans* to the tooth surface producing a copious reservoir of bioavailable calcium ions [67]. Cheese and yogurt CPP have the ability to stabilize calcium phosphate in solution, forming small CPP-ACP nanocomplexes. The calcium-binding ability of CPP has been applied by clinical dentists to show that CPP stabilize high concentrations of calcium, phosphate, and fluoride ions on the tooth surface by binding them to pellicle and plaque [66,123]. In dental plaque, CPP-ACP binds onto the surface components of the intercellular plaque matrix. Incorporation of CPP-ACP into the plaque will increase the calcium and phosphate content by forming a stable supersaturated solution of calcium phosphate. Thus, the availability of calcium in plaque provides a natural anticaries protective effect, either by suppression of demineralization promoted by fermentative acids in mouth, through an increased remineralization by binding calcium ions to teeth enamel, or possibly a combination of both. An inverse relation between plaque calcium and caries incidence has been evidenced [124-125]. Reynolds (1997) [126] has demonstrated that CPP-ACP can actually remineralize subsurface lesions in human enamel, and this is indeed the basis of one claim of his patent [127]. The diffusion of available CPP-ACP in the mouth is controlled by two main factors: i) the molecular weight of the diffusing species, the square of which is inversely proportional to the diffusion coefficient; and ii) the binding characteristics of the diffusing species, which dictate how much CPP-ACP is free to diffuse at a given time [128]. At neutral pH, calcium diffusion is limited by the quantity of bound calcium, reducing the effective diffusion coefficient (De) and creating a measurable restricted effective diffusion coefficient (rDe), where $rDe = De/(R + 1)$ and R is the ratio of bound to free calcium. A large number of potential binding sites for calcium can have significant effects on the calcium diffusion coefficient; this effect is maintained at low pH, although overall diffusion is slightly faster [129-130]. Conversely, one may infer that ACP may also bind to dental plaque and tooth enamel, thus having beneficial effects on teeth remineralization [67]. In addition, SerP residues of the CCP-HA complex are exposed to intestinal alkaline phosphatase, which favors metal ion bioavailability by releasing inorganic phosphate. Milk, ice cream, and cheese have been observed to lower the incidence of dental caries in rats [131]. Elderly people that eat cheese several times per week had a lower incidence of root surface caries development [132]. CPP of yogurt have been observed to have an inhibitory effect on demineralization and are able to promote remineralization of dental enamel [133]. Moreover, anticariogenic activity has also been reported for egg phosphopeptides (viz. phosvitin and phosphophorin) [134-135]. The heating process affects the bioavailability of CPP; for example, milk sterilization can induce dephosphorylation of phosphoseryl residues and dehydroalanine residue formation [136].

6. CPP as nutraceutical ingredients in functional foods

6.1. Examples of commercial CPP preparation

Some commercial products have been developed containing moderately hydrolyzed milk proteins as the sole protein source. Hypoallergenic formulas are based on partially or extensively hydrolyzed proteins. Both formulae are better tolerated by small premature infants than native cow milk protein. Ultrafiltered micellar casein and microfiltered whey protein concentrate are known to slow down the digestive process. Typical composition data of commercial phosphopeptide preparations (m/m) include 91.3% (TN x 6.47) or 94.8 (dry basis) protein, 16.0% CPP, and 6.0% free amino acids. Compared with the expected CPP composition, the commercial preparation contained ~5% undigested protein, ~30% peptides in the intermediate molecular mass 5000-20000 Da range, and ~48% of molecular mass in the 500-5000 Da range (Tatua, New Zealand). The preparation is generally very complex and dependent on the procedure used to perform casein hydrolysis. Therefore, commercial products can be considered to be an enriched-CCP preparation containing 16% CPP without specification of the peptide size and phosphorylation degree. Preliminary analysis performed by MALDI-TOF analysis indicated that the signal in the mass spectra originated exclusively from the β-CN digestion. The intensity of non-CPP such as β-CN (f191-209), β-CN (f184-209), β-CN (f177-209) and β-CN (f170-209) was sufficiently high to obscure other CN peptides (MALDI spectrum not shown). As reported above, a number of available techniques allow the separation of CPP and non-CPP. To reduce the large dynamic range of non-CPP, HA was used for CPP enrichment [90,94,102]. CPP included multiply phosphorylated peptides (up to 4 phosphorylation residues). CPP β-CN (f1-28)4P and 3P, β-CN (f1-25)4P, 3P and 2P and β-CN (f2-25)4P, 3P and 2P were found and may be indicative of the progressive CPP dephosphorylation (MALDI spectrum not shown). Interestingly, in addition to β-CN, the commercial CPP (Tatua, NZ) also displayed α_{s1}- or α_{s2}-CN-derived CPP, invisible or weakly visible before the sample was treated with HA (Table 6).

Finally, a method for reducing the complexity of peptide mixtures was the separation of non-CPP and CPP by trapping CPP on HA under neutral conditions. This could be the principle for an industrially based production of CPP.

Parent Protein	Molecular Mass (Da) Measured	Theoretical	Start	End	Peptide Sequence	Peptide Modifications
α_{s1}-CN	1138.4	1138.4	115	123	(N)SAEERLHSM(K)	1P
	1526.7	1526.7	35	47	(K)EKVNELSKDIGSE(S)	1P
	1831.9	1831.8	75	89	(N)SVEQKHIQKEDVPSE(R)	1P
	1859.9	1859.9	75	89	(N)SVEKHIQKEDVPSER(Y)*	1P
	1926.5	1926.7	43	58	(K)DIGSESTEDQAMEDIK(Q)	2P
	1950.9	1950.9	104	119	(K)YKVPQLEIVPNSAEER(L)	1P
	1988.0	1987.9	75	90	(N)SVEQKHIQKEDVPSER(Y)	1P
	2054.7	2054.7	43	59	(K)DIGSESTEDQAMEDIKQ(M)	2P
	2185.8	2185.8	43	60	(K)DIGSESTEDQAMEDIKQM(E)	2P
	2400.9	2400.9	41	60	(L)SKDIGSESTEDQAMEDIKQM(E)	2P
	2597.0	2597.0	37	58	(K)VNELSKDIGSESTEDQAMEDIK(Q)	2P
	2856.1	2856.1	37	60	(K)VNELSKDIGSESTEDQAMEDIKQM(E)	2P
	3113.2	3113.3	35	60	(K)EKVNELSKDIGSESTEDQAMEDIKQM(E)	2P
	3193.2	3193.3	35	60	(K)EKVNELSKDIGSESTEDQAMEDIKQM(E)	3P
	3699.6	3699.5	61	90	(M)EAESISSSEEIVPNSVEQKHIQKEDVPSER(Y)	4P
	3779.6	3779.5	61	90	(M)EAESISSSEEIVPNSVEQKHIQKEDVPSER(Y)	5P
	3974.5	3974.6	59	90	(K)QMEAESISSSEEIVPNSVEQKHIQKEDVPSER(Y)	4P; Oxidation (Met)
	4479.0	4479.0	23	60	(R)FFVAPFPEVFGKEKVNELSKDIGSESTEDQAMEDIKQM(E)	2P
α_{s2}-CN	1432.6	1432.6	135	146	(N)SKKTVDMESTEV(F)	1P
	1538.6	1538.6	126	137	(R)EQLSTSEENSKK(T)	2P
	1578.6	1578.6	123	134	(T)LNREQLSTSEEN(S)	2P
	1694.7	1694.7	125	137	(N)REQLSTSEENSKK(T)	2P
	1793.7	1793.7	123	136	(T)LNREQLSTSEENSK(K)	2P
	1921.8	1921.8	123	137	(T)LNREQLSTSEENSKK(T)	2P
	2715.3	2715.3	115	137	(R)NAVPITPTLNREQLSTSEENSKK(T)	2P
	3051.3	3051.2	1	24	KNTMEHVSSSEESIISQETYKQEK(N)	3P
	3131.2	3131.2	1	24	KNTMEHVSSSEESIISQETYKQEK(N)	4P
	3786.6	3786.6	115	146	(R)NAVPITPTLNREQLSTSEENSKKTVDMESTEV(F)	3P
	4034.6	4034.8	115	148	(R)NAVPITPTLNREQLSTSEENSKKTVDMESTEVFT(K)	3P
β-CN	1600.7	1600.7	29	40	(K)IEKFQSEEQQQ(T)	1P
	1785.7	1785.7	35	48	(Q)SEEQQQTEDELQDK(I)	1P
	1945.8	1945.8	29	43	(K)IEKFQSEEQQQTED(E)	1P
	2559.2	2559.1	29	48	(K)IEKFQSEEQQQTEDELQDK(I)	1P
	2805.2	2805.2	1	24	RELEELNVPGEIVESLSSSEESIT(R)	2P
	2885.2	2885.2	1	24	RELEELNVPGEIVESLSSSEESIT(R)	3P
	2906.3	2906.3	29	51	(K)IEKFQSEEQQQTEDELQDKIHP(F)	1P
	2961.3	2961.3	1	25	RELEELNVPGEIVESLSSSEESITR(I)	2P
	2965.3	2965.2	1	24	RELEELNVPGEIVESLSSSEESIT(R)	4P
	3041.2	3041.3	1	25	RELEELNVPGEIVESLSSSEESITR(I)	3P
	3053.4	3053.4	29	52	(K)IEKFQSEEQQQTEDELQDKIHPF(A)	1P
	3121.1	3121.3	1	25	RELEELNVPGEIVESLSSSEESITR(I)	4P
	3396.5	3396.5	1	28	RELEELNVPGEIVESLSSSEESITRINK(K)	3P
	3476.5	3476.5	1	28	RELEELNVPGEIVESLSSSEESITRINK(K)	4P

* indicates a CPP containing the alternative non-allelic deletion of Gln[78] [137]; Phosphoserine residues are coloured red.

Table 6. CPP identified in a commercial CPP preparation (Tatua Co-operative Dairy Company Ltd) by tandem MS sequencing.

6.2. Examples of industrial methods for CPP preparation

Starting from CN, the overall preparation process gave 16% CPP (Figure 10), less than the theoretical yield of 23% (Y_{theor} =22.8%), which means approximately 20% yield on the basis of weight, a yield higher than that obtained by other researchers [138-140]. In their production experiments, the authors obtained a CPP preparation as high as 18.8% degree of hydrolysis (DH).

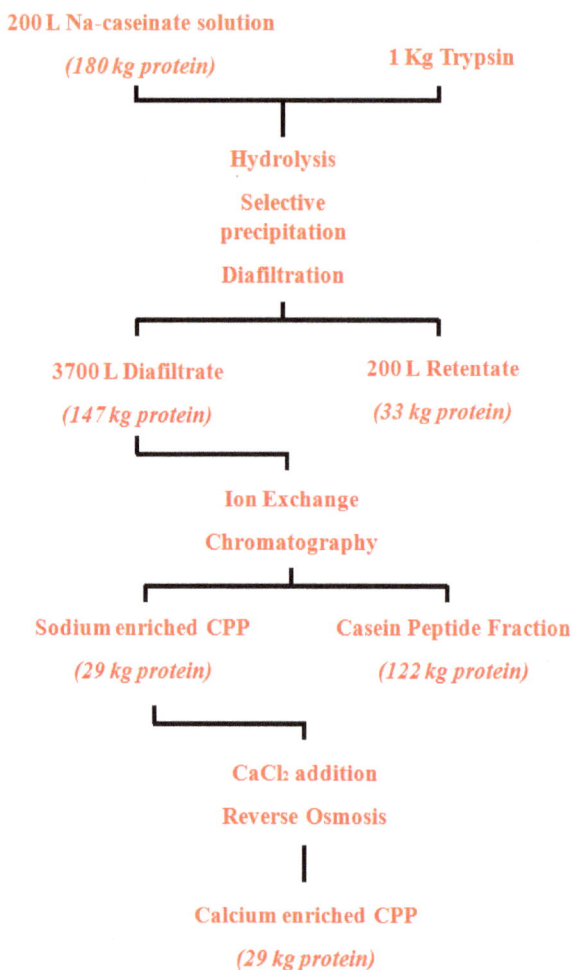

Figure 10. Schematic representation of the process-scale isolation of tryptic CPP from caseinate, showing the protein flow through the process (source: [141]).

The example of 2000 L Na-caseinate solution containing 180 kg protein and 1 kg trypsin yielded 29 kg of calcium-enriched CPP, corresponding to a yield of ~16% (w/w) [141]. A variety of raw materials such as acid casein, sodium caseinate, and calcium caseinate, as well as skimmed raw milk, milk concentrate by ultrafiltration, pasteurized and UHT milk near the limit date for consumption, may be used as the substrate for CPP production. The optimal parameters for the hydrolysis with trypsin were 37 °C and pH between 7.5 and 8.5. The casein hydrolysate solution was roughly fractioned by ultrafiltration with appropriate membranes to obtain soluble CPP both in the "permeate" and "retentate". Lower-molecular peptides/phosphopeptides occurred in the "permeate" and larger in the "retentate". The diafiltrate containing the tryptic casein digest was loaded on the ion exchange resin, and the non-CPP flowed through and were recovered for further use, *e.g.*, as substrate for bacterial culture. Bound CPP, free of non-CPP, are eluted using sodium hydroxide, *e.g.*, 0.2 M, and conductivity and absorbance monitored at 280 nm. The eluate containing CPP is collected and then concentrated, typically by reverse osmosis. Peptides were pasteurized (85 °C, 15 s) and spray dried, yielding sodium enriched CPP (Na-CPP). In other cases, to obtain calcium-enriched CPP, the concentrate is added with CaCl$_2$ in excess, diafiltered, and then CPP solution concentrated by reverse osmosis until the filtrate conductivity was negligible (less than 3 mS cm^{-1}). The concentrate is spray dried, and the product is labeled as calcium-enriched CPP (Ca-CPP).

6.3. Traditional and new processes for the use of CPP in alimentary products

A Ca^{2+}/ethanol selective precipitation procedure was used to produce a CPP and non-CPP concurrently from an alcalase digest of whole casein in which the traditional and new processes for CPP production were reported [142]. CN is trypsinized, and the pH of the solution is adjusted to 4.6 to separate the non-peptide material. The CPP-Ca^{2+} aggregation was induced by ethanol addition in the supernatant and recovered as precipitate for freeze drying. In the novel process, the step of non-peptide material removal was omitted, and non-CPP (CNPP) was recovered as supernatant for use in alimentary products. For casein, the use of alcalase, a cheap enzyme suitable for industrial application, for hydrolysis was suggested [142]. The CN hydrolysates were separated into the two types of peptides using combined treatment with CaCl$_2$ and ethanol. CPP and non-CPP comprised components with molecular weight lower than 2509 Da and 2254 Da, respectively, as determined using size exclusion HPLC. A DH of 20% for the CN hydrolysate was achieved. At the end, the recovery of CPP reached 24%. The phosphorus component of CPP was 3.08%, and nitrogen recovery was approximately 76% [142]. CPP generally had an improved solubility and transparency even under acid conditions and could be used as ingredient for beverages such as sport drinks, soft drinks, health drinks, fermented products, vitamin concentrates, fruit or fruit fractions.

6.4. Patented methods for CPP production as ingredients for alimentary products

A method for the preparation of selected anticariogenic CPP comprised the steps of complete digestion of CN as soluble monovalent cation salt with a proteolytic enzyme: the addition of a mineral acid to the solution to adjust the pH to approximately 4.7; the removal of any produced precipitate; the addition of CaCl$_2$ to a concentration of approximately 1.0% (w/v) to cause the aggregation of CPP; the separation of the aggregated CPP from the solution through a filter with a molecular weight exclusion limit within the range 10000 to

20000 while passing the bulk of the remaining CPP in solution; the diafiltration of the separated CPP with water through a filter; the concentration of the solution; and the drying the retentate [143]. Peptides not included in the aggregation were removed by ultrafiltration/diafiltration. By this means, anticariogenic CPP at purity greater than 90% were obtained [143]. CPP including calcium, magnesium or both salts (or zinc, ferric or other salts) are produced by submitting CN to proteolytic enzyme hydrolysis, ultrafiltering the resulting hydrolyzate to produce a permeate containing CPP, adding a bivalent cation salt to the peptides to form CPP aggregates, and separating by ultrafiltration the CPP aggregates and non-CPP [144]. When CPP salts need to be converted to free phosphopeptides, they can be restored by acidification with HCl; the solution is then diafiltered extensively through a 1000 molecular weight cut-off membrane to remove excess calcium chloride.

6.5. Example of CPP applications as nutraceuticals in functional foods

The CPP-salts complex can be added to different foods. A stable acidic beverage or other alimentary products can be obtained by digesting casein with trypsin, precipitating the insoluble components at acid pH, adjusting the pH of the obtained supernatant to approximately 6.0, then adding calcium chloride and ethanol to recover an acid-soluble calcium complex of CPP and the reaction product and adding them to a soft drink [145]. The acid-soluble calcium complex of CPP enhances calcium absorption from food because calcium may be absorbed by the body in the form of soluble calcium. Acid-soluble CPP is a mixture of α_{s1}-, α_{s2}- and β-CPP, forming essentially no turbidity in solution at pH of 3.0 or less, and having purity greater than 90%, molecular weights between approximately 2500 and 4600 Da, and the ability to solubilize at least 100 ppm calcium at a concentration of 0.5 mg/ml CPP. The acid-soluble CPP produced *in vitro* also have a solubilizing capability on iron. It is widely believed that iron must be solubilized for absorption through the small intestine. Accordingly, health may also be enhanced by the absorption of the soluble iron in the drink by the human body. Similarly, magnesium may be solubilized in a drink or edible product of this type [145]. Therefore, in addition to drinks, preparation of CPP with high negative charge could be used as additive for healthy foods or for dietetic or pharmaceutical compositions, as they are capable of increasing the *in vivo* absorption of calcium or other ions [146]. Another interesting invention relates to processes and the compositions that are useful to remineralize the teeth of mammals, particularly humans, and impart acid resistance thereto. These compositions included a gum base or carrier, sweetening agents, CPP-ACP preparation and food-grade acids [147]. Because many chewing gum and confectionery products usually contain acids, many consumers enjoying chewing gum and confectionery products ingest acids causing demineralization of the tooth surface. CPP-calcium phosphate complexes are known to have anticariogenic teeth strengthening effects that could be used address the problem of dissolution or demineralization of tooth enamel and the resultant formation of dental caries. Exogenous CPP-ACP preparations have also been added to milk in 2.0-5.0 g/L amounts to remineralize enamel subsurface lesions, which actually increased with respect to the control [148].

These inventions create functional foods with undoubted beneficial effects on human health, possibly promoting recalcification of bones, protecting the tooth enamel from decay and other possible health benefits.

7. Conclusion

Proteins are no longer considered merely nutritional components because they possess encrypted peptides with possible biological properties [149-153]. The cited literature has highlighted that bioactive peptides may be released *in vitro* or *in vivo* by digestive and bacterial enzymes starting from casein or generally inactive precursors [151,154-155]. We have examined the case of casein-derived phosphopeptides that can be applied as dietary supplements in "functional foods" and produced on an industrial scale. With this, it has been demonstrated that, each time a health-enhancing nutraceutical is required for a functional food, an appropriate enzymatic hydrolysis of casein needs to be designed [156-157]. In addition to the anticariogenic activity, the most important function displayed by CPP is that soluble and encrypted casein phosphopeptides can arrive without modification of the phosphorylated sequence to the brush border membrane. It is postulated that short-sequence peptides reach their putative receptors in many tissues without modification. Whether they enter blood circulation or whether their action is restricted to a peripheral circle are current questions that await response. In conclusion, this review has considered milk and cheese CPP for specific food ingredients. The contribution of *in vitro* casein digests by gastric proteases to potential biologically active substances in the intestine must be simultaneously considered. Scientists are currently involved in investigations to define the *in vivo* fate of all of the bioactive peptides. The *in vitro* studies have allowed the scientists to compare the predicted and the experimental sequence of CPP. This step is preliminary to the clinical investigations designed to determine the bioactivity of the milk hydrolysates. These findings open an industrial perspective that will permit the unrestricted use of CPP in healthy promoting food application.

Author details

Gabriella Pinto, Marina Cuollo, Sergio Lilla, Lina Chianese and Francesco Addeo
Department of Food Science, University of Naples "Federico II", Parco Gussone, Portici (Naples), Italy

Simonetta Caira
Food Science Institute of the National Research Council (C.N.R.), Via Roma, Avellino, Italy

Acknowledgments

Publication in partial fulfilment of the requirements for PhD in Sciences and Technologies of Food Productions - XXV Cycle, University of Naples 'Federico II'. The Authors gratefully acknowledge the American Journal Experts Association for the text revision (http://www.journalexperts.com/). This work was partly supported by the financial aid to C.L. from MIUR, Program PRIN-2008 HNHAT7-004.

8. References

[1] Lambert JD, Hong J, Yang GY, Liao J, Yang CS. Inhibition of Carcinogenesis by Polyphenols: Evidence from Laboratory Investigations. American Society for Clinical Nutrition 2005;81(1) 284S-291S.

[2] Aggarwal BA, Shishodia S. Molecular Targets of Dietary Agents for Prevention and Therapy of Cancer. Biochemical Pharmacology 2006;71(10) 1397-1421.

[3] Chang DS. Preparations for Sustained Release of Nutraceuticals and Methods of Controllably Releasing Nutraceuticals. US Patent 07115283.

[4] Cartwright R, Hendricks LE. Nutraceutical Composition. US Patent 20030091552.

[5] Torres-Giner S, Martinez-Abad A, Ocio MJ, Lagaron JM. Stabilization of a Nutraceutical Omega-3 Fatty Acid by Encapsulation in Ultrathin Electrosprayed Zein Prolamine. Journal of Food Science 2010;75(6) N69-79.

[6] Mora-Gutierrez A, Gurin MH. Bioactive Complexes Compositions and Methods of Use Thereof. US Patent 7780873.

[7] Policker Y. Nutraceutical Sweetener Composition. US Patent 2010/0068347.

[8] Schanbacher FL, Talhouk RS, Murray FA, Gherman LI, Willett LB. Milk-Borne Bioactive Peptides. International Dairy Journal 1998;8(5) 393-403.

[9] Husson SJ, Landuyt B, Nys T, Baggerman G, Boonen K, Clynen E, Lindemans M, Janssen T, Schoofs L. Comparative Peptidomics of Caenorhabditis Elegans versus C. Briggsae by LC-MALDI-TOF MS. Peptides 2009;30(3) 449-457.

[10] Fosset S, Tomé D. Nutraceuticals from Milk. In: Roginski H, Fuquay JW, Fox PF. (ed.) Enciclopedia Dairy Science. London: Accasemy Press; 2003. p2108-2112.

[11] Silva SV, Malcata FX. Caseins as Source of Bioactive Peptides. International Dairy Journal 2005;15(1) 1-15.

[12] Recio Sanchez I. Bioactive Peptides Identified in Enzymatic Hydrolyzates of Milk Caseins and Method of Obtaining Same. US 2010/0048464.

[13] FitzGerald RJ. Potential Uses of Caseinophosphoeptides. International Dairy Journal 1998;8(5-6) 451-457.

[14] Luo SJ, Wong LL. Oral Care Confections and Method of Using. US Patent 6733818.

[15] Reynolds EC, Cai F, Shen P, Walker GD. Retention in Plaque and Remineralization of Enamel Lesions by Various Forms of Calcium in a Mouthrinse or Sugar-free Chewing Gum. Journal of Dental Research 2003;82(3) 206-211.

[16] Korhonen H, Pihlanto A. Bioactive Peptides: Production and Functionality. International Dairy Journal 2006;16(9) 945-960.

[17] Law J, Haandrikman A. Proteolytic Enzymes of Lactic Acid Bacteria. International Dairy Journal 1997;7(1) 1-11.

[18] Gatti M, Fornasari ME, Mucchetti G, Addeo F, Neviani E. Presence of Peptidase Activities in Different Varieties of Cheese. Letters in Applied Microbiology 1999;28(5) 368-372.

[19] Léonila J, Gagnaire V, Mollé D, Pezennec S, Bouhallab S. Application of Chromatography and Mass Spectrometry to the Characterization of Food Proteins and Derived Peptides. Journal of Chromatography A 2000;881(1-2) 1-21.

[20] Christensen JE, Dudley EG, Pederson JA, Steele JL. Peptidases and Amino Acid Catabolism in Lactic Acid Bacteria. Antonie Van Leeuwenhoek 1999;76(1-4) 217-246.

[21] Kunji ERS, Mierau I, Hagting A, Poolman B, Konings WN. The Proteolytic Systems of Lactic Acid Bacteria. Antonie Van Leeuwenhoek 1996;70(2-4) 187-221.

[22] Doeven MK, Kok J, Poolman B. Specificity and Selectivity Determinants of Peptide Transport in Lactococcus Lactis and Other Microorganisms. Molecular Microbiology 2005;57(3) 640-649.

[23] Fernandez-Espla MD, Rul F. PepS from Streptococcus thermophilus. A New Member of the Aminopeptidase T Family of Thermophilic Bacteria. European Journal of Biochemistry 1999;263(2) 502-510.

[24] Gagnaire V, Piot M, Camier B, Vissers JP, Jan G, Léonil J. Survey of Bacterial Proteins Released in Cheese: a Proteomic Approach. International Journal of Food Microbiology 2004;94(2) 185-201.

[25] Saz JM, Marina ML. Application of Micro- and Nano-HPLC to the Determination and Characterization of Bioactive and Biomarker Peptides. Journal of Separation Science 2008;31(3) 446-458.

[26] Miclo L, Roux É, Genay M, Brusseaux É, Poirson C, Jameh N, Perrin C, Dary A. Variability of Hydrolysis of β-, αs1-, and αs2-Caseins by 10 Strains of Streptococcus thermophilus and Resulting Bioactive Peptides. Journal of Agricultural and Food Chemistry 2012;60(2) 554-565.

[27] Pihlanto-Leppälä A, Rokka T, Korhonen H. Angiotensin I Converting Enzyme Inhibitory Peptides Derived from Bovine Milk Proteins. International Dairy Journal 1998;8(4) 325-331.

[28] Pélissier JP. Protéolyse des Caséines. Sciences des Aliments 1984;4(1) 1-35.

[29] Teschemacher H, Koch G, Brantl V. Milk Protein-Derived Opioid Receptor Ligands. Biopolymers 1997;43(2) 99-117.

[30] Meisel H. Chemical Characterization and Opioid Activity of an Exorphin Isolated from in Vivo Digests of Casein. FEBS Letters 1986;196(2) 223-227.

[31] Ondetti MA, Cushman DW. Enzymes of the Renin-Angiotensin System and their Inhibitors. Annual Review of Biochemistry 1982;51 283-308

[32] Bruneval P, Hinglais N, Alhenc-Gelas F. Angiotensin I Converting Enzyme in Human Intestine and Kidney. Ultrastructural Immunohistochemical Localization. Histochemistry 1986;85(1) 73-80.

[33] Meisel H. Casokinins as Bioactive Peptides in the Primary Structure of Casein. In Schwenke KD, Mothes R. (ed.) Food Proteins: Functionality. New York: VCH-Weinheim; 1993. p67-75.

[34] Sato R, Noguchi T, Naito H. Casein phosphopeptide (CPP) Enhances Calcium Absorption from the Ligated Segment of Rat Small Intestine. Journal of Nutritional Science and Vitaminology 1986;32(1) 67-76.

[35] Reynolds EC. Remineralization of Enamel Subsurface Lesions by Casein Phosphopeptide-Stabilized Calcium Phosphate Solutions. Journal of Dental Research 1997;76(9) 1587-1595.

[36] Meisel H, Olieman C. Estimation of Calcium binding Constants of Casein Phosphopeptides by Capillary Zone Electrophoresis. Analytica Chimica Acta 1998;372(1-2) 291-297.

[37] Hala I, Higashiyama S, Otani H. Identification of a Phosphopeptide in Bovine α_{s1}-Casein Digest as a Factor Influencing Proliferation and Immunoglobulin Production in Lymphocyte Cultures. Journal of Dairy Research 1998;65(4) 569-578.

[38] Hala I, Ueda J, Otani H. Immunostimulatory Action of a Commercially Available Casein Phosphopeptide Preparation, CPP-III. in Cell Cultures. Milchwissenschaft 1999; 54(1) 3-7.

[39] Loukas S, Varoucha D, Zioudrou C, Streaty R, Klee WA. Opioid Activities and Structures of alpha-Casein-Derived Exorphins. Biochemistry 1983;22(19) 4567-4573.

[40] Loukas S, Panetsos F, Donga E, Zioudrou C. Selective δ-Antagonist Peptides, Analogs of α-Casein Exorphin, as Probes for the Opioid Receptor In: Nyberg F. Brantl V. (ed.) β-Casomorphins and Related Peptides. Uppsala: Fyris-Tryck AB; 1990. p143-149.

[41] Pihlanto-Leppälä A, Antila P, Mäntsälä P, Hellman J. Opioid Peptides Produced by in Vitro Proteolysis of Bovine Caseins. International Dairy Journal 1994;4(4) 291-301.

[42] Miquel E, Gómez JÁ, Alegría A, Barberá R, Farré R Recio I. Identification of Casein Phosphopeptides after Simulated Gastrointestinal Digestion by Tandem Mass Spectrometry. European Food Research and Technology 2006;222(1-2) 48-53.

[43] Picariello G, Ferranti P, Fierro O, Mamone G, Caira S, Di Luccia A, Monica S, Addeo F. Peptides Surviving the Simulated Gastrointestinal Digestion of Milk Proteins: Biological and Toxicological Implications. Journal of Chromatography B. Analytical Technologies in the Biomedical and Life Sciences 2010;878(3-4) 295-308.

[44] Farrell HM Jr, Jimenez-Flores R, Bleck GT, Brown EM, Butler JE, Creamer LK, Hicks CL, Hollar CM, Ng-Kwai-Hang KF, Swaisgood HE. Nomenclature of the Proteins of Cows' Milk—Sixth Revision. Journal of Dairy Science 2004;87(6) 1641-1674.

[45] Kenny AJ, Maroux S. Topology of Microvillar Membrane Hydrolases of Kidney and Intestine. Physiological Reviews 1982;62(1) 91-128.

[46] Yoshioka M, Erickson RH, Woodley JF, Gulli R, Guan D, Kim YS. Role of Rat Intestinal Brush-Border Membrane Angiotensin-Converting Enzyme in Dietary Protein Digestion. American Journal of Physiology-Gastrointestinal and Liver Physiology 1987;253(6 Pt 1) G781-G786.

[47] Erickson RH, Song IS, Yoshioka M, Gulli R, Miura S, Kim YS. Identification of Proline-Specific Carboxypeptidase Localized to Brush Border Membrane of Rat Small Intestine and its Possible Role in Protein Digestion. Digestive Diseases and Sciences 1989;34(3) 400-406.

[48] Mellander O, Folsch G. Enzyme Resistance and Metal Binding of Phosphorylated Peptides. In: Bigwood EJ. (ed.) Protein and Amino Acid Function. New York: Pergamoon Press Inc; 1972. p569-79.

[49] Boutrou, R, Coirre E, Jardin J, Léonil J. Phosphorylation and Coordination Bond of Mineral Inhibit the Hydrolysis of the β-Casein (1–25) Peptide by Intestinal Brush-Border Membrane Enzymes. Journal of Agricultural and Food Chemistry 2010;58(13) 7955-7961.

[50] Kibangou IB, Bouhallab S, Henry G, Bureau F, Allouche S, Blais A, Guérin P, Arhan P, Bouglé DL. Milk Proteins and Iron Absorption: Contrasting Effects of Different Caseinophosphopeptides. Pediatric Research 2005;58(4) 731-734.

[51] Pérès JM, Bouhallab S, Petit C, Bureau F, Maubois JL, Arhan P, Bougle D. Improvement of Zinc Intestinal Absorption and Reduction of Zinc/Iron Interaction Using Metal Bound to the Caseinophosphopeptide 1-25 of β-casein. Reproduction, Nutrition & Development 1998;38(4) 465-472.

[52] Kasai T, Iwasaki R, Tanaka M, Kiriyama S. Caseinphosphopeptides (CPP) in Feces and Contents in Digestive Tract of Rats Fed Casein and CPP Preparations. Bioscience, Biotechnology and Biochemistry 1995;59(1) 26-30.

[53] Ferraretto A, Signorile A, Gravaghi C, Fiorilli A, Tettamanti G. Casein Phosphopeptides Influence Calcium Uptake by Cultured Human Intestinal HT-29 Tumor Cells. Journal of Nutrition 2001;131(6) 1655-1661.

[54] Richter S, Bergmann R, Pietzsch J, Ramenda T, Steinbach J, Wuest F. Fluorine-18 Labeling of Phosphopeptides: A Potential Approach for the Evaluation of Phosphopeptide Metabolism In Vivo. Biopolymers 2009;92(6) 479-488.

[55] Erba D, Ciappellano S, Testolin G. Effect of Caseinphosphopeptides on Inhibition of Calcium Intestinal Absorption due to Phosphate. Nutrition Research 2001;21(4) 649-656.

[56] Marcus CS, Lengemann FW. Absorption of Ca^{45} and Sr^{85} from Solid and Liquid Food at Various Levels of the Alimentary Tract of the Rat. Journal of Nutrition 1962;77(2) 155-160.

[57] Marcus CS, Lengemann FW. Use of Radioyttrium to Study Food Movement in the Small Intestine of the Rat. Journal of Nutrition 1962;76(2) 179-182.

[58] Meisel H, Bernard H, Fairweather-Tait S, FitzGerald RJ, Hartmann R, Lane CN, McDonagh D, Teucher B, Wal JM. Detection of Caseinophosphopeptides in the Distal Ileostomy Fluid of Human Subjects. British Journal of Nutrition 2003;89(3) 351-359.

[59] Meisel H, Frister H. Chemical Characterization of a Caseinophosphopeptide Isolated from in Vivo Digests of a Casein Diet. Biological Chemistry Hoppe-Seyler 1988;369(12) 1275-1279.

[60] Chabance B, Marteau P, Rambaud JC, Migliore-Samour D, Boynard M, Perrotin P, Guillet R, Jollès P, Fiat AM. Casein Peptide Release and Passage to the Blood in Humans during Digestion of Milk or Yogurt. Biochimie 1998;80(2) 155-165.

[61] Meisel H. Biochemical Properties of Regulatory Peptides Derived from Milk Proteins. Biopolymers 1997;43(2) 119-128.

[62] Sato R, Noguchi T, Naito H. Casein phosphopeptide (CPP) Enhances Calcium Absorption from the Ligated Segment of Rat Small Intestine. Journal of Nutritional Science and Vitaminology 1986;32(1) 67-76.

[63] Ferraretto A, Gravaghi C, Fiorilli A, Tettamanti G. Casein-Derived Bioactive Phosphopeptides: Role of Phosphorylation and Primary Structure in Promoting Calcium Uptake by HT-29 Tumor Cells. FEBS Letters 2003;551(1-3) 92-98.

[64] Cosentino S, Gravaghi C, Donetti E, Donida BM, Lombardi G, Bedoni M, Fiorilli A, Tettamanti G, Ferraretto A. Caseinphosphopeptide-Induced Calcium Uptake in Human Intestinal Cell Lines HT-29 and Caco2 is Correlated to Cellular Differentiation. Journal of Nutritional Biochemistry 2010;21(3) 247-254.

[65] Donida BM, Mrak E, Gravaghi C, Villa I, Cosentino S, Zacchi E, Perego S, Rubinacci A, Fiorilli A, Tettamanti G, Ferraretto A. Casein Phosphopeptides Promote Calcium Uptake and Modulate the Differentiation Pathway in Human Primary Osteoblast-like Cells. Peptides 2009;30(12) 2233-2241.

[66] Reynolds EC. Casein Phosphopeptide-Amorphous Calcium Phosphate: The Scientific Evidence. Advances in Dental Research 2009;21(1) 25-29.

[67] Rose RK. Binding Characteristics of *Streptococcus mutans* for Calcium and Casein Phosphopeptide. Caries Res 2000;34(5) 427-431.

[68] McCormack AL, Schieltz DM, Goode B, Yang S, Barnes G, Drubin D, Yates JR III. Direct Analysis and Identification of Proteins in Mixtures by LC/MS/MS and Database Searching at the Low-femtomole Level Analytical Chemistry 1997;69(4):767-776.

[69] Eng JK, McCormack AL, Yates JR III. An Approach to Correlate Tandem Mass Spectral Data of Peptides with Amino Acid Sequences in a Protein Database. Journal of The American Society for Mass Spectrometry 1994;5(11) 976-989.

[70] Stensballe A, Andersen S, Jensen ON. Characterization of Phosphoproteins from Electrophoretic Gels by Nanoscale Fe(III) Affinity Chromatography with Off-line Mass Spectrometry Analysis. Proteomics 2001;1(2) 207-222.

[71] Craig AG, Hoeger CA, Miller CL, Goedken T, Rivier JE, Fischer WH. Monitoring Protein Kinase and Phosphatase Reactions with Matrix-Assisted Laser Desorption/Ionization Mass Spectrometry and Capillary Zone Electrophoresis: Comparison of the Detection Efficiency of Peptide-Phosphopeptide Mixtures. Biological Mass Spectrometry 1994;23(8) 519-528.

[72] Annan RS, Huddleston MJ, Verma R, Deshaies RJ, Carr SA. A Multidimensional Electrospray MS-Based Approach to Phosphopeptide Mapping. Analytical Chemistry 2001;73(3) 393-404.

[73] Manson W, Annan WD. The Structure of a Phosphopeptide Derived from β-casein. Archives of Biochemistry and Biophysics 1971;145(1) 16-26.

[74] Gygi SP, Corthals GL, Zhang Y, Rochon Y, Aebersold R. Evaluation of two-Dimensional Gel Electrophoresis-Based Proteome Analysis Technology. PNAS 2000;97(17) 9390-9395.

[75] Oda Y, Nagasu T, Chait, BT. Enrichment Analysis of Phosphorylated Proteins as a Tool for Probing the Phosphoproteome. Nature Biotechnology 2001;19(4) 379-382.

[76] Zhou H, Watts, JD, Aebersold R. A Systematic Approach to the Analysis of Protein Phosphorylation. Nature Biotechnology 2001;19(4) 375-378.

[77] Posewitz MC, Tempst P. Immobilized Gallium(III) Affinity Chromatography of Phosphopeptides. Analytical Chemistry 1999;71(14) 2883-2892.

[78] Kokubu M, Ishihama Y, Sato T, Nagasu T, Oda Y. Specificity of Immobilized Metal Affinity-Based IMAC/C18 Tip Enrichment of Phosphopeptides for Protein Phosphorylation Analysis. Analytical Chemistry 2005;77(16) 5144-5154.

[79] Machida M, Kosako H, Shirakabe K, Kobayashi M, Ushiyama M, Inagawa J, Hiran J, Nakano T, Bando Y, Nishida E, Hattori S. Purification of Phosphoproteins by Immobilized Metal Affinity Chromatography and its Application to Phosphoproteome Analysis. FEBS Journal 2007;274(6) 1576-1587.

[80] Thingholm TE, Jensen ON. Enrichment and Characterization of Phosphopeptides by Immobilized Metal Affinity Chromatography (IMAC) and Mass Spectrometry. Methods in Molecular Biology 2009;527 47-56.

[81] Pinkse MWH, Uitto PM, Hilhorst MJ, Ooms B, Heck AJ Selective Isolation at the Femtomole Level of Phosphopeptides from Proteolytic Digests Using 2D-NanoLC-ESI-MS/MS and Titanium Oxide Precolumns. Analytical Chemistry 2004;76(14) 3935-3943.

[82] Rinalducci S, Larsen MR, Mohammed S, Zolla L. Novel Protein Phosphorylation Site Identification in Spinach Stroma Membranes by Titanium Dioxide Microcolumns and Tandem Mass Spectrometry. Journal of Proteome Research 2006;5(4) 973-982.

[83] Thingholm TE, Jørgensen TJ, Jensen ON, Larsen MR. Highly Selective Enrichment of Phosphorylated Peptides using Titanium Dioxide. Nature Protocols 2006;1(4) 1929-1935.

[84] Dunn JD, Reid GE, Bruening ML. Techniques for Phosphopeptide Enrichment prior to Analysis by Mass Spectrometry. Mass Spectrometry Reviews 2010;29(1) 29-54.

[85] Moser K, White FM. Phosphoproteomic Analysis of Rat Liver by High Capacity IMAC and LC-MS/MS. Journal of Proteome Research 2006;5(1) 98-104.

[86] Larsen MR, Thingholm TE, Jensen ON, Roepstorff P, Jorgensen TJD. Highly Selective Enrichment of Phosphorylated Peptides from Peptide Mixtures Using Titanium Dioxide Microcolumns. Molecular & Cellular Proteomics 2005;4(7) 873-886.

[87] Bodenmiller B, Mueller LN, Mueller M, Domon B, Aebersold R. Reproducible Isolation of Distinct, Overlapping Segments of the Phosphoproteome. Nature Methods 2007;4(3) 231-237.

[88] Sugiyama N, Masuda T, Shinoda K, Nakamura A, Tomita M, Ishihama Y. Phosphopeptide Enrichment by Aliphatic Hydroxyl Acid-Modified Metal Oxide Chromatography for Nano-LC-MS/MS in Proteomics Applications. Molecular & Cellular Proteomics 2007;6(6) 1103-1109.

[89] McLachlin DT, Chait BT. Analysis of Phosphorylated Proteins and Peptides by Mass Spectrometry. Current Opinion in Chemical Biology 2001;5(5) 591-602.

[90] Pinto G, Caira S, Cuollo M, Lilla S, Fierro O, Addeo F. Hydroxyapatite as a Concentrating Probe for Phosphoproteomic Analyses. Journal of Chromatography B 2010;878(28) 2669-2678.

[91] Zhou HY, Salih E, Glimcher MJ. Isolation of a Novel Bone Glycosylated Phosphoprotein with Disulphide Cross-Links to Osteonectin. Biochemical Journal 1998;330(3) 1423-1431.

[92] Schmidt SR, Schweikart F, Andersson ME. Current Methods for Phosphoprotein Isolation and Enrichment. Journal of Chromatography B 2007;849(1-2) 154-162.

[93] Kawasaki T. Hydroxyapatite as a Liquid Chromatographic Packing. Journal of Chromatography A 544(17) 147-184.

[94] Mamone G, Picariello G, Ferranti P, Addeo F. Hydroxyapatite Affinity Chromatography for the Highly Selective Enrichment of Mono- and Multi-Phosphorylated Peptides in Phosphoproteome Analysis. Proteomics 2010;10(3) 380-393.

[95] Cirulli C, Chiappetta G, Marino G, Mauri P, Amoresano A. Identification of Free Phosphopeptides in Different Biological Fluids by a Mass Spectrometry Approach. Analytical and Bioanalytical Chemistry 2008;392(1-2) 147-159.

[96] Hu L, Zhou H, Li Y, Sun S, Guo L, Ye M, Tian X, Gu J, Yang S, Zou H. Profiling of Endogenous Serum Phosphorylated Peptides by Titanium (IV) Immobilized Mesoporous Silica Particles Enrichment and MALDI-TOFMS Detection. Analytical Chemistry 2009;81(1) 94-104.

[97] Li Y, Qi D, Deng C, Yang P, Zhang X. Cerium Ion-Chelated Magnetic Silica Microspheres for Enrichment and Direct Determination of Phosphopeptides by Matrix-Assisted Laser Desorption Ionization Mass Spectrometry. Journal of Proteome Research 2008;7(4) 1767-1777.

[98] Maurer MC, Peng JL, An SS, Trosset JY, Henschen-Edman A, Scheraga, HA. Structural Examination of the Influence of Phosphorylation on the Binding of Fibrinopeptide A to Bovine Thrombin. Biochemistry 1998, 37(17) 5888–5902.

[99] Matrisian LM, Sledge GW Jr, Mohla S. Extracellular Proteolysis and Cancer: Meeting Summary and Future Directions Cancer Research 2003;63(19) 6105-6109.

[100] Resmini P, Pellegrino L, Battelli G. Accurate Quantification of Furosine in Milk and Dairy Products by a Direct HPLC Method. Italian Journal of Food Science 1990;2(3) 173-183.

[101] Corzo N, López-Fandiño R, Delgado T, Ramos M, Olano A. Changes in Furosine and Proteins of UHT-Treated Milks Stored at High Ambient Temperatures. Zeitschrift fur Lebensmittel-Untersuchung und-Forschung 1994;198(4) 302-306.

[102] Pinto G, Caira S, Cuollo M, Fierro O, Nicolai MA, Chianese L, Addeo F. Lactosylated Casein Phosphopeptides as Specific Indicators of Heated Milks. Analytical and Bioanalytical Chemistry 2012 2012;402(5) 1961-1972.

[103] Mendoza MR, Olano A, Villamiel M. Chemical Indicators of Heat Treatment in Fortified and Special Milks. Journal of Agriculture and Food Chemistry 2005;53(8) 2995-2999.

[104] Mottar J, Bassier A, Joniau M, Baert J. Effect of Heat-Induced Association of Whey Proteins and Casein Micelles on Yogurt Texture. Journal of Dairy Science 1989;72(9) 2247-2256.

[105] Tokusoglu Ö, Akalin AS, Unal K. Rapid High Performance Liquid Chromatographic Detection of Furosine (epsilon- N-2-furoylmethyl-L-lysine) in Yogurt and Cheese Marketed in Turkey. Journal of Food Quality 2006;29(1) 38-46.

[106] Bouhallab S, Oukhatar NA, Molle D, Henry G, Maubois JL, Arhan P, Bougle D. Sensitivity of Beta-Casein Phosphopeptide-Iron Complex to Digestive Enzymes in Ligated Segment of Rat Duodenum. The Journal of Nutritional Biochemistry 1999;10(12) 723-727.

[107] Pérès JM, Bouhallab S, Bureau P, Neuville D, Maubois JL, Devroede G, Arhan P, Bouglé D. Mechanisms of Absorption of Caseinophosphopeptide Bound Iron. The Journal of Nutritional Biochemistry 1999;10(4) 215-222.

[108] Adamson NJ, Reynolds EC. Characterization of Multiply Phosphorylated Peptides Selectively Precipitated from a Pancreatic Casein Digest. Journal of Dairy Science 1995;78(12) 2653-2659.

[109] Klaver FAM, Kingma F, Weerkamp AH. Growth and Survival of Bifidobacteria in Milk. Netherlands Milk Dairy Journal 1993;47(3-4) 151-164.

[110] Roudot-Algaron F, Le Bars D, Kerhoas L, Einhorn J, Gripon JC. Phosphopeptides from Comté Cheese. Nature and Origin. Journal of Food Science 1994; 59(3) 544-547.

[111] Ferranti P, Barone F, Chianese L, Addeo F, Scaloni A, Pellegrino L, Resmini P. Phosphopeptides from Grana Padano: Nature, Origin and Changes during Ripening. Journal of Dairy Research 1997;64(4) 601-615.

[112] Lund M, Ardö Y. Purification and Identification of Water Soluble Phosphopeptides from Cheese using Fe(III) Affinity Chromatography and Mass Spectrometry. Journal of Agricultural and Food Chemistry 2004;52(21) 6616-6622.

[113] Pellegrino L, Battelli G, Resmini P, Ferranti P, Barane F, Addeo F. Effects of Heat Load Gradient Occurring in Moulding on Characterization and Ripening of Grana Padano. Lait 1997;77(2) 217-228.

[114] Walstra P, Jenness R. Dairy Chemistry and Physics. New York: John Wiley and Sons; 1984.

[115] Pirisi A, Pinna G, Addis M, Piredda G, Mauriello R, De Pascale S, Addeo F, Chianese L. Relationship Between the Enzymatic Composition of Lamb Paste Rennet and Proteolytic and Lypolitic Pattern and Texture of PDO Fiore Sardo Cheese. International Dairy Journal 2007;17(2) 143-156.

[116] Battistotti B, Corradini C. Italian Cheese. In: Fox PF. (ed.) Cheese: Chemistry, Physics and Microbiology. Volume II. Major Cheese Groups. London: Elsevier Applied Science; 1993. p221-243.

[117] Voigt DD, Chevalier F, Qian MC, Kelly AL. Effect of High-Pressure Treatment on Microbiology, Proteolysis, Lipolysis and Levels of Flavour Compounds in Mature Blue-Veined Cheese. Innovative Food Science & Emerging Technologies 2010;11(1) 68-77.

[118] Mykkanen HM, Wasserman RH. Enhanced Absorption of Calcium by Casein Phosphopeptides in Rachitic and Normal Chicks. Journal of Nutrition 1980;110(11) 2141-2148.

[119] Meisel H. Overview on Milk Protein-Derived Peptides. International Dairy Journal 1998;8(5-6) 363-373.

[120] Reynolds EC, Johnson IH. Effect of Milk on Caries Incidence and Bacterial Composition of Dental Plaque in the Rat. Archives of Oral Biology 1981;26 (5) 445-451.

[121] Rosen S, Min DB, Harper DS, Harper WJ, Beck EX, Beck FM. Effect of Cheese with and without Sucrose, on Dental Caries and Recovery of Streptococcus mutans in Rats. Journal of Dental Research 1984;63(6) 894-896.

[122] Harper DS, Osborne JC, Hefferren JJ, Clayton R. Cariostatic Evaluation of Cheeses with Diverse Physical and Compositional Characteristics. Caries Research 1986;20 (2) 123-130.

[123] Walker GD, Cai F, Shen P, Adams GG, Reynolds C, Reynolds EC. Casein Phosphopeptide Amorphous Calcium Phosphate Incorporated into Sugar Confections Inhibits the Progression of Enamel Subsurface Lesions in Situ. Caries Research 2010;44(1) 3-40.

[124] Shaw L, Murray JJ, Burchell CK, Best JS. Calcium and Phosphorus Contents of Plaque and Saliva in Relation to Dental Caries. Caries Research 1983;17 (6) 543-548.

[125] Margolis HC, Moreno EC. Composition of Pooled Plaque Fluid from Caries-Free and Caries-Positve Individuals Following Sucrose Exposure. Journal of Dental Research 1992;71(11) 1776-1784.

[126] Reynolds EC. Remineralization of Enamel Subsurface Lesions by Casein Phosphopeptide-Stabilized Calcium Phosphate Solutions. Journal of Dental Research 1997;76(9) 1587-1595.

[127] Reynolds EC. Anticariogenic Phosphopeptides. US Patent 5015628.

[128] Rose RK, Turner SJ. Fluoride-Induced Enhancement of Diffusion in Streptococcal Model Plaque Biofilms. Caries Research 1998; 32(3) 227-232.

[129] Rose RK, Dibdin GH. Calcium and Water Diffusion in Streptococcal Model Plaques. Archives of Oral Biology 1995;40(5) 385-391.

[130] Rose RK, Turner SJ, Dibdin GH. Effect of pH and Calcium Concentration on Calcium Diffusion in Streptococcal Model Plaque Biofilms. Archives of Oral Biology 1997;42(12) 795-800.

[131] Shaw JH, Ensfield BJ, Wollman DH. Studies on the Relation of Dairy Products to Dental Caries in Caries-Susceptible Rats. Journal of Nutrition 1959;67(2) 253-273.

[132] Papas AS, Joshi A, Belanger AJ, Kent RL, Palmer CA, De Paola PF. Dietary Models for Root Caries. American Journal of Clinical Nutrition 1995;61(2) 417S-422S.

[133] Ferrazzano GF, Cantile T, Quarto M, Ingenito A, Chianese L, Addeo F. Protective Effect of Yogurt Extract on Dental Enamel Demineralization in Vitro. Australian Dental Journal 2008;53(4) 314-319.

[134] Reynolds EC, Riley PF, Adamson N. A Selective Precipitation Purification Procedure for Multiple Phosphoseryl-Containing Peptides and Methods for their Indentification. Analytical Biochemistry 1994;217(2) 277-284.

[135] Tirelli A, De Noni I, Resmini P. Bioactive Peptides in Milk Products. Italian Journal of Food Science 1997;9(2) 91-98.

[136] Meisel H, Andersson HB, Buhl K, Erbersdobler HF, Schlimme E. Heat-Induced Changes in Casein-Derived Phosphopeptides. Zeitschrift Fur Ernahrungswissenschaft 1991;30(3) 227-232.

[137] Ferranti P, Lilla S, Chianese L, Addeo F. Alternative Nonallelic Deletion is Constitutive of Ruminant α_{s1}-Casein. Journal of Protein Chemistry 1999;18(5) 595-602.

[138] Juillerat MA, Baechler R, Berrocal R, Chanton S, Scherz JC, Jost R. Tryptic Phosphopeptides from Whole Casein. I. Preparation and Analysis by Fast Protein Liquid Chromatography. Journal of Dairy Research 1989;56(4) 603-611.

[139] Adamson NJ, Reynolds EC. Characterization of Tryptic Casein Phosphopeptides Prepared under Industrially Relevant Conditions. Biotechnology and Bioengineering 1995;45(3) 196-204.

[140] Goepfert A, Meisel H. Semi-Preparative Isolation of Phosphopeptides Derived from Bovine Casein and Dephosphorylation of Casein Phosphopeptides. Nahrung 1996;40(5) 245-248.

[141] Ellegard KH, Gammelgard-Larsen C, Sørensen ES, Fedosovc S. Process Scale Chromatographic Isolation, Characterization and Identification of Tryptic Bioactive Casein Phosphopeptides. International Dairy Journal 1999;9(9) 639-652.

[142] Zhao L, Wang Z, Xu S. Preparation of Casein Phosphorylated Peptides and Casein non-Phosphorylated Peptides using Alcalase. European Food Research and Technology 2007;225(3-4) 579-584.

[143] Reynolds EC. Production of Phosphopeptides from Casein. US Patent 6448374.

[144] Brule G, Roger L, Fauquant J, Piot M. Casein Phosphopeptide Salts. US Patent 5028589.

[145] Naito H, Noguchi T, Sato R, Tsuji K, Hidaka H. Transparent Acid Drink Containing Acid-Soluble Casein Phosphopeptides. US Patent 5405756.

[146] Ramalingam L, Messer LB, Reynolds EC. Adding Casein Phosphopeptide-Amorphous Calcium Phosphate to Sports Drinks to Eliminate in Vitro Erosion. Pediatric Dentistry 2005;27(1) 61-67.

[147] Tancredi D, Ming D, Holme S. Calcium Phosphate Complex in Acid Containing Chewing Gum. US Patent 8133475.

[148] Walker G, Cai F, Shen P, Reynolds C, Ward B, Fone C, Honda S, Koganei M, Oda M, Reynolds E. Increased Remineralization of Tooth Enamel by Milk Containing Added Casein Phosphopeptide-Amorphous Calcium Phosphate. Journal of Dairy Research 2006;73(1) 74-78.

[149] Kostyra H, Kostyra E. Biologically Active Peptides Derived from Food Proteins. Polish Journal of Food and Nutrition Sciences 1992;1/42(4) 5-17.

[150] Meisel H, Schlimme E. Milk Proteins: Precursors of Bioactive Peptides. Trends in Food Science and Technology 1990;1 41-43.

[151] Rokka T, Syväoja EL, Tuominen J, Korhonen H. Release of Bioactive Peptides by Enzymatic Proteolysis of Lactobacillus GG Fermented UHT Milk. Milchwissensch 1997;52 675-678.

[152] Jarmolowska B, Kostyra E, Krawczuk S, Kostyra H. β-Casomorphin-7 Isolated from Brie Cheese. Journal of the Science of Food and Agriculture 1999;79(13) 1788-1792.

[153] Muehlenkamp MR, Warthesen JJ. β-Casomorphins: Analysis in Cheese and Susceptibility to Proteolytic Enzymes from Lactococcus lactic ssp. cremoris. Journal of Dairy Science 1996;79(1) 20-26.

[154] Smacchi E, Gobbetti M. Bioactive Peptides in Dairy Products: Synthesis and Interaction with Proteolytic Enzymes. Food Microbiology 2000;17(2) 129-141.

[155] Ferranti P, Traisci MV, Picariello G, Nasi A, Boschi MS, Falconi C, Chianese L, Addeo F. Casein Proteolysis in Human Milk Tracing the Pattern of Casein Breakdown and the Formation of Potential Bioactive Peptides. Journal of Dairy Research 2004;71(1) 74-87.

[156] Pihlanto-Leppälä A. Bioactive Peptides Derived from Bovine Whey Proteins: Opioid and Ace-Inhibitory Peptides. Trends in Food Science and Technology 2000;11(9-10) 347-356.

[157] Bitri L. Optimization Study for the Production of an Opioid-like Preparation from Bovine Casein by Mild Acidic Hydrolysis. International Dairy Journal 2004;14(6) 535-539.

Novel Functions of Milk Proteins

Chapter 3

Alpha-Casein as a Molecular Chaperone

Teresa Treweek

Additional information is available at the end of the chapter

1. Introduction

The caseins are a heterogeneous group of dairy proteins constituting 80% of the protein content of bovine milk. The operational definition of casein is that proportion of total milk protein which precipitates on acidification of milk to a pH value of 4.6 [1]. The remaining dairy proteins, known collectively as whey proteins, do not precipitate. Caseins are synthesised in the mammary gland and are found nowhere else among the plant and animal kingdoms [2]. The casein family of proteins comprises α- , β- and κ-caseins, all with little sequence homology [3]. As their primary function is nutritional, binding large amounts of calcium, zinc and other biologically important metals, amino acid substitutions or deletions have little impact on function. The caseins also lack well-defined structure and as a result their amino acid sequence is less critical to function than in many 'classic' globular proteins. As a result, the caseins are one of the most evolutionarily divergent protein families characterised in mammals [2]. Alpha-casein, also known as α_S-casein, is in fact two distinct gene products, α_{S1}- and α_{S2}-casein, with the 'S' denoting a sensitivity to calcium. Of all the caseins, α_{S1}- and β-casein are predominant in bovine milk, representing 37 and 35% of whole casein respectively, whereas α_{S2}- and κ-casein make up 10 and 12% of whole casein, respectively [2].

1.1. Key structural features of the casein proteins

The casein proteins are important nutritionally not just because of their high phosphate content which allows them to bind significant quantities of calcium, but because they are high in lysine. The constituent proteins of α_S-casein, α_{S1}- and α_{S2}-casein, possess 14 and 24 lysines, respectively [2]. Lysine is an essential amino acid in humans and one in which many plant sources are lacking, therefore casein extracts form an effective nutritional supplement for cereals [2].

In addition to the variability inherent in their amino acid sequences, each of the caseins exhibit significant variability as a result of their degree of post-translational modification, disulfide bonding, genetic polymorphism and the manner in which they are hydrolysed by the milk protease, plasmin. In terms of the extent of phosphorylation, each of the four caseins may have various numbers of phosphate groups attached via their serine or threonine residues [4]. For example, α_{s1}-casein may have 8 or 9, α_{s2}-casein 10, 11, 12 or 13, β-casein may have 4 or 5 and κ-casein, 1, 2 or 3 [5]. It is not known whether this variability results from the action of casein kinases phosphorylating to different degrees or by phosphatases dephosphorylating to a greater or lesser extent [5]. The predominant caseins, α_{s1}- and β-casein, contain no disulfide bonds; however α_{s2}- and κ-casein contain two cysteine residues which form intra- or intermolecular disulfide bonds under normal conditions. In the absence of a reducing agent, α_{s2}-casein exists as a disulfide-linked dimer and κ-casein can adopt dimeric to decameric forms, depending on the pattern of intermolecular disulfide bonding. κ-Casein is also the only casein which is glycosylated and the degree of glycosylation varies so that ten different molecular forms of κ-casein are possible on this basis alone [4].

Genetic polymorphism is yet another source of variability in the caseins. This phenomenon was first described in 1955 in relation to β-lactoglobulin [6] and exists when the same protein exists in a number of forms, differing from one another in just a few amino acids. This has since been shown to occur in all dairy proteins. The milk of one animal may contain one polymorphic form alone, or both, and the occurrence of particular polymorphs is breed-specific [7, 8] . Genetic variants are indicated by a latin letter i.e. α_{s1}-casein has been shown to be present in bovine milk as α_{s1}-caseinA – D; α_{s2}- caseinA – D; β-caseinA^1, A^2, A^3 – E and κ-caseinA and B [9]. With the combined variability between the caseins themselves, contributed by low sequence homology, glycosylation and disulfide bonding and within individual caseins due to the degree of phosphorylation and genetic polymorphism, this is a very interesting family of proteins. As will be described in greater detail in this chapter, the caseins have created even greater intrigue with the recent discovery of their chaperone abilities.

Structurally, the caseins are classified as 'intrinsically or natively disordered' proteins under physiological conditions [10, 11]. This disordered structure, which is present to some extent even in globular proteins, is different to random coil conformation. In natively disordered proteins, conformations of these regions are still relatively fixed with respect to the ϕ and ψ angles of the peptide bonds, as opposed to true 'random coil' polypeptide chains, which exhibit greater and more rapid fluctuation in bond angles [4]. The lack of well-defined structure in the casein proteins is believed to facilitate proteolysis and therefore ready absorption of amino acids and small peptides in the gut [2], but is another likely factor in the unwillingness of the caseins to crystallise to provide a 3D crystal structure [12]. Physical characterisations of caseins in solution and predicted 3D models have shown that the caseins have relatively little tertiary structure, but possess some secondary structure, similar to the classic 'molten globule' states described in [13]. The greatest degree of secondary structure exists in α_{s2}- and κ-casein, mainly in the form of β-sheets and β-turns rather than α-helix [14-16]. The formation of higher proportions of secondary and tertiary structural elements in the caseins is likely to be inhibited by high numbers of proline residues which distort protein folding into α-helices and β-sheets [2].

Each of the casein proteins has a high degree of hydrophobicity as a result of containing approximately 35-45% non-polar amino acids (e.g. Val, Leu, Phe, Tyr, Pro), but this does not preclude them from being quite soluble in aqueous solvents due to the presence of high numbers of phosphate and sulfur-containing amino acids, and in the case of κ-caseins, carbohydrates [2]. These hydrophobic regions are likely exposed in the caseins as a result of their flexible and relatively unfolded structure. The hydrophobicity tends to occur in patches along the sequence of the caseins, however, and is interspersed with hydrophilic regions. It is this feature that is credited with making the caseins good emulsifying agents – a property exploited in the food industry. The clustered exposed hydrophobicity is also thought to be a major feature of the molecular chaperone action of the caseins [17] as discussed later in this chapter.

1.2. Self-association, fibril formation and micellar arrangement of the caseins

Although relatively small in size with molecular masses of 23.6 and 25.2 kDa for α_{s1}- and α_{s2}-casein, respectively, α-caseins readily associate with one another and with the other caseins (β- and κ-casein) to form large aggregates up to 1.4 MDa in size [18]. This tendency to form multimeric assemblies is likely to be another reason why it has not been possible to obtain crystal structures for the caseins thus far. In the presence of calcium, associations between the various caseins can lead to the formation of micelles [2]. These micelles are composed of approx. 94% protein and approx. 6% low molecular weight species such as calcium, phosphate, citrate and magnesium which together form 'colloidal calcium phosphate' [2] or amorphous calcium phosphate; APC [19]. Evolutionarily, it is thought that the formation of micelles has served as a means by which to increase the calcium concentration in milk over many millennia to satisfy its nutritional function without compromising physical stability [19]. The makeup of the micelle, which is roughly spherical in shape and has a radius of approx. 600 nm or less, comprises the amorphous calcium phosphate, more recently referred to as 'nanoclusters', bound to specific phosphorylated sequences in the surrounding α_{s1}- , α_{s2}- and β-casein chains [19]. The major protein constituent of casein micelles, accounting for 65% of protein is α_s-casein [4]. The function of κ-casein, present at the surface of the micelle, appears to be related to limiting the size of the micelle [19].

The C-terminal region of κ-casein is strongly hydrophilic, whereas the N-terminal region is strongly hydrophobic [4]. Such amphipathic qualities have been shown to be of great importance for molecules residing at the interface between hydrophobic and hydrophilic environments in various biological contexts (e.g. the phospholipid cell membrane, the assembly of lipoprotein particles) and are no doubt also important in the stabilisation of the micelle in the aqueous environment of milk. Further evidence for the localisation of κ-casein at the surface of the micelle forming a diffuse outer region [20] was provided by the discovery that in the formation of cheese, the more hydrophilic C-terminal portion is the one cleaved by the action of rennet [2]. Recent Cryo-TEM and TEM studies have shown that the small electron-dense regions consistent with calcium phosphate nanoclusters are evenly distributed throughout the micellar structure rather than being sequestered within the core of the micelle [21-24] and that these structures, linked together by chains of caseins, were continuous throughout the entire micelle [25]. The sequestration of calcium phosphate, which accounts for 7% of the solute mass of bovine casein micelles, within a phosphoprotein matrix in this way is critical to maintaining the stability of these potentially very insoluble minerals in milk which would otherwise precipitate, compromising lactation [19]. It is the light scattering caused by these large (10^3 to 3×10^6 kDa) casein micelles in a colloidal dispersion that is thought to give milk its characteristic white colour [2, 25, 26]. The caseins are also very stable at high temperatures, a property thought to be associated with their high phosphate content [2] and lack of well-defined structure [11] which makes them resistant to denaturation by heat and chemical agents. Milk may be heated at 100°C at its native pH (~6.7) for 24 hours without coagulating and will withstand 140°C temperatures for 20 min. The current ultra high heat treatments applied to milk and milk products are made possible by the extreme thermostability of the caseins [2]. Our studies have shown that solutions of individual α_{s1}- and α_{s2}-caseins are stable at 70°C for a period of at least 8 hours at a concentration of 5 mg/mL [18].

Interestingly, both of the disulfide bond-containing caseins, κ- and α_{s2}-casein, have been shown to form amyloid fibrils under physiological conditions [27-30]. It has been suggested that casein proteins may have a propensity to form amyloid fibrils because they possessed similar structural features to the amyloid forming proteins tau, α-synuclein and amyloid β [31]. The similarity lies in the tendency of all four casein proteins to adopt a flexible and relatively unfolded conformation but also their possession of significant amounts of poly-L-proline (PPII) helix structure which likely arose from the relatively uniform distribution of proline residues [2, 31]. In contrast to α-helix, PPII helix is more open, flexible and extended and the conversion of this to antiparallel β-sheets, the precursor to amyloid fibril, is a highly energetically-favourable one [32]. As suggested in [33], other factors in the formation of amyloid fibrils must also be important , as only two of the four caseins (namely κ- and α_{s2}-casein) form fibrils under physiological conditions. Whole casein does not form fibrils under the same conditions as κ- and α_{s2}-casein and this is thought to be related to the inhibitory action of other caseins present. It has been shown previously that both α_{s1}- and β-casein are able to inhibit fibril formation by κ-casein [28].

1.3. Separation and purification of α_{s1}- and α_{s2}-casein

Caseins can be separated from whey proteins in milk by a variety of methods that are effective for large scale applications such as the acid precipitation already described [2]. The individual casein proteins can also be separated from one another by classical methods based on solubility differences and more recently via gel chromatography (summarised in [4]). In order to study α_{s1}- and α_{s2}-caseins separately, these proteins were purified from their associated form, α_s-casein, by successfully employing the method previously described in reference [34] using a Q-Sepharose column with some minor modifications [17]. Subsequent investigation of the chaperone activities of individual α_{S1}- and α_{S2}-casein proteins was then possible. As shown in Figure 1, purification gave an initial smaller peak corresponding to the predominant form of α_{S2}-casein, α_{S2}A-casein 11P (25.2 kDa with 11 phosphate groups attached), and a second, much larger peak corresponding to the predominant form of α_{S1}-casein, α_{S1}B-casein 8P (23.6 kDa; 8 phosphate groups attached). It would be logical to expect that the areas under these peaks to be roughly representative of the 4:1 ratio of α_{S1}:α_{S2}-caseins present in bovine milk, however, due to their differing amino acid compositions α_{S1}- and α_{S2}-casein have quite different specific absorbances of 10.1 and 14.0 A1%1 cm, respectively [2]. The smallest peak visible in Figure 1 is representative of a small amount of dimeric α_{S2}-casein eluting first from the column. This species had an approximate mass of 50 kDa due to the presence of intermolecular disulfide bonds between cysteine residues at positions 36 and 40 [17].

Successful separation and purification of the constituent α_{S}-casein proteins was confirmed by SDS-PAGE and ESI-MS, which gave a major peak at 23,619 kDa (Figure 2). This closely corresponded to the major variant of α_{S}-casein, α_{S1}B-casein 8P. Unfortunately it was not possible to obtain a similar spectrum for α_{S2}A-casein 11P, most likely as a result of the greater hydrophobicity of this casein and its propensity for amyloid fibril formation [17].

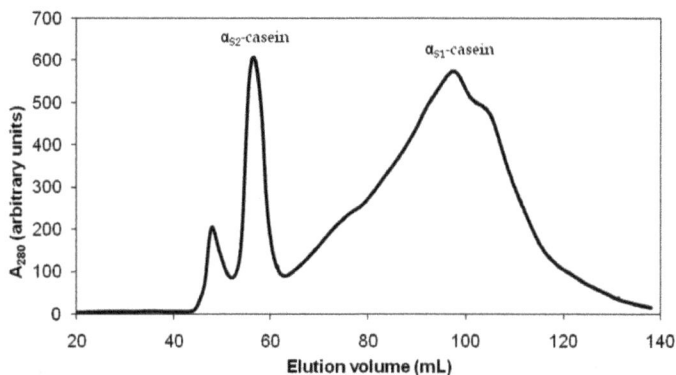

Figure 1. Purification of α_{S1}- and α_{S2}-casein from total α_{S}-casein. Reprinted with permission from [17]. Copyright (2011) Elsevier Inc.

1.4. Characterisation of purified α_{S1}- and α_{S2}-casein proteins

Biophysical characterisation of the purified α_{S}-casein proteins showed that the proteins possessed a similar degree of secondary structure to that expected from literature values [3, 35-38]. The far-UV CD spectra of purified α_{S1}-, α_{S2}- and α_{S}-casein proteins (Figure 3) show a minimum at approximately 202 nm for α_{S1}-casein, 203 nm for α_{S}-casein and 205 nm for α_{S2}-casein and a second minimum for all three proteins at 222 nm. Deconvolution of the far-UV CD data for α_{S1}- and α_{S2}-casein shown in Figure 3 was performed using SELCON software via the DICHROWEB database [39-43]. These data are summarised in Table 1. Deconvolution of data for α_{S}-casein could not be performed due to the presence of a 4:1 ratio of α_{S1}:α_{S2}-casein, each with a different number of amino acid residues.

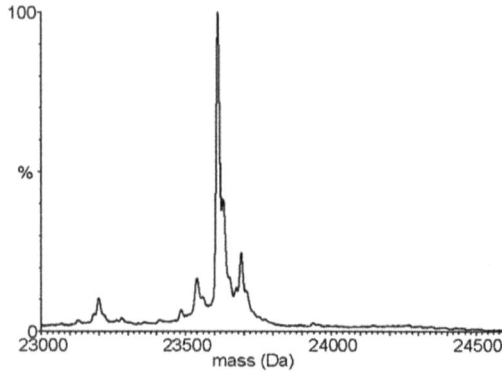

Figure 2. ESI-MS spectrum of purified αs₁-casein with a major peak occurring at 23, 619 kDa corresponding to αs₁B-casein 8P. Reprinted with permission from [17]. Copyright (2011) Elsevier Inc.

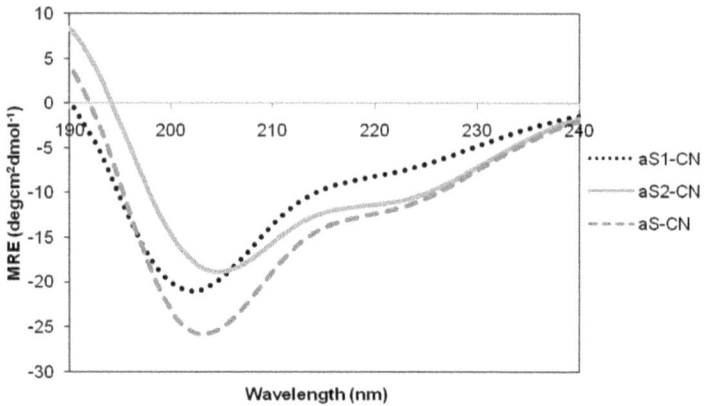

Figure 3. Far-UV CD spectrum of purified αs₁-, αs₂- and αs-casein at 0.20 – 0.28 mg/mL in prefiltered 10 mM sodium phosphate buffer, pH 7.4. Reprinted with permission from [17]. Copyright (2011) Elsevier Inc.

Protein	α-helix (%)	β-sheet (%)	β-turns (%)	Disordered structure (%)
αs₁-casein	14	28	22	35
αs₂-casein	18	26	22	33

Table 1. Secondary structural elements of αs₁- and αs₂-casein as determined by deconvolution of far-UV CD spectra [39-43].

2. Introduction to molecular chaperones

2.1. The perils of protein folding

Early experiments on the folding of ribonuclease *in vitro* revealed that all of the information required for a protein to fold into its native 3D conformation was contained within the primary sequence of that protein i.e. from the characteristics of amino acids and their positions in the polypeptide chain [44]. However, it also became evident that the folding of large, multi-domain proteins was complicated by incorrect intermolecular interactions involving the folding of the polypeptide chain [45] which prevented a proportion of proteins from reaching their native state, both *in vitro* [46, 47] and *in vivo* [48]. The process of protein folding, especially in the case of large proteins, appears to occur via a limited number of pathways. These pathways involve distinct intermediately-folded states known as 'molten globule' states [49]. This term is used to refer to a partially folded but compact state of a protein that has substantial amounts of secondary structure, but little or no tertiary structure [50]. In contrast to the natively folded state which is rigid and constrained, the molten globule state is not a single conformation but is instead a range of multiple conformations. These conformations are dynamic and rapidly interconvert with one another in response to the external environment [51] and they are also in equilibrium with more unfolded states [52]. Theoretical models have predicted a major loss of hydrophobic contacts in the molten globule compared to the natively folded state [50]. These intermediately folded states transiently expose previously buried hydrophobic areas on their surfaces and it is this characteristic that makes them prone to intermolecular association with other partially folded proteins, leading to aggregation and potential precipitation into insoluble aggregates [53, 54].

The currently accepted theory of protein folding is that natively folded proteins exist in equilibrium with less 'ordered' molten globule states. This 'folding/unfolding' pathway is reversible and slow, therefore it is possible for a protein that has partially unfolded but remains soluble to adopt a native state again provided it does not begin the process of aggregation. Once a molten globule state has begun to aggregate (either through interactions with other proteins or other molecules of the same protein), it has entered the fast and irreversible 'off-folding' pathway which ends with precipitation [55]. Within the cell, where proteins are present in high concentrations (e.g. 340 g/L in *E. coli*) and rates of protein production can be extremely high, folding of nascent proteins is further challenged by molecular crowding. Under such conditions, there are a great number of opportunities for inappropriate protein interactions. As the three-dimensional structure of a protein largely determines its function, the incorrect folding of proteins has the consequence of loss of function. It is not surprising therefore that abnormalities in protein folding form the basis of many human pathologies such as prion diseases and amyloidoses [56, 57].

2.2. The role of molecular chaperones

Molecular chaperones are a diverse group of proteins that act to prevent 'improper' interactions between other proteins that may result in aggregation and precipitation [58]. They ensure high fidelity protein folding and assembly without becoming part of the natively folded structure. In doing so, chaperones perform important roles in the stabilisation of many other proteins both intra- and extracellularly. Proteins from many unrelated families have been identified as possessing molecular chaperone function [59]. There are four key features that must be exhibited by a protein in order for it to be classified as a molecular chaperone: 1) suppression of aggregation during protein folding, 2) suppression of aggregation during protein unfolding, 3) influence on the yield and kinetics of folding and 4) effects exerted at near stoichiometric levels [60]. Many molecular chaperones have been identified in the eukaryotic cell, particularly in the endoplasmic reticulum and cytosol where they ensure correct folding, transport and biological activity of countless proteins [61]. Some chaperones act in sequence with others, passing on intermediately folded proteins to continue the folding process [62, 63]. Other chaperones, such as the small heat-shock proteins which are explained in greater detail later in this chapter, specifically interact with proteins on the off-folding pathway only. These intermediates, more prone to aggregation, are the ones recognised by the chaperone and stabilised against precipitation [58, 64].

2.3. Molecular chaperones are also heat-shock proteins

The expression of molecular chaperones is markedly increased (10 to 20 fold) under conditions of physiological stress (e.g. heat, reduction, oxidation stresses) - a feature which explains their other moniker as 'heat-shock' proteins (Hsps). It has been demonstrated that both the protein coding sequences [65, 66] and regulatory sequences [67] of some heat-shock genes have been highly conserved [68, 69].

2.4. The sHsps and clusterin

There are several classes of heat shock proteins, and the accepted nomenclature is based on their approximate molecular mass on SDS-PAGE i.e. Hsp60 and Hsp70 are 60 and 70 kDa, respectively. A subset of the Hsps is called the small heat-shock proteins, or sHsps, with molecular masses of monomers ranging from 15-30 kDa. These chaperones act in an ATP-independent manner, which unlike Hsp60 and Hsp70 do not actively refold in the presence of ATP. Instead, they interact with partially unfolded or 'stressed' proteins, stabilising them in a soluble, high molecular weight complex to prevent their precipitation from solution [70]. They do not interact with natively folded proteins, nor with those that have already aggregated [71]. Rather than simply serving as a one-way 'sink' for denatured proteins, however, previous studies have shown that in addition to their ability to interact with and stabilise stressed proteins, some sHsps such as Hsp25 [58], α-crystallin [72] and clusterin [73] act co-operatively with ATP-dependent chaperones (e.g. Hsp70) to refold the stressed protein when the stress is removed and normal cellular conditions are restored. Studies on α-crystallin have shown that the presence of ATP causes the sHsp to undergo a conformational change whereby the stressed target protein is released facilitating refolding by chaperones such as Hsp70 [74]. Members of the sHsp family have several features in common, including size and amino acid sequence homology within the C-terminal domain [55]. There have been 10 human sHsps identified thus far, including αA- and αB-crystallin (discussed below), Hsp27 and its murine equivalent, Hsp25 [75, 76].

The sHsps are also able to exist as monomeric and dimeric species which associate with one another to form large multimeric complexes, somewhat akin to the behaviour of the casein proteins both within and without the casein micelle. Our studies have shown that α_s-casein, and more specifically, α_{s1}- and α_{s2}-casein, also act as molecular chaperones and do so in a manner that is similar to the sHsps and an extracellular chaperone, clusterin. Clusterin is a secreted mammalian chaperone present in bodily fluids such as blood and semen. Like the sHsps, clusterin is highly conserved and is upregulated in many cell types under conditions of stress and in protein misfolding diseases such as Alzheimer's disease [77]. It is a disulfide-linked dimer which is 75 – 80 kDa in size and highly glycosylated [73, 77]. Clusterin has been shown to aggregate in aqueous solution and at physiological pH values is present in monomeric, dimeric and multimeric states [78]. Like the sHsps, clusterin is an ATP-independent molecular chaperone capable of stabilising a wide array of target proteins [73, 77]. Stoichiometrically, clusterin has been shown to protect stressed target proteins at levels consistent with the concentration range of clusterin normally found in extracellular fluids (50 – 370 µg/mL in human serum and 2.1 – 15.0 mg/mL in human seminal fluid). In addition, clusterin displays greater chaperone activity with smaller target proteins versus larger ones [77] indicating that it has the ability to interact differently with various target proteins depending on the conditions. This variability of chaperone action has also been demonstrated extensively in studies on sHsps such as α-crystallin which have been shown to exhibit increased or decreased chaperone activity depending on the mode of aggregation of a target protein [79-81].

α-Crystallin is a member of the small heat-shock protein family and has for some time been known to play a major role in stabilising other crystallin proteins in the eye lens. It was identified some time ago that α-crystallin performed an important structural role in the lens, as a member of the crystallin family of proteins which also comprises β- and γ-crystallin [82]. In its normal state, the lens is transparent despite the high concentrations of these proteins in the cell cytoplasm (33% in the human lens and 50% in rat and bovine lenses [83]). Such a high concentration of proteins would ordinarily cause a significant degree of light scatter, but these highly homologous proteins adopt a critical short-range order that allows them to exist in a dense glass-like liquid resulting in unimpeded transmission of light through the lens [84, 85]. Their structural integrity therefore is of prime importance and disruption to their three-dimensional arrangement as a result of chemical modification, for example, has been shown to result in increased light scattering manifesting as cataract [86]. Due to the lack of protein turnover in the lens over the lifetime of an organism, the occurrence of α-crystallin in the lens allows it to perform a second, equally important role as a molecular chaperone. Numerous studies have shown that under conditions of cellular stress, α-crystallin interacts with not only other crystallins but with a plethora of other proteins, stabilising them against precipitation. Its wide tissue distribution and localisation in various disease states associated with protein misfolding also provides strong evidence for the role of α-crystallin as a molecular chaperone outside the lens. Like α_s-casein, α-crystallin is made up of two distinct gene products designated α_A- and α_B-crystallin, so named because of their relatively acidic and basic properties. The individual subunits of α-crystallin are present in a 3:1 ratio (α_A:α_B) in the human lens, are each ~20 kDa in size and these proteins readily associate with one another to form dimeric and multimeric species. These multimers can be up to 1.2 MDa in size and exhibit dynamic subunit exchange. As such, α-crystallin, like αs-casein has been resistant to crystallisation and a precise picture of its mechanism of action is still to be elucidated. Almost 100 years after being identified in the lens, individual αA- and αB-crystallin subunits were found in non-lenticular tissues [87-91]. These proteins and other sHsps have been identified in the brain where they are associated with neurodegenerative diseases such as Alzheimer's, Creutzfeldt-Jakob and Parkinson's diseases (summarised in [92]).

2.5. Putative mode of action of sHsps and clusterin

As reviewed in [93], current theories of the chaperone action of both sHsps and clusterin centre around their ability to expose hydrophobic regions that interact with partially folded target proteins, also known as 'disordered' molten globules, forming soluble, high molecular weight complexes and preventing them from precipitation. It has been shown that sHsps do not interact with target proteins that are natively folded, completely denatured, or those in stable molten globule states [55, 94-97]. The stoichiometry of sHsp and clusterin interaction with target proteins suggests that one oligomer of chaperone can bind to and stabilise many molecules of stressed target protein, in fact, CryoEM and X-ray solution studies on stressed α-lactalbumin and αB-crystallin showed that the target protein coated the exterior surface of the αB-crystallin oligomer upon formation of the chaperone-target protein complex [98, 99].

The relatively hydrophobic nature of the more globular portion of the sHsp is balanced by the adjoining flexible, dynamic C-terminal extension which is solvent exposed and hydrophilic and is thought to play a major role in ensuring the solubility of the huge complex formed upon chaperone interaction [100, 101]. It has also been demonstrated that the C-terminal extension in α-crystallin is critical for oligomeric assembly [79]. Altered 'spacing' of chaperone molecules resulting from modification of the C-terminal sequence results in abnormally sized oligomers with perturbed structure, physical stability and chaperone function [79]. The large, oligomeric forms of sHsps are thought to exist in dynamic equilibrium with smaller species which rapidly interchange with the oligomer [102, 103]. This equilibrium of subunit exchange is believed to be key to sHsps broad target protein specificity [97, 103-105], and although clusterin shares many features with the sHsps in terms of its chaperone action, the potential importance of subunit exchange for it has not been described to date [106].

3. αs-Casein as a molecular chaperone

The casein proteins and their derivatives have been used by the food industry as important nutritional and stabilising proteins for many years [4]. Early studies showed that whole casein (i.e. αs-, β- and κ-casein) prevented heat induced aggregation of whey proteins, even in calcium-containing systems [107, 108]. This stabilising action of the caseins on heat-denatured target proteins was proposed to occur through non-specific interactions and opened up a new avenue of uses for casein proteins in stabilising both milk and non-milk proteins and thereby contributing to novel properties of milk products. It was then demonstrated in 1999 that individual αs-casein possessed molecular chaperone activity [109]. Since then, β- and κ-casein have both been shown to also act as molecular chaperones [18, 33, 110-112]. The presence of high numbers of phosphate groups in the casein proteins appears to be important for chaperone action against amorphously aggregating target proteins under both reduction and heat stress, with studies showing that removal of these in αs- and β-casein reduced their ability to prevent the aggregation of target proteins [113, 114]. Previous work on αs-casein showed that it prevented the stress-induced aggregation of natural target proteins such as the whey proteins β-lactoglobulin and bovine serum albumin, but also of unrelated proteins such as alcohol dehydrogenase and carbonic anhydrase [109]. In acting as a molecular chaperone under these conditions, αs-casein was able to interact with partially unfolded target proteins and prevent their incorporation into insoluble aggregates which would then have formed precipitates. Furthermore, when added to partially aggregated reduction-stressed insulin, αs-casein not only prevented further aggregation of insulin but facilitated its re-solubilisation when present at a 2:1 (w:w) ratio [109].

As shown in Figure 4, the putative mode of action of αs-casein is based on a similar model proposed for sHsps [70] where a natively folded protein (N) unfolds via a number of intermediately folded states (I₁, I₂ etc.) or 'molten globule' states on its way to the unfolded state (U). This folding and unfolding is fast and reversible and involves the exposure of hydrophobic regions normally buried in the interior of the protein.

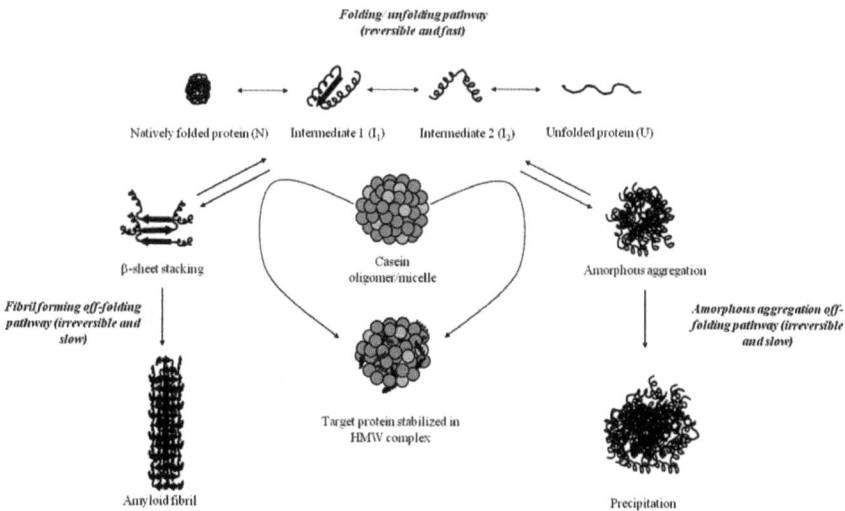

Figure 4. Putative mechanism of action of αs-casein showing its interaction with target proteins on the folding and off-folding pathways. See text for explanation. Reprinted with permission from [33]. Copyright (2009) Dairy Industry Association of Australia.

Under conditions of cellular stress, when intermediately folded states are present for longer periods, self-association is promoted by the prolonged exposure of hydrophobic surfaces. When self-association occurs, the intermediately folded states enter the off-folding pathways which are slow and irreversible and may lead to either amorphous aggregation as shown on the right hand side of the figure or to fibril formation as shown on the left. Amorphous aggregates result from disordered aggregation and lead to the formation of insoluble protein precipitates. Conversely, the ordered amyloid pathway leads to highly ordered β-sheet stacking giving cross β-sheet fibrils. Casein micelles or oligomeric forms of αs- or β-casein are able to interact with partially folded proteins and stabilise them against aggregation and precipitation by forming a soluble high molecular weight (HMW) complex [33].

3.1. Assessment of molecular chaperone activity

Assessment of molecular chaperone activity traditionally involves an *in vitro* assay in which a target protein is subjected to a form of stress similar to what would be encountered under physiological conditions (e.g. heat, oxidation, reduction etc.) in the presence and absence of different amounts of chaperone. These assays are not complicated by the aggregation of the chaperone itself which is stable under these conditions, and suppression of aggregation has been shown to be specific to the action of molecular chaperones as substitution of these with non-chaperones (e.g. ovalbumin) has been shown to have no effect on the extent of target protein precipitation [73]. For the assessment of amorphous aggregation, light scattering of the proteins in solution at 360 nm is monitored over time, whereas for fibril forming proteins, amyloid formation is monitored by increasing Thioflavin T (ThT) fluorescence. These values increase to a maximum over the timeframe of the experiment and are used to estimate % protection provided by the chaperone as per the following formula:

$$\% \text{ Protection} = \frac{100 * \left(\Delta I - \Delta I_{\text{chaperone}} \right)}{\Delta I} \tag{1}$$

(1) Calculation of percentage protection of a target protein by molecular chaperone. In this formula, ΔI and $\Delta I_{\text{chaperone}}$ represent the change in light scattering (for amorphously aggregating target proteins) or in Thioflavin T (ThT) fluorescence (for amyloid forming target proteins) in the absence and presence of the chaperone, respectively [79].

3.2. αs-Casein stabilises proteins *via* formation of soluble high molecular weight complexes

Consistent with the well-characterised properties of a molecular chaperone, it has been shown that αs-casein (and β-casein) form soluble, high molecular weight (HMW) complexes with stressed target proteins that can be identified by size-exclusion chromatography. These complexes, formed between αs-casein and heat-stressed β-lactoglobulin or apo-α-lactalbumin eluted from the column at a retention time correlating to an approximate mass of 1.8 MDa [18]. Control experiments showed that in the absence of a chaperone, heat stressed β-lactoglobulin also formed large, multimeric species but these were largely insoluble [18]. Mixtures of the individual whey proteins and αs-casein that were not exposed to heat stress eluted as separate peaks with elution times corresponding to the individual proteins i.e. there was an absence of a target protein-chaperone or HMW complex. Additional experiments showed that the elution time of αs-casein was not altered in any discernible way by the addition of heat stress [18].

The molten globule states of a natural target protein of αs-casein, α-lactalbumin, have been well characterised [13, 55, 115-122]. Intermediately folded states of the apo- form of α-lactalbumin provide an ideal model for the investigation of protein folding and unfolding and therefore the action of molecular chaperones when present in solution. The molten globule states of apo-α-lactalbumin exhibit a relatively compact structure in which a secondary structure is largely preserved, but tertiary structure is lost [123]. A characteristic of these states is that they expose significant amounts of hydrophobicity to solution as a result of their being 'uncovered' from the interior of the previously natively folded protein and it is these exposed hydrophobic areas that appear key to their interaction with molecular chaperones [124, 125]. Destabilisation of α-lactalbumin upon removal of its calcium ion by a chelating agent such as EDTA induces a conformational change that further exposes the disulfide bonds of the protein to reduction with DTT [122]. Under this reduction stress, α-lactalbumin adopts a molten globule state that is structurally unstable and similar to that formed at pH 2.0 [13]. This state, in the absence of a molecular chaperone, such partially unfolded proteins will readily aggregate and precipitate. When monitored via real-time ^{1}H NMR spectroscopy, the formation of the molten globule state, its aggregation and eventual precipitation can be visualised. The aromatic protons in the region of the spectrum from 6-8 ppm can be attributed almost exclusively to the signals of protons in the target protein, which is small (14.4 kDa) and monomeric. Resonances arising from aromatic protons in αs-casein are relatively broad by comparison, so that even in the presence of added chaperone, the structural alterations in the target protein are easily observed. Isolated resonances arising from tyrosine 3,5 ring protons at 6.8 ppm are therefore a reliable indicator of molten globule formation and stabilisation in α-lactalbumin [18, 122, 126]. The well-resolved resonances visible at Time 0 are representative of the native state and are quickly lost with the addition of DTT. In the absence of αs-casein (Figure 5A), an initial increase in resonance arising from the Tyr (3,5) protons can be observed from ~ 0-200 s following the addition of DTT. This period represents the reduction of the disulfide bonds in α-lactalbumin by DTT, giving the molten globule state which in the absence of chaperone is prone to aggregation then precipitation [122] after a period of approximately 1000 s [18]. In the presence of a 2:1 (w:w) ratio of αs-casein to apo-α-lactalbumin (Figure 5B) , signals arising from the aromatic protons are preserved, indicating that the molten globule state of α-lactalbumin is stabilised by the interaction between the two proteins. As shown in Figure 5C, in the absence of αs-casein the rate of decay of the Tyr (3,5) resonance of apo-α-lactalbumin is rapid, occurring at rate of $2.70 \ (\pm 0.11) \times 10^{-3} \ s^{-1}$ whereas in the presence of αs-casein it is approximately 50% slower at $1.25 \ (\pm 0.08) \times 10^{-3} \ s^{-1}$ [18].

Extrinsic fluorescence studies have shown that upon formation of a high-molecular weight complex between αS-casein and reduction stressed insulin, a conformational change in one or both proteins resulted in an increase in clustered, exposed hydrophobicity. The other casein proteins, β- and κ-casein also exhibited similar increases in hydrophobicity in the presence of reduction stressed insulin, implying that similar interactions were occurring during formation of the chaperone-target protein complex [18]. A mixture of the three caseins (αs-, β- and κ-casein) combined according to their approximate proportions in bovine milk (60%, 25% and 15%, respectively), exhibited considerably increased exposure of clustered hydrophobic areas, indicating a synergy between the various subunits during stabilisation of stressed insulin [18]. Intrinsic (tryptophan) fluorescence studies showed that the tryptophan residues in the C-terminal region of αs-casein were exposed to a more non-polar environment as a result of the interaction with the reduced insulin B-chain, and may indicate involvement of this region of the chaperone with the hydrophobic (bound) target protein [18].

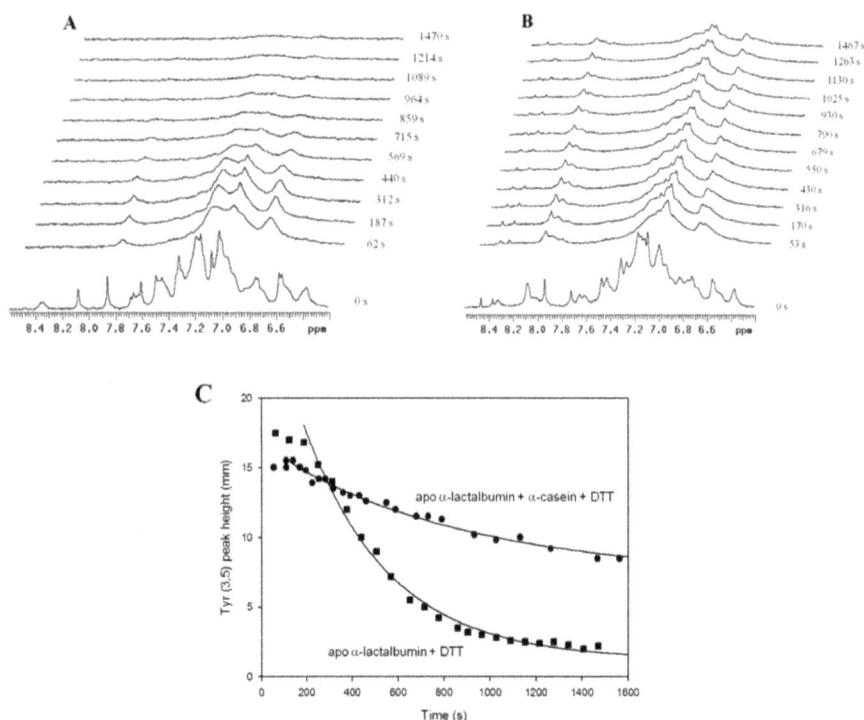

Figure 5. Real-time 1H NMR spectroscopy of apo-α-lactalbumin under reduction stress at 37°C in the presence and absence of αs-casein. Stacked plots of 1D 1H NMR spectra show molten globule formation by reduced apo-α-lactalbumin in A) the absence and B) the presence of αs-casein at a 2:1 (w:w) ratio. First-order decay of the resonance arising from the tyrosine (3,5) ring protons at 3.5 ppm in both cases (C) was used to calculate rates i.e. peak height (mm) v time (s). Reprinted with permission from [18]. Copyright (2005) American Chemical Society.

3.3. Effects of temperature, pH and molecular crowding

As shown in [18], αs-casein displayed temperature dependence in preventing aggregation of the reduced insulin B-chain, with comparatively better chaperone ability observed at 25°C compared to that at 37°C. This was in keeping with previous studies that showed that αs-casein was a considerably better chaperone at lower temperatures, giving 52% protection of insulin aggregation at 27°C compared to almost complete protection at 18°C [109]. This 'cold shock protein' characteristic is quite different to the temperature effects of other molecular chaperones such as α-crystallin and tubulin, which display enhanced chaperone activity at increased temperatures [127-129]. At temperatures above 30°C, α-crystallin undergoes structural transitions that involve rearranging and/or increasing hydrophobic surfaces, enhancing hydrophobic interactions between the chaperone and the target protein [124, 125]. Conversely, clusterin's chaperone ability appears to be quite independent of temperature effects [130]. The properties of αs-casein that allow it to chaperone at lower temperatures are reflected in the fact that greater amounts of αs-casein have been identified in mammals inhabiting low temperature zones compared to those in more temperate climates [131, 132] which may indicate an important stabilising role in physiological adaptation to low temperatures. Both β- and κ-casein also display chaperone activity, but in contrast to αs-casein, the chaperone action of these proteins does not appear to be as dependent on temperature as αs-casein [18]. In stabilising insulin, apo- and holo-α-lactalbumin as target proteins, β- and κ-casein did not provide as much protection against precipitation as αs-casein on a w:w and molar basis, but were similar to each other in their efficacy [18]. This was in contrast to previous reports which described greater chaperone action in β-casein compared to αs-casein in stabilising heat stressed catalase and reduction stressed lysozyme [110], whereas others found that αs-casein was better than β-casein at preventing the aggregation of heat-stressed ovotransferrin [113]. This apparent disparity in the relative chaperone performances of the casein proteins is likely to be related to their broad target protein specificities. The same observations have been made in relation to the chaperone action of sHsps and chaperone efficiency has been shown to be dependent upon the size of the target protein and the rate and mode of aggregation, as previously discussed [79-81]. A combined form of the casein proteins (60% αs-, 25% β- and 15% κ-casein) was shown to possess similar chaperone activity to that of αs-casein alone [18].

In the investigation into the mechanism of chaperone action of αs-casein, experiments with its natural target protein, β-lactoglobulin, have been performed under heat stress at 70°C [18, 109]. Obviously this level of heat stress is not physiologically appropriate, but the ability of αs-casein to stabilise other dairy proteins under extreme heat stress is of importance to the dairy industry which employs treatment processes such as pasteurisation and ultra-high temperature treatment [18]. In these studies, it was found that pH had a major effect on the chaperone's ability to suppress the aggregation of the target protein. The ability of αs-casein to suppress the aggregation of β-lactoglobulin at a 0.25:1 (w:w) ratio decreased from 73% at pH 7.0 to 33% at pH 7.5 and 19% at pH 8.0 after 450 min [18]. This is likely to be related to the rate at which β-lactoglobulin aggregates at higher pH values and may also be related to changes in the chaperone ability of αs-casein at more alkaline pH values. At slightly alkaline pH values, the histidine residues of αs-casein deprotonate ($\alpha s1$-casein has five histidines and $\alpha s2$-casein, three) and it is feasible that this results in a loss of electrostatic contacts between the subunits, disrupting their ability to effectively bind the target protein [18].

A pH dependence is also exhibited by clusterin, which is in fact a better chaperone at more acidic pH values [130]. This is believed to be due to the increased dissociation of clusterin from its larger multimeric state at acidic pH values [130]. Conversely, α-crystallin is less effective as a chaperone at slightly acidic pH values [130, 133] but this is thought to be due to subtle structural changes in the protein with a change in pH rather than a change in aggregation state [134]. It may also be that at increased pH values αs-casein is less effective as a chaperone against β-lactoglobulin aggregation because the nature of the aggregation changes at more alkaline pH values. Greater intermolecular disulfide bond formation produces oligomers of β-lactoglobulin which then polymerise to form an aggregate. It is likely that αs-casein is similar to α-crystallin in that it cannot interact with and stabilise these forms of aggregated proteins as effectively as those that aggregate in a nucleation-dependent manner [135]. Other factors that have been shown to influence the nature and rate of target protein aggregation also have an impact on αs-casein's chaperone action. One of these is molecular crowding. Experiments were conducted in the presence of 10% dextran in order to simulate the protein-rich environment of milk and a greater rate of target protein (insulin, apo- and holo-α-lactalbumin) precipitation was observed under these conditions [18]. This increased rate of aggregation affects αs-casein's chaperone ability in the same way as described previously for α-crystallin, other sHsps and clusterin, which are all poorer chaperones with rapidly aggregating target proteins [94, 96, 122, 136, 137].

3.4. αs-Casein is unable to protect enzymes from loss of function due to heat stress

As previously discussed, it is known that molecular chaperones such as sHsps interact with and bind partially unfolded target proteins in molten globule-like states, but not once they have begun to aggregate [71]. The chaperone behaviour of α- and β-casein appears different in this regard, with both having been shown to re-solubilise aggregated DTT-treated insulin [109, 110]. According to reference [138], one of the features of a protein used to classify it as a molecular chaperone is that it be able to aid in the recovery of lost biological activity. Like the sHsps and clusterin, however, αs-casein is unable to prevent the loss of activity in enzymes induced by heat stress. Our experiments with catalase, GST and ADH show that the presence of αs-casein at a 1:1 (w:w) ratio does not protect these enzymes to any significant extent against heat-induced loss of function (Figure 6).

3.5. αs-Casein is ATP-independent in its chaperone action

In order to obtain further insight into the molecular mechanism of action of αs-casein, the effect of ATP on its chaperone function was also investigated [17; Treweek, Price & Carver, unpublished work]. Previous studies have shown that αs-casein acts in a similar manner to small heat shock proteins (sHsps) and clusterin [17, 18, 28, 29, 33]. A major characteristic of

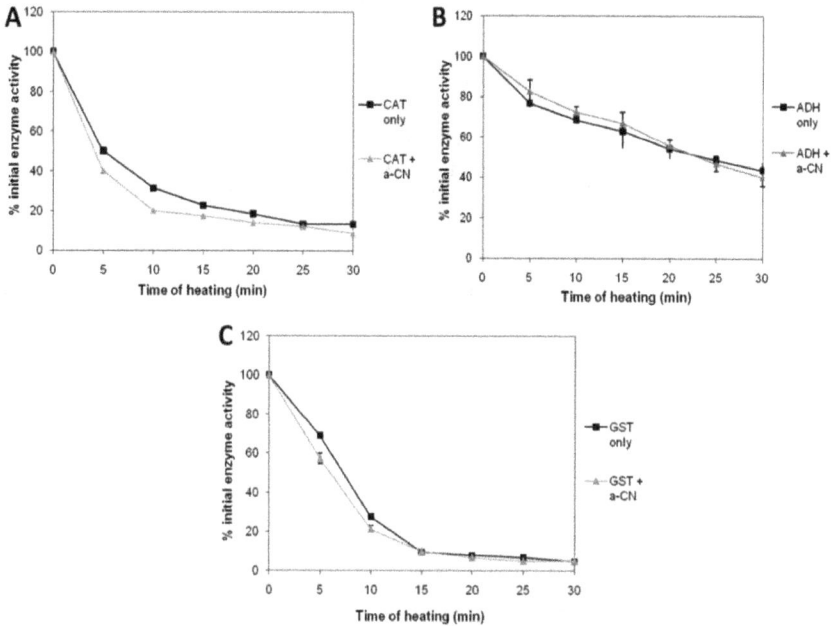

Figure 6. Loss of enzyme activity (catalase, CAT; ADH, alcohol dehydrogenase; GST, glutathione-S-transferase) with heat stress at 55°C and the effect of addition of αs-casein at a 1:1 (w:w) ratio. Values represented are means of independent triplicate measurements and error bars shown are standard deviations of the mean which in some cases are too small to be visible. Reprinted with permission from [17]. Copyright (2011) Elsevier Inc.

sHsps and clusterin is that they function in an ATP-independent manner consistent with their inability to refold stressed proteins [139-141]. ATP levels in milk are relatively low (5 μM) and as such it is likely to be non-essential to αs-casein's chaperone activity, however, it is an important mechanistic tool. A similar study on clusterin (which is also extracellular and as such experiences low physiological ATP levels) provided valuable insight into the ATP-independent action of the chaperone in binding the stressed enzymes catalase, ADH and GST [73]. Subsequent refolding of bound target proteins to clusterin and sHsps (e.g. Hsp25 and α-crystallin) is achieved via ATP-dependent chaperones such as Hsp70 [58, 72]. The presence of physiologically relevant levels of ATP on the ability of αs-casein to suppress aggregation of stressed target proteins (specifically catalase, ADH and insulin) was investigated and it was found that the chaperone action of αs-casein was unaffected by the presence of ATP [17; Treweek, Price & Carver, unpublished work]. In addition, the ATPase activity of αs-casein was assessed and it was found that αs-casein had no detectable ATPase activity either on its own or during chaperone action i.e. when interacting with heat stressed β-lactoglobulin in a chaperone complex (Figure 7) [Treweek, Price & Carver, unpublished work].

3.6. αs-Casein does not bind target proteins in a way that allows refolding by Hsp70

As previously mentioned, some sHsps such as Hsp25 [58], α-crystallin [72] and clusterin [73] act co-operatively with ATP-dependent chaperones such as Hsp70 to refold stressed proteins when the stress is removed and cellular conditions are restored [139]. Studies on α-crystallin have revealed that a region in the conserved α-crystallin domain of αB-crystallin undergoes structural modification upon binding of ATP [142] and that this conformational change causes α-crystallin to release stressed target proteins, facilitating their refolding by chaperones such as Hsp70 [74]. Although not a natural component of milk, Hsp70 was used to probe the mechanism of αs-casein's chaperone action and to allow comparisons with better characterised chaperones to be made.

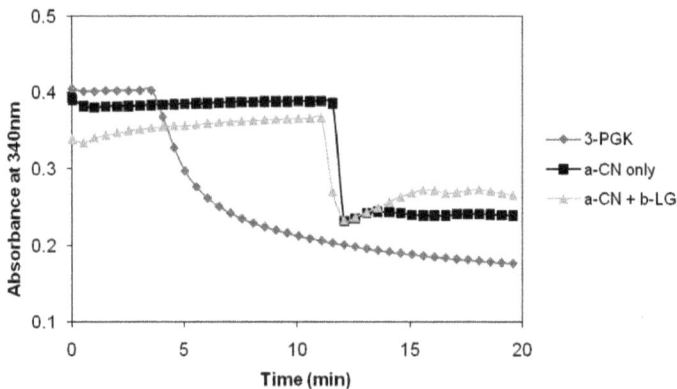

Figure 7. ATPase activity of αs-casein. The ability of αs-casein (a-CN) to hydrolyse 170 µM ATP was examined in the presence and absence of heat-stressed β-lactoglobulin (b-LG) at a ratio of 2:1 (w:w) in 50 mM sodium phosphate buffer containing 0.2 M NaCl, 2.5 mM EDTA and 0.02% NaN₃ at pH 7.2. Generation of ADP was monitored *via* NADH oxidation giving a decrease in absorbance at 340 nm. This can be seen with the addition of 21 nmol exogenous ADP at 11 min (black arrow) and the addition of 1.1 µmol 3-phosphoglycerate to 0.76 units of 3-phosphoglycerate kinase (3-PGK) at 4 min (grey arrow). All experiments were performed in triplicate at 37°C.

Recovery of enzyme activity was assessed using heat-stressed catalase and ADH [17, 73] in the presence and absence of 1:1 (w:w) ratios of αs-casein similar to the assays shown in Figure 6, with the addition of Hsp70 and ATP after a recommended 30 min 'recovery period' [17, 58]. In the presence of αs-casein, neither catalase nor ADH showed any significant recovery from heat stress at 55°C, with only 1(±5)% activity remaining 5 hours after the addition of Hsp70 and ATP. In contrast, control experiments showed 27(±5)% recovery of catalase activity over the same time period and under the same conditions, but with the addition of clusterin instead of αs-casein [17]. This effect of clusterin has been shown to be specific for the chaperone, as other proteins added in its place (e.g. lysozyme, myoglobin) have been shown to be unable to facilitate enzyme recovery to the same extent

[73]. It could be concluded, therefore, that the binding of Hsp70 to its target protein to achieve refolding and release, coupled with ATP binding and hydrolysis, is impeded by the manner in which αs-casein binds the destabilised target protein [17]. In stabilising the heat-stressed enzymes, αs-casein may hold its target more tightly or incorporate them into the complex in such a way that they are not accessible to Hsp70 to allow for the refolding that has been seen for target proteins stabilised by sHsps and clusterin [58, 72, 73, 139]. It is possible that the αs-casein-target protein complex is to some extent incorporated the casein micellar structure and this arrangement influences the accessibility of Hsp70 [17]. CryoEM and X-ray solution studies on sHsps have revealed that stressed α-lactalbumin molecules coat the exterior surface of the oligomeric form of αB-crystallin when the chaperone-target protein complex is formed [99, 143]. Further studies with spin-labelled melittin peptides have shown that there is a stoichiometry of approximately 1:1 [144] in the binding of the peptides to each monomer of α-crystallin and that these are relatively evenly spaced [145]. This regular arrangement of target protein binding to α-crystallin may be an important factor in the ability of Hsp70 to subsequently refold target proteins [17].

3.7. αs1- and αs2-Casein also act as molecular chaperones

The constituent proteins of αs-casein, αs1- and αs2-casein (separated and purified as described earlier in this chapter) have been shown to exhibit chaperone action independently of one another. Like αs-casein, the chaperone action of αs1- and αs2-casein has been described for a range of target proteins under various stress conditions [17]. Studies with reduced insulin as the target protein showed that at ratios of 0.25:1 and 0.5:1 (w:w) casein: insulin) both αs1- and αs2-casein had comparable chaperone activity to αs-casein (summarised in Table 2).

α-CN:insulin (w:w)	αs-CN	αs1-CN	αs2-CN
0.1:1	66 (± 2)%	49 (± 2)%	64 (± 2)%
0.25:1	74 (± 0.6)%	69 (± 2)%	73 (± 1)%
0.5:1	96 (± 0.4)%	91 (± 1)%	91 (± 0.3)%
1:1	98 (± 0.9)%	98 (± 0.1)%	97 (± 0.4)%

Table 2. Summary of chaperone assay data with insulin under reduction stress in the presence of increasing amounts of αs-, αs1- or αs2-CN. Figures shown are % protection of stressed target protein by the chaperone. Percentage protection is calculated as previously described. Reprinted with permission from [17]. Copyright (2011) Elsevier Inc.

When assessed with catalase under heat stress, a 0.5:1 ratio of αs-casein to catalase provided 88 (± 2)% protection after 50 min. Under these conditions, αs2-casein was the better chaperone, giving 84 (± 4)% protection at the same ratio and time point, whereas αs1-casein provided only 64 (± 1)% at the same ratio and time point [17]. In another set of experiments that included 0.1 M NaCl in order to more accurately simulate the high salt conditions in milk, catalase aggregation occurred more rapidly and as a result, all of the α-casein proteins tested were less effective in preventing catalase aggregation and precipitation. These results are summarised in Table 3 and visible spectroscopy spectra are shown in Figure 8.

α-CN:cat (w:w)	αs-CN	αs1-CN	αs2-CN	αs-CN + salt	αs1-CN + salt	αs2-CN + salt
0.25:1	76 (± 1)%	44 (± 9)%	77 (± 3)%	15 (± 0.1)%	3 (± 0.8)%	16 (± 2)%
0.5:1	88 (± 2)%	64 (± 1)%	84 (± 4)%	27 (± 1)%	16 (± 2)%	26 (± 2)%
1:1	96 (± 2)%	80 (± 3)%	85 (± 6)%	57 (± 1)%	16 (± 3)%	18 (± 2)%

Table 3. Summary of chaperone assay data with catalase (cat) under heat stress in the presence of increasing amounts of αs-, αs1- or αs2-CN and in the presence and absence of 0.1 M NaCl. Figures shown are % protection of stressed target protein by the chaperone. Percentage protection is calculated as previously described. Reprinted with permission from [17]. Copyright (2011) Elsevier Inc.

Under conditions of added salt, the chaperone abilities of αs-, αs1- and αs2-casein were all greatly reduced, with a 0.5:1 ratio of αs-casein: catalase providing only 27 (± 1)% protection, compared with 88 (± 2)% for the same ratio in the absence of salt. At the same ratio and time point αs2-casein showed similar chaperone ability to αs-casein (26 (± 2)% protection), with the most profoundly affected protein being αs1-casein. Under these conditions, even a 1:1 ratio of αs1-casein: catalase provided only 16 (± 2)% protection. Analysis of the samples from these assays by SDS-PAGE showed that αs1-casein had actually formed high molecular weight complexes that failed to migrate through the gel matrix to any extent, but like the other chaperones, remained soluble [17]. Despite remaining in solution, the data show that this aggregated form of αs1-casein was no longer an effective chaperone. The self-association of αs1-casein in the presence of salt most likely occurs as a result of neutralisation of charged residues on the protein ('charge screening' [146]) by the interaction with sodium and chloride ions i.e. the early stages of isoelectric precipitation [12]. Aromatic residues in αs1-casein are also thought to play a major role in the hydrophobic interactions between αs1-casein molecules, which at increasing ionic strengths go from monomers to dimers, tetramers, hexamers, octamers, then higher order aggregates [147].

Molecular chaperones are known to stabilise amorphously aggregating proteins like those described above, but they also interact with and stabilise proteins destined to form fibrillar aggregates (refer to Figure 4). This property has been described for sHsps in suppressing amyloid fibril formation in β-amyloid peptide [148, 149], apolipoprotein C-II [150] and in α-synuclein, the protein present in the Lewy bodies of Parkinson's disease [151]. A form of κ-casein which is destabilised as a result of being reduced and carboxymethylated (RCM-κ-casein) has been shown to form fibrils at 37°C in the presence of DTT and has been widely used as a model for investigating chaperone action against fibrillar proteins [28, 33, 79, 152].

Studies have shown that the presence of a ~0.6:1 w:w ratio of αs-casein:RCM-κ-casein reduced the Thioflavin T fluorescence (an indicator of the probe's binding to forming fibrils) by 65%, and in the presence of a 2.5:1 w:w ratio of αs-casein:RCM-κ-casein, fibril formation was completely abrogated [17, 28]. In the presence of αs1- or αs2-casein at the same ratio, fibril formation was also completely suppressed. At lower ratios however i.e. ~0.6:1 w:w α-casein:RCM-κ-casein, αs1-casein was comparable to αs-casein in that it suppressed fibril formation by 96%, but αs2-casein was able to only provide 56% protection under the same conditions as shown in Figure 9 [17].

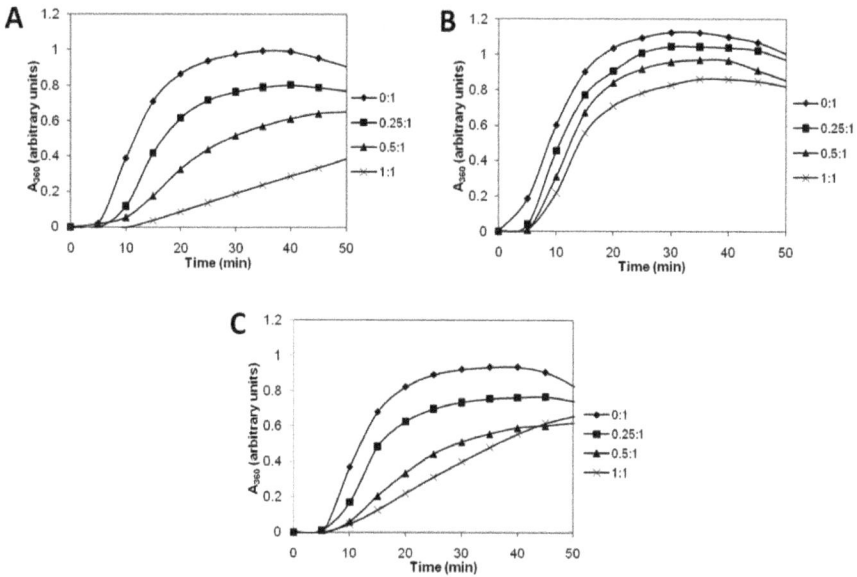

Figure 8. Chaperone activity of A) αs-, B) αs1- and C) αs2-casein against amorphously aggregating catalase, in the presence of 0.1 M NaCl and various w:w ratios as determined by light scattering at 360 nm. The 0:1 ratio in each assay represents catalase aggregation in the absence of α-casein i.e. no chaperone present. Reprinted with permission from [17]. Copyright (2011) Elsevier Inc.

The results of the fibril-forming experiments discussed above were further confirmed by TEM studies (Figure 10) which showed that the presence of a 1.25:1 ratio of αs1-casein resulted in reduced numbers of fibrils being formed by RCM-κ-casein (Figure 10B). In addition, those fibrils that were formed were shorter in length than those observed for RCM-κ-casein in the absence of chaperone (Figure 10A). Conversely, in the presence of the same ratio of αs2-casein, RCM-κ-casein fibrils were abundant and were associated with rounded aggregates 50–100 nm in diameter (Figure 10C) that may have contained one or both proteins. αs2-Casein, which forms characteristic twisted fibrils, was included as a control at the same concentration but in the absence of chaperone [17].

Previous studies have shown that both β- and αs-casein have the ability to inhibit the formation of fibrils by κ-casein [28], and this observation has led to the conclusion that amyloid formation in mixtures of casein (i.e. whole casein), namely by αs2- and κ-casein, is prevented by the action of the casein chaperones, β- and αs1-casein. As αs1-casein was a more potent inhibitor of fibril formation than αs2-casein under these conditions, it would be reasonable to assume that a large proportion of the fibril-preventing action of αs-casein is provided by αs1-casein. As previously discussed, the chaperone activity of αs1-casein against amorphously aggregating catalase in the presence of salt was a stark contrast, but provides the rationale that perhaps in milk, where salt concentrations are high, αs1-casein has an important role in preventing fibrillar aggregation rather than amorphous aggregation [17].

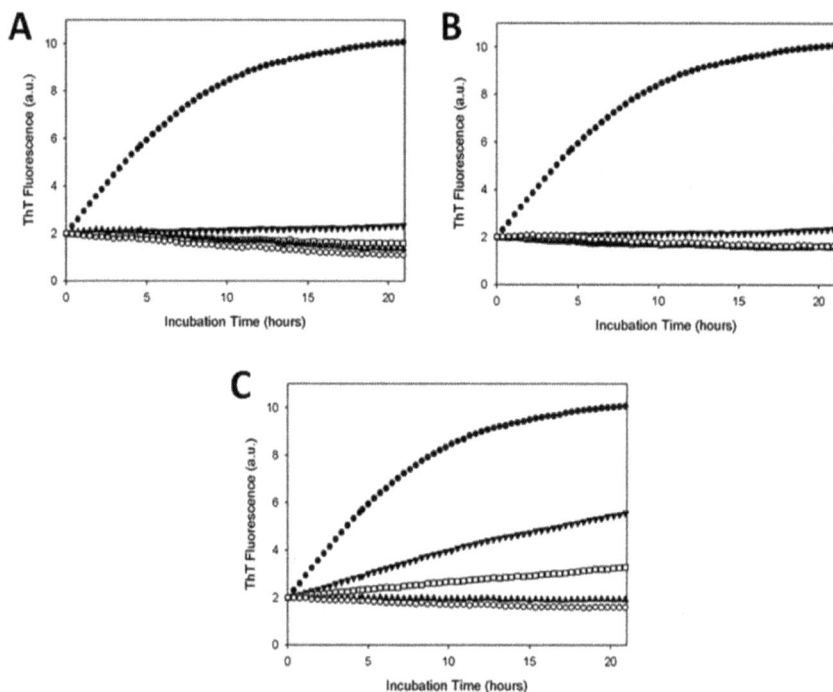

Figure 9. Chaperone activity of A) αs-, B) αs1- and C) αs2-casein against amyloid forming RCM-κ-casein as determined by Thioflavin T (ThT) fluorescence. Ratios of chaperone:RCM-κ-casein equivalent to ~0.6:1, 1.25:1 and 2.5:1 (w:w) are represented by ▼, □ and ▲, respectively. Reprinted with permission from [17]. Copyright (2011) Elsevier Inc.

Investigations into the mechanism of αs-casein's chaperone action are only in their infancy, and the precise nature of its chaperone function remains largely an enigma. Despite this, comparisons with the chaperone action of sHsps have provided several important insights. In sHsps, binding of a stressed target protein is thought to occur primarily via hydrophobic interactions between exposed hydrophobic regions on the chaperone and on the partially unfolded protein [139]. It has been well described that under conditions of heat stress, some sHsps undergo a conformational change which increases the extent of exposed hydrophobicity and these structural changes are accompanied by functional ones i.e. increased chaperone action [94, 125, 127, 153, 154]. Solubility of the large sHsp aggregates is maintained by hydrophilic regions of the chaperone that are dynamic and solvent-exposed, such as the flexible C-terminal extensions in α-crystallin [70, 93]. The presence of these polar regions is also believed to be important in maintaining the solubility of the target-protein-chaperone complex once interaction between the two proteins has taken place, resulting in a high molecular weight complex [155]. It is plausible that α_s-casein has a similar mode of action. The distribution of hydrophobic and hydrophilic residues in the caseins is not uniform and their lack of well-defined structure encourages self-association but also likely aids in chaperone action with partially unfolded target proteins. Within the predominant α_s-casein subunit, α_{s1}-casein, hydrophobic residues are clustered into three distinct regions (residues 1-44, 90-113 and 132-199) and the hydrophilic phosphoserine residues are also clustered in the polar domains (residues 41-80) [5]. It is possible that the structure of α_{s1}-casein is similar to that of the sHsps whereby it has a predominant, relatively globular hydrophobic domain linked to a highly polar region akin to the flexible polar C-terminal extension of the sHsps [155-157]. Hypothetical structures for α_s-casein subunits obtained through energy minimisation calculations are consistent with this model [15]. The most hydrophilic of the caseins is α_{s2}-casein (present at a ratio of 4:1 α_{s1}:α_{s2} in bovine milk) with only two areas of hydrophobicity arising from residues 160-207 and 90-120 [17]. The C-terminal region of α_{s2}-casein possesses a high net charge despite being relatively hydrophobic [5].

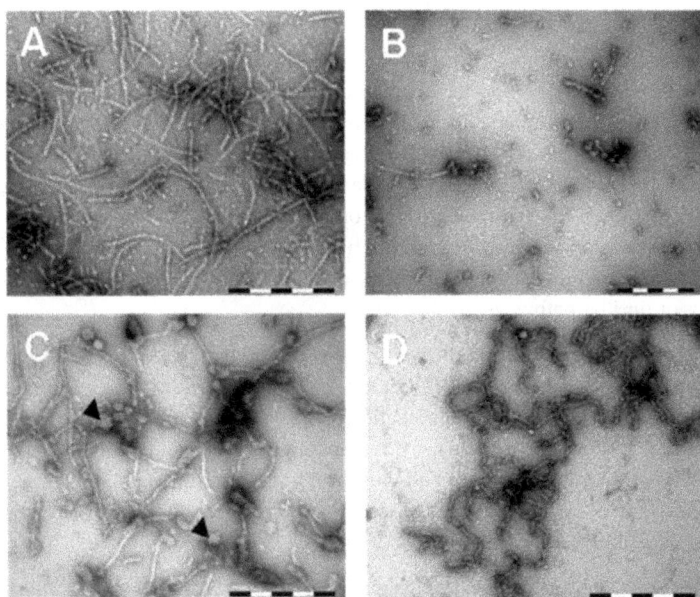

Figure 10. Electron micrographs of RCM-κ-casein incubated for 50 h at 37°C with and without α_s-casein proteins. RCM-κ-casein incubated in the absence (A) or presence of ~1.0 mol:mol ratio (3.75 mg/mL) of either α_{s1}-casein (B) or α_{s2}-casein (C). α_{s2}-Casein alone is also shown (D; 3.75 mg/mL). In (C), small, rounded aggregates are indicated by ◄. Scale bars represent 500 nm. Reprinted with permission from [17]. Copyright (2011) Elsevier Inc.

Size-exclusion HPLC studies have shown that α_s-casein exists as a polydisperse aggregate [18] and it is likely that this heterogeneity arises from association of the two subunits. It is apparent from Figure 4 and from the previous discussion of the chaperone mechanism of sHsps that a crucial part of the process is the dynamic interaction between large, heterogeneous aggregates formed by α_s-casein and smaller oligomers, which, in the case of the sHsps, are believed to be the active form of the chaperone [18, 110]. As the other casein proteins present in milk (β-, and κ-casein) have also been shown to possess chaperone activity [18, 33, 110-112] it is likely that these subunits also play an important role in dynamic subunit exchange and stabilisation of the casein aggregate. The ability of α_s-casein to resolubilise aggregated target protein species is especially interesting in the context of the mechanism shown in Figure 4. The data on insulin resolubilisation by α_s-casein indicates that an equilibrium certainly exists between intermediately folded states of target proteins (e.g. I_2) and their amorphous aggregates, but that in addition, α_s-casein is capable of pushing this equilibrium back toward a more soluble and therefore more stable state (I_1 or I_2) following formation of the aggregate. Dynamic equilibrium between the α_s-casein aggregate and smaller, dissociated species is likely to be important in this aspect of its chaperone function, but this remains to be elucidated.

4. Conclusion and future directions

The predominant milk protein, α_s-casein, has been shown to possess molecular chaperone abilities with a range of target proteins, under different stress conditions. Like the sHsps, α_s-casein is also ATP-independent in its chaperone action. As described in this chapter, α_s-casein and its two constituent proteins, α_{s1}- and α_{s2}-casein are capable of interacting with and stabilising a range of physiological and non-physiological target proteins. This 'promiscuous' nature is a feature of many other chaperones. Under a variety of stress conditions, α_s-, α_{s1}- and α_{s2}-casein form high molecular weight complexes with partially unfolded target proteins, stabilising them against precipitation, whether this be via amorphous or fibrillar pathways.

Like the sHsps and clusterin, the α-casein proteins exhibit different degrees of chaperone activity depending on the mode of target protein aggregation (i.e. amorphous versus fibrillar), the rate of target protein aggregation, the size of the target protein, the conditions of stress applied and the presence of competing ions (e.g. salt). Unlike the sHsps (specifically Hsp25 and α-crystallin) and clusterin, however, α_s-casein binds target proteins in a manner that does not allow subsequent interaction and reactivation by the ATP-dependent Hsp70.

The mechanism/s by which α_s-casein stabilises and prevents the precipitation of other proteins in milk (such as the other caseins and whey proteins such as α-lactoglobulin and β-casein) is of interest to the dairy industry as it may provide an alternative method for long-life milk treatment [18]. It has been demonstrated that α_s-casein, and indeed other caseins such as β-casein, interact with 'molten globule' states or folding intermediates of proteins. As suggested by others, processing treatments in dairy foods have the potential to transform previously native structures into denatured or partially denatured states and that the presence of these states may either present a problem or offer opportunities for novel foods to be developed [123]. This is where the action of molecular chaperones may play an important role. Thus, a better understanding of the aggregation processes in milk and how these can be modified opens up potential avenues for new milk based products with novel textures and other organoleptic properties to be developed.

Author details

Teresa Treweek

Graduate School of Medicine, University of Wollongong, Australia

Acknowledgement

The author gratefully acknowledges the financial support of Dairy Australia for much of the work presented here. Sincere thanks also go to Prof. Will Price, Prof. John Carver, Dr Heath Ecroyd and Dr David Thorn for their assistance with the preparation of this manuscript.

5. References

[1] Swaisgood HE. Chemistry of the caseins. In: Fox PF. (ed.) Advanced Dairy Chemistry- 1: Proteins. London: Elsevier Applied Science; 1992. p63-110.

[2] Fox PF, McSweeney PLH. Dairy chemistry and biochemistry. London: Blackie Academic and Professional; 1998.

[3] Holt C, Sawyer L. Caseins as rheomorphic proteins: Interpretations of the primary and secondary structures of the alphaS1-, beta- and kappa-caseins. Journal of the Chemical Society, Faraday Transactions 1993; 89 2683-2692.

[4] Swaisgood HE. Chemistry of the caseins. In: Fox PF, McSweeney PLH (eds.) Advanced Dairy Chemistry – 1: Proteins. 3rd ed., Part B. New York: Kluwer Academic/Plenum Publishers; 2003. p. 139-201.

[5] Fox PF. Milk Proteins: General and historical aspects. In: Fox PF, McSweeney PLH. (eds.) Advanced Dairy Chemistry – 1: Proteins. 3rd ed., Part B. New York: Kluwer Academic/Plenum Publishers; 2003. p1-48.

[6] Aschaffenburg R, Drewry J. Occurrence of different beta-lactoglobulins in cow's milk. Nature 1955;176(4474) 1309-1314.

[7] Ng Kwai-Hang KF, Grosclaude F. Genetic polymorphism of milk protein. In: Fox PF. (ed.) Advanced Dairy Chemistry - 1: Proteins. London: Elsevier Applied Science; 1992. p405-455.

[8] Jakob E, Puhan Z. Technological properties of milk as influenced by genetic polymorphism of dairy proteins. A review. International Dairy Journal 1992;2 157-178.

[9] Eigel WN, Butler JE, Ernstrom CA, Farrell HMJ, Harwalkar VR, Jenness R, McLWhitney R. Nomenclature of proteins of cow's milk: Fifth revision. Journal of Dairy Science 1984;67 1599-1631.

[10] Dunker AK, Brown CJ, Lawson JD, Iakoucheva LM, Obradovic Z. Intrinsic disorder and protein function. Biochemistry 2002;41(21) 6573-6582.

[11] Uversky VN. What does it mean to be natively unfolded? European Journal of Biochemistry 2002;269(1) 2-12.

[12] Alaimo MH, Farrell HMJ, Germann MW. Conformational analysis of the hydrophobic peptide αs1-casein(136-196). Biochimica et Biophysica Acta 1999;1431 410-420.

[13] Kuwajima K. The molten globule state of α-lactalbumin. The FASEB Journal 1996;10 102-109.

[14] Kumosinski TF, Brown EM, Farrell HMJ. Three-dimensional molecular modeling of bovine caseins: a refined, energy-minimized κ-casein structure. Journal of Dairy Science 1993;76(9) 2507-2520.

[15] Kumosinski TF, Brown FM, Farrell HMJ. Three-dimensional molecular modeling of bovine caseins: αs1-casein. Journal of Dairy Science 1991;74 2889-2895.

[16] Kumosinski TF, King G, Farrell HMJ. An energy-minimized casein submicelle working model. Journal of Protein Chemistry 1994;13(8) 681-700.

[17] Treweek TM, Thorn DC, Price WE, Carver JA. The chaperone action of bovine milk αs1- and αs2-caseins and their associated form αs-casein. Archives of Biochemistry and Biophysics 2011;510 42-52.

[18] Morgan PE, Treweek TM, Lindner RA, Price WE, Carver JA. Casein proteins as molecular chaperones. Journal of Agricultural and Food Chemistry 2005;53(7) 2670-2683.

[19] Holt C, Carver JA. Darwinian transformation of a 'scarcely nutritious fluid' into milk. Journal of Evolution Biology 2012;In press.

[20] Holt C, Dalgleish DG. Electrophoretic and hydrodynamic properties of bovine casein micelles interpreted in terms of particles with an outer hairy layer. Journal of Colloid Interface Science 1986;114 513-524.

[21] Marchin S, Putaux JL, Pignon F, Leonil J. Effects of the environmental factors on the casein micelle structure studied by cryo transmission electron microscopy and small-angle x-ray scattering/ultrasmall-angle x-ray scattering. The Journal of Chemical Physics 2007;126(4) 045101.

[22] McMahon DJ, McManus WR. Rethinking casein micelle structure using electron microscopy. Journal of Dairy Science 1998;81 2985-2993.

[23] McMahon DJ, Oommen BS. Supramolecular structure of the casein micelle. Journal of Dairy Science 2008;91(5) 1709-1721.

[24] Knudsen JC, Skibsted LH. High pressure effects on the structure of the casein micelles in milk as studied by cryo-transmission electron microscopy. Food Chemistry 2010;119 202-208.

[25] Dalgleish DG. On the structural models of bovine casein micelles - review and possible improvements. Soft Matter 2011;7 2265-2272.

[26] DeKruif CG, Holt C. Casein micelle structure, functions and interactions. In: Fox PF, McSweeney PLH. (eds.). Advanced Dairy Chemistry – 1: Proteins. 3rd ed., Part B. New York: Kluwer Academic/Plenum Publishers; 2003. p233-276.

[27] Farrell HMJ, Cooke PH, Wickham ED, Piotrowski EG, Hoagland PD. Environmental influences on bovine κ-casein: reduction and conversion to fibrillar (amyloid) structures. Journal of Protein Chemistry 2003;22(3) 259-273.

[28] Thorn DC, Meehan S, Sunde M, Rekas A, Gras SL, MacPhee CE, Dobson CM, Wilson MR, Carver JA. Amyloid fibril formation by bovine milk kappa-casein and its inhibition by the molecular chaperones αs- and β-casein. Biochemistry 2005;44(51) 17027-17036.

[29] Thorn DC, Ecroyd H, Sunde M, Poon S, Carver JA. Amyloid fibril formation by bovine milk α s2-casein occurs under physiological conditions yet is prevented by its natural counterpart, αs1-casein. Biochemistry 2008;47(12) 3926-3936.

[30] Leonil J, Henry G, Jouanneau D, Delage MM, Forge V, Putaux JL. Kinetics of fibril formation of bovine κ-casein indicate a conformational rearrangment as a critical step in the process. Journal of Molecular Biology 2008;381 1267-1280.

[31] Syme CD, Blanch EW, Holt C, Jakes R, Goedert M, Hecht L, Barron L. A Raman optical activity study of rheomorphism in caseins, synuclein and tau. New insight into the structure and behaviour of natively unfolded proteins. European Journal of Biochemistry 2002;269 148-156.

[32] Blanch EW, Morozova-Roche LA, Cochran DA, Doig AJ, Hecht L, Barron LD. Is polyproline II helix the killer conformation? A raman optical activity study of the amyloidogenic perfibrillar intermediate of human lysozyme. Journal of Molecular Biology 2000;301(2) 553-563.

[33] Thorn DC, Ecroyd H, Carver JA. The two-faced nature of milk casein proteins: amyloid fibril formation and chaperone-like activity. Australian Journal of Dairy Technology 2009;64(1) 34-40.

[34] Thompson MP. DEAE-cellulose-urea chromatography of casein in the presence of 2-mercaptoethanol. Journal of Dairy Science 1966;49(7) 792-795.

[35] Creamer LK, Richardson T, Parry DAD. Secondary structure of bovine αs1- and β-casein in solution. Archives of Biochemistry and Biophysics 1981;211 689-693.

[36] Haga M, Yamauchi K, Aoyagi S. Conformation and some properties of bovine αs2-group casein. Agricultural and Biological Chemistry 1983;47(7) 1467-1471.

[37] Hoagland PD, Unruh JJ, Wickham ED, Farrell HMJ. Secondary structure of bovine αs2-casein: Theoretical and experimental approaches. Journal of Dairy Science 2001;84 1944-1949.

[38] Chakraborty A, Basak S. pH-induced structural transitions of caseins. Journal of Photochemistry and Photobiology 2007;87 191-199.

[39] Sreerama N, Woody RW. A self-consistent method for the analysis of protein secondary structure from circular dichroism. Analytical Biochemistry 1993;209(1) 32-44.

[40] Sreerama N, Woody RW. Estimation of protein secondary structure from circular dichroism spectra: Comparison of CONTIN, SELCON, and CDSSTR methods with an expanded reference set. Analytical Biochemistry 2000;287 252-260.

[41] Lobley A, Whitmore L, Wallace BA. DICHROWEB: A website for the analysis of protein secondary structure from circular dichroism spectra. Biophysical Journal 2001;80 373a.

[42] Lobley A, Whitmore L, Wallace BA. DICHROWEB: an interactive website for the analysis of protein secondary structure from circular dichroism spectra. Bioinformatics 2002;18(1) 211-212.

[43] Whitmore L, Wallace BA. DICHROWEB: an online server for protein secondary structure analyses from circular dichroism spectroscopic data. Nucleic Acids Research 2004;32 W668-673.

[44] Anfinsen CB, Haber E, Sela M, White FHJ. The kinetics of formation of native ribonuclease during oxidation of the reduced polypeptide chain. Proceedings of the National Academy of Sciences, U.S.A. 1961;47(9) 1309-1314.

[45] Buchner J. Supervising the fold: functional principles of molecular chaperones. The FASEB Journal 1996;10 10-19.

[46] Anfinsen CB. The formation and stabilization of protein structure. Biochemical Journal 1972;128(4) 737-749.

[47] Anfinsen CB. Principles that govern the folding of protein chains. Science 1973;181(96) 223-230.

[48] Rudolph R, Lilie H. *In vitro* folding of inclusion body proteins. The FASEB Journal 1996;10(1) 49-56.

[49] Ptitsyn OB. How does protein synthesis give rise to the 3D-structure? FEBS Letters 1991;285(2) 176-181.

[50] Fink AL. Molten globules. In: Shirley BA. (ed.) Methods in Molecular Biology. New Jersey: Humana Press Inc.; 1995.

[51] Jaenicke R, Creighton TE. Junior chaperones. Current Biology 1993;3(4) 234-235.

[52] Ewbank JJ, Creighton TE, Hayer-Hartl M-K, Hartl FU. What is the molten globule? Nature Structural and Molecular Biology 1995;2(1) 10-11.

[53] Ellis RJ, Hartl FU. Protein folding in the cell: competing models of chaperonin function. The FASEB Journal 1996;10 20-26.

[54] Wright PE, Dyson HJ. Insights into the structure and dynamics of unfolded proteins from nuclear magnetic resonance. Advances in Protein Chemistry 2002;62 311-340.

[55] Lindner RA, Kapur A, Carver JA. The interaction of the molecular chaperone, α-crystallin, with molten globule states of bovine α-lactalbumin. Journal of Biological Chemistry 1998;272 27722-27729.

[56] Thomas PJ, Qu B-H, Pedersen PL. Defective protein folding as a basis of human disease. Trends in Biochemical Sciences 1995;20 456-459.

[57] Welch WJ, Brown CR. Influence of molecular and chemical chaperones on protein folding. Cell Stress and Chaperones 1996 April 1996;1(2) 109-115.

[58] Ehrnsperger M, Graber S, Gaestel M, Buchner J. Binding of non-native protein to Hsp25 during heat shock creates a reservoir of folding intermediates for reactivation. EMBO Journal 1997;16(2) 221-229.

[59] Ellis RJ, van der Vies SM. Molecular Chaperones. Annual Review of Biochemistry 1991;60 321-347.

[60] Jakob U, Buchner J. Assisting spontaneity: the role of Hsp90 and small Hsps as molecular chaperones. Trends in Biochemical Sciences 1994;19(5) 205-211.

[61] Ruddon RW, Bedows E. Assisted protein folding. Journal of Biological Chemistry 1997;272(6) 3125-3128.

[62] Kim PS, Arvan P. Calnexin and BiP act as sequential molecular chaperones during thyroglobulin folding in the endoplasmic reticulum. Journal of Cell Biology 1995;128(1-2) 29-38.

[63] Frydman J, Hartl FU. Principles of chaperone-assisted protein folding: differences between *in vitro* and *in vivo* mechanisms. Science 1996;272(526) 1497-1502.

[64] Martin J, Langer T, Boteva R, Schramel A, Horwich AL, Hartl FU. Chaperonin-mediated protein folding at the surface of groEL through a 'molten globule'-like intermediate. Nature 1991;352(6330) 17-18.

[65] Ingolia TD, Craig EA, McCarthy BJ. Sequence of three copies of the gene for the major *Drosophila* heat shock induced protein and their flanking regions. Cell 1980;21(3) 669-676.

[66] Hunt C, Morimoto R. Conserved features of eukaryotic hsp70 genes revealed by comparison with the nucleotide sequence of human hsp70. Proceedings of the National Academy of Sciences, U.S.A. 1985;82(19) 6455-6459.

[67] Corces V, Holmgren R, Freund R, Morimoto R, Meselson M. Four heat shock proteins of *Drosophila melanogaster* coded within a 12-kilobase region in chromosome subdivision 67B. Proceedings of the National Academy of Sciences, U.S.A. 1980;77 5390-5393.

[68] Lindquist S. The Heat Shock Response. Annual Review of Biochemistry 1986;55 1151-1191.

[69] Lindquist S, Craig EA. The heat-shock proteins. Annual Review of Genetics 1988;22 631-677.

[70] Treweek TM, Morris AM, Carver JA. Intracellular protein unfolding and aggregation: The role of small heat-shock chaperone proteins. Australian Journal of Chemistry 2003;56 357-367.

[71] Wang K, Spector A. The chaperone activity of bovine α-crystallin. Interaction with other lens crystallins in native and denatured states. Journal of Biological Chemistry 1994;269(18) 13601-13608.

[72] Wang K, Spector A. α-Crystallin prevents irreversible protein denaturation and acts cooperatively with other heat-shock proteins to renature the stabilized partially denatured protein in an ATP-dependent manner. European Journal of Biochemistry 2000;267 4705-4712.

[73] Poon S, Easterbrook-Smith SB, Rybchyn MS, Carver JA, Wilson MR. Clusterin is an ATP-independent chaperone with very broad substrate specificity that stabilizes stressed proteins in a folding-competent state. Biochemistry 2000;39 15953-15960.

[74] Wang K, Spector A. ATP causes small heat shock proteins to release denatured protein. European Journal of Biochemistry 2001;268 6335-6345.

[75] Kappe G, Franck E, Verschuure P, Boelens WC, Leunissen JA, de Jong WW. The human genome encodes 10 α-crystallin-related small heat shock proteins: HspB1-10. Cell Stress and Chaperones 2003;8(1) 53-61.

[76] Fontaine JM, Rest JS, Welsh MJ, Benndorf R. The sperm outer dense fiber protein is the 10th member of the superfamily of mammalian small stress proteins. Cell Stress and Chaperones 2003;8(1) 62-69.

[77] Humphreys DT, Carver JA, Easterbrook-Smith SB, Wilson MR. Clusterin has chaperone-like activity similar to that of small heat shock proteins. Journal of Biological Chemistry 1999;274(11) 6875-6881.

[78] Kapron JT, Hilliard GM, Lakins JN, Tenniswood MPR, West KA, Carr SA, Crabbe JW. Identification and characterization of glycosylation sites in human serum clusterin. Protein Science 1997;6(10) 2120 - 2133.

[79] Treweek TM, Ecroyd H, Williams DM, Meehan S, Carver JA, Walker MJ. Site-directed mutations in the C-terminal extension of human αB-crystallin affect chaperone function and block amyloid fibril formation. Public Library of Science One 2007;e1046(10) 1-10.

[80] Ecroyd H, Meehan SM, Horwitz J, Aquilina JA, Benesch JLP, Robinson CV, MacPhee CE, Carver JA. Mimicking phosphorylation of αB-crystallin affects its chaperone activity. Biochemical Journal 2007; 401(1) 129-141.

[81] Ecroyd H, Carver JA. Crystallin proteins and amyloid fibrils. Cell and Molecular Life Sciences 2009;66(1) 62-81.

[82] Boelens WC, de Jong WW. α-Crystallins, versatile stress proteins. Molecular Biology Reports 1995;21 75-80.

[83] Harding JJ. Cataract: Biochemistry, epidemiology and pharmacology. London: Chapman and Hall; 1991.

[84] Delaye M, Tardieu A. Short-range order of crystallin proteins accounts for eye lens transparency. Nature 1983;302 415-417.

[85] Tardieu A, Delaye M. Eye lens proteins and transparency: from light transmission theory to solution X-ray structural analysis. Annual Review of Biophysics and Biophysical Chemistry 1988;17 47-70.

[86] Veretout F, Delaye M, Tardieu A. Molecular basis of eye lens transparency. Osmotic pressure and X-ray analysis of α-crystallin solutions. Journal of Molecular Biology 1989;205 713-728.

[87] Bhat SP, Nagimeni CN. αB-subunit of lens specific α-crystallin is present in other ocular and non-ocular tissue. Biophysical and Biochemical Research Communications 1989;158(1) 319-325.

[88] Iwaki T, Kume-Iwaki A, Goldman JE. Cellular distribution of αB-crystallin in non-lenticular tissues. Journal of Histochemistry and Cytochemistry 1990;38(1) 31-39.

[89] Kato K, Shinohara H, Kurobe N, Goto S, Inaguma Y, Ohshima K. Immunoreactive αA-crystallin in rat non-lenticular tissues is detected with a sensitive immunoassay system. Biochimica et Biophysica Acta 1991;1080 173-180.

[90] Kato K, Shinohara H, Kurobe N, Inaguma Y, Shimizu K, Ohshima K. Tissue distribution and developmental profiles of immunoreactive αB-crystallin in the rat determined with a sensitive immunoassay system. Biochimica et Biophysica Acta 1991;1074(1) 201-208.

[91] Srinivasan AN, Nagimeni CN, Bhat SP. αA-Crystallin is expressed in non-ocular tissues. Journal of Biological Chemistry 1992;267(32) 23337-23341.

[92] van Rijk AF, Bloemendal H. Alpha-B-crystallin in neuropathology. Ophthalmologica 2000;214 7-12.

[93] Carver JA, Rekas A, Thorn DC, Wilson MR. Small heat-shock proteins and clusterin: intra- and extracellular molecular chaperones with a common mechanism of action and function? IUBMB Life 2003;55(12) 661-668.

[94] Carver JA, Guerreiro N, Nicholls KA, Truscott RJW. On the interaction of α-crystallin with unfolded proteins. Biochimica et Biophysica Acta 1995 June 28, 1995;1252 251-260.

[95] Treweek TM, Lindner RA, Mariani M, Carver JA. The small heat-shock chaperone protein, α-crystallin, does not recognise stable molten globule states of cytosolic proteins. Biochimica et Biophysica Acta 2000 May 31, 2000;1481 175-188.

[96] Lindner RA, Treweek TM, Carver JA. The molecular chaperone α-crystallin is in kinetic competition with aggregation to stabilize a monomeric molten-globule form of α-lactalbumin. Biochemical Journal 2001;354(1) 79-87.

[97] van Montfort RLM, Slingsby C, Vierling E. Structure and function of the small heat shock protein/alpha-crystallin family of molecular chaperones. Advances in Protein Chemistry 2001;59 105-156.

[98] Haley DA, Bova MP, Huang Q-L, Mchaourab HS, Stewart PL. Small heat-shock protein structures reveal a continuum from symmetric to variable assemblies. Journal of Molecular Biology 2000;298 261-272.

[99] Horwitz J. Review: Alpha-crystallin. Experimental Eye Research 2003;76 145-153.

[100] Lindner RA, Kapur A, Mariani M, Titmuss SJ, Carver JA. Structural alterations of α-crystallin during its chaperone action. European Journal of Biochemistry 1998;258 170-183.

[101] Lindner RA, Carver JA, Ehrnsperger M, Buchner J, Esposito G, Behlke J, Lutsch G, Kotlyarov A, Gaestel M. Mouse Hsp25, a small heat shock protein. The role of its C-terminal extension in oligomerization and chaperone action. European Journal of Biochemistry 2000 28 Jan 2000;267 1923-1932.

[102] Sobott F, Benesch JLP, Vierling E, Robinson CV. Subunit exchange of multimeric protein complexes. Journal of Biological Chemistry 2002;277(41) 38921-38929.

[103] Treweek TM, Morris AM, Carver JA. Intracellular protein unfolding and aggregation: The role of small heat-shock chaperone proteins. Australian Journal of Chemistry 2003;56 357-367.

[104] Bova MP, Mchaourab HS, Han Y, Fung BK-K. Subunit exchange of small heat shock proteins. Journal of Biological Chemistry 2000;275(2) 1035-1042.

[105] Bova MP, Huang Q-L, Ding L, Horwitz J. Subunit exchange, conformational stability, and chaperone-like function of the small heat shock protein 16.5 from *Methanococcus jannaschii*. Journal of Biological Chemistry 2002;277(41) 38468-38475.

[106] Carver JA, Rekas A, Thorn DC, Wilson MR. Small heat shock proteins and clusterin: intra- and extracellular molecular chaperones with a common mechanism of action and function? IUBMB Life 2003;55 661-668.

[107] Morr CV, Josephson RV. Effect of calcium, N-ethylmaleimide and casein upon heat-induced whey protein aggregation. Journal of Dairy Science 1968;51 1349-1355.

[108] Kenkare DB, Morr CV, Gould IA. Factors affecting the heat aggregation of proteins in selected skim milk sera. Journal of Dairy Science 1964;47 947-952.

[109] Bhattacharyya J, Das KP. Molecular chaperone-like properties of an unfolded protein, αs-casein. Journal of Biological Chemistry 1999;274(22) 15505-15509.

[110] Zhang X, Fu X, Zhang H, Liu C, Jiao W, Chang Z. Chaperone-like activity of β-casein. International Journal of Biochemistry and Cell Biology 2005;37 1232-1240.

[111] Barzegar A, Yousefi R, Sharifzadeh A, Dalgalarrondo M, Saboury AA, Haertle T, Moosavi-Movahedi AA. Chaperone activities of bovine and camel β-caseins: Importance of their surface hydrophobicity in protection against alcohol dehydrogenase aggregation. International Journal of Biological Macromolecules 2008;42(4) 392-399.

[112] Yong YH, Foegeding EA. Effects of caseins on thermal stability of bovine β-lactoglobulin. Journal of Agricultural and Food Chemistry 2008;56(21) 10352-10558.

[113] Matsudomi N, Kanda Y, Yoshika Y, Moriwaki H. Ability of αS-casein to suppress the heat aggregation of ovotransferrin. Journal of Agricultural and Food Chemistry 2004;52 4882-4886.

[114] Koudelka T, Hoffmann P, Carver JA. Dephosphorylation of αs- and β-caseins and its effect on chaperone activity: a structural and functional investigation. Journal of Agricultural and Food Chemistry 2009;57(13) 5956-5964.

[115] Ptitsyn OB. Protein folding: Hypotheses and experiments. Journal of Protein Chemistry 1987;6 273-293.

[116] Gilmanshin RI, Ptitsyn OB. An early intermediate of refolding α-lactalbumin forms within 20 ms. FEBS Letters 1987;223(2) 327-329.

[117] Baum J, Dobson CM, Evans MC, Hanley C. Characterization of a partly folded protein by NMR methods: studies on the molten globule state of guinea pig α-lactalbumin. Biochemistry 1989;28(1) 7-13.

[118] Lala AK, Kaul P. Increased exposure of hydrophobic surfaces in molten globule state of α-lactalbumin: Fluorescence and hydrophobic photolabeling studies. Journal of Biological Chemistry 1992;267(28) 19914-19918.

[119] Alexandrescu AT, Evans MC, Pitkeathly M, Baum J, Dobson CM. Structure and dynamics of the acid-denatured molten globule state of α-lactalbumin: A two-dimensional NMR study. Biochemistry 1993;32 1707-1718.

[120] Chyan C-L, Wormald C, Dobson CM, Evans MC, Baum J. Structure and stability of the molten globule state of guinea-pig α-lactalbumin: A hydrogen exchange study. Biochemistry 1993;32 5681-5691.

[121] Dobson CM. Protein folding. Solid evidence for molten globules. Current Biology 1994;4(7) 636-640.

[122] Carver JA, Lindner RA, Lyon C, Canet D, Hernandez H, Dobson CM, Redfield C. The interaction of the molecular chaperone α-crystallin with unfolding a-lactalbumin: A structural and kinetic spectroscopic study. Journal of Molecular Biology 2002;318 815-827.

[123] Farrell HMJ, Qi PX, Brown EM, Cooke PH, Tunick MH, Wickham ED, Unruh JJ. Molten globule structures in milk proteins: implications for potential new structure-function relationships. Journal of Dairy Science 2002;85 459-471.

[124] Raman B, Rao CM. Chaperone-like activity and quaternary structure of α-crystallin. Journal of Biological Chemistry 1994;269(44) 27264-27268.

[125] Raman B, Ramakrishna T, Rao CM. Temperature dependent chaperone-like activity of α-crystallin. FEBS Letters 1995;365(2-3) 133-136.

[126] Treweek TM, Rekas A, Lindner RA, Walker MJ, Aquilina JA, Robinson CV, Horwitz J, Perng MD, Quinlan RA, Carver JA. R120G αB-crystallin promotes the unfolding

of reduced α-lactalbumin and is inherently unstable. FEBS Journal 2005;272(3) 711-724.

[127] Das BK, Surewicz WK. Temperature-induced exposure of hydrophobic surfaces and its effect on the chaperone activity of α-crystallin. FEBS Letters 1995;369(2-3) 321-325.

[128] Bhattacharyya J, Das KP. Alpha-crystallin does not require temperature activation for its chaperone-like activity. Biochemistry and Molecular Biology International. 1998;46(2) 249-258.

[129] Guha S, Manna TK, Das KP, Bhattacharyya J. Chaperone-like activity of tubulin. Journal of Biological Chemistry 1998;273 30077-30080.

[130] Poon S, Rybchyn MS, Easterbrook-Smith SB, Carver JA, Pankhurst GJ, Wilson MR. Mildly acidic pH activates the extracellular molecular chaperone clusterin. Journal of Biological Chemistry 2002;277(42) 39532-39540.

[131] McMeekin TL, Polis BD. Milk Proteins. Advances in Protein Chemistry. 1949;5 202-228.

[132] Waugh DF. Formation and structure of casein micelles. In: McKenzie HA (ed.) Milk Proteins: Chemistry and Molecular Biology. New York: Academic Press; 1971. p17-40.

[133] Koretz JF, Doss EW, LaButti JN. Environmental factors influencing the chaperone-like activity of a-crystallin. International Journal of Biological Macromolecules 1998;22(3-4) 283-294.

[134] Brockwell CH. The effect of pH on the structure and function of α-crystallin and cyclodextrins as artificial molecular chaperones. PhD thesis. University of Adelaide; 2009.

[135] Devlin GL, Carver JA, Bottomley SP. The selective inhibition of serpin aggregation by the molecular chaperone, α-crystallin, indicates a nucleation-dependent specificity. Journal of Biological Chemistry 2003;278(49) 48644-48650.

[136] Poon S, Treweek TM, Wilson MR, Easterbrook-Smith SB, Carver JA. Clusterin is an extracellular chaperone that specifically interacts with slowly aggregating proteins on their off-folding pathway. FEBS Letters 2002;513(2-3) 259-266.

[137] Ghahghaei A. The chaperone action of α-crystallin. PhD thesis. University of Wollongong; 2006.

[138] Manna T, Sarkar T, Poddar A, Roychowdhury M, Das KP, Bhattacharyya B. Chaperone-like activity of tubulin. Binding and reactivation of unfolded substrate enzymes. Journal of Biological Chemistry 2001;276 39742-39747.

[139] Haslbeck M, Buchner J. Chaperone function of sHsps. In: Arrigo AP, Muller WEG. (eds.) Small stress proteins. Berlin: Springer-Verlag; 2002.

[140] Stromer T, Ehrnsperger M, Gaestel M, Buchner J. Analysis of the interaction of small heat shock proteins with unfolding proteins. Journal of Biological Chemistry 2003;278(20) 18015-18021.

[141] Friedrich KL, Giese KC, Buan NR, Vierling E. Interactions between small heat shock protein subunits and substrate in small heat shock protein-substrate complexes. Journal of Biological Chemistry 2004;279(2) 1080-1089.

[142] Muchowski PJ, Hays LG, Yates JRI, Clark JI. ATP and the core "α-crystallin" domain on the small heat-shock protein αB-crystallin. Journal of Biological Chemistry 1999;274(42) 30190-30195.

[143] Haley DA, Bova MP, Huang Q-L, Mchaourab HS, Stewart PL. Small heat-shock protein structures reveal a continuum from symmetric to variable assemblies. Journal of Molecular Biology 2000;298 261-272.

[144] Regini JW, Ecroyd H, Meehan S, Bremmell K, Clarke MJ, Lammie D, Wess T, Carver JA. The interaction of unfolding α-lactalbumin and malate dehydrogenase with the molecular chaperone αB-crystallin: a light and X-ray scattering investigation. Molecular Vision 2010;16 2446-2456.

[145] Farahbakhsh ZT, Huang QL, Ding LL, Altenbach C, Steinhoff HJ, Horwitz J, Hubbell WL. Interaction of α-crystallin with spin-labeled peptides. Biochemistry 1995;34(2) 509-516.

[146] Alaimo MH, Wickham ED, Farrell HMJ. Effect of self-association of αS1-casein and its cleavage fractions αS1-casein(136-196) and αS1-casein(1-197), on aromatic circular dischroism spectra: comparison with predicted models. Biochimica et Biophysica Acta 1999;1431 395-409.

[147] Schmidt DG. Association of caseins and casein micelle structures. In: Fox PF. (ed.) Developments in Dairy Chemistry - 1. Proteins. London: Elsevier Applied Science Publishers; 1982. p161-86.

[148] Kudva YC, Hiddinga HJ, Butler PC, Mueske CS, Eberhardt NL. Small heat shock proteins inhibit in vitro A beta(1-42) amyloidogenesis. FEBS Letters 1997;416(1) 117-121.

[149] Stege GJ, Renkawek K, Overkamp PS, Verschuure P, van Rijk AF, Reijnen-Aalbers A, Boelens WC, Bosman GJ, de Jong WW. The molecular chaperone αB-crystallin enhances amyloid β neurotoxicity. Biophysical and Biochemical Research Communications 1999;262(1) 152-156.

[150] Hatters DM, Lindner RA, Carver JA, Howlett GJ. The molecular chaperone, α-crystallin, inhibits amyloid formation by Apolipoprotein C-II. Journal of Biological Chemistry 2001;276 33755-33761.

[151] Rekas A, Adda CG, Aquilina JA, Barnham KJ, Sunde M, Galatis D, Williamson NA, Masters CL, Anders RF, Robinson CV, Cappai R, Carver J. Interaction of the molecular chaperone αB-crystallin with a-synuclein: Effects on amyloid fibril formation and chaperone activity. Journal of Molecular Biology 2004;340 1167-1183.

[152] Ecroyd H, Koudelka T, Thorn DC, Williams DM, Devlin G, Hoffmann P, Carver JA. Dissociation from the oligomeric state is the rate-limiting step in fibril formation by κ-casein. Journal of Biological Chemistry 2008;283(14) 9012-9022.

[153] Das BK, Surewicz WK. On the substrate specificity of α-crystallin as a molecular chaperone. Biochemical Journal 1995;311(2) 367-370.

[154] Haslbeck M, Walke S, Stromer T, Ehrnsperger M, White HE, Chen S, Saibil HR, Buchner J. Hsp26: a temperature-regulated chaperone. EMBO Journal 1999;18(23) 6744-6751.

[155] Carver JA, Lindner RA. NMR spectroscopy of α-crystallin. Insights into the structure, interactions and chaperone action of small heat-shock proteins. International Journal of Biological Macromolecules 1998;22 197-209.

[156] Jakob U, Gaestel M, Engel K, Buchner J. Small heat shock proteins are molecular chaperones. Journal of Biological Chemistry 1993;268(3) 1517-1520.

[157] Merck KB, Groenen PJ, Voorter CE, de Haard-Hoekman WA, Horwitz J, Bloemendal H, de Jong WW. Structural and functional similarities of bovine α-crystallin and mouse small heat-shock protein. A family of chaperones. Journal of Biological Chemistry 1993;268(2) 1046-1052.

Chapter 4

The Alpha-Lactalbumin/Oleic Acid Complex and Its Cytotoxic Activity

Marcel Jøhnke and Torben E. Petersen

Additional information is available at the end of the chapter

1. Introduction

The protein alpha-lactalbumin (α-LA) is present in the milk of nearly all mammals. Here it has a well-described function of controlling the specificity of a galatosyltransferase towards the formation of lactose, the common sugar present in milk. In recent years an entirely different activity has been associated with the protein. By forming a complex with oleic acid (OA) it shows cytotoxic activity against cells with an apparent selectivity for cancer cells. The complex is called HAMLET (Human α-lactalbumin Made LEthal to Tumor cells) and was originally discovered by serendipity. The complex was revealed during studies of the effect of anti-adhesive molecules from human milk on bacterial attachment to alveolar lung carcinoma cells [1]. A casein fraction, obtained after low pH precipitation of human milk at pH 4.3, was shown to exhibit cytotoxic activity. The dying cells displayed changes in morphology, cytoplasmic blebbing, nuclear condensation, and formed apoptotic bodies, comparable to cells that undergo classical apoptosis [2]. The tumoricidal component of the casein was retained in a DEAE-Trisacryl M column upon ion exchange chromatography, but eluted at high salt concentration (1M).

The active component was identified as multimeric α-LA (MAL) by its oligomeric appearance in SDS-PAGE analysis with bands at 14, 28, and 100 kDa and was subsequently confirmed by Edman degradation and mass spectrometry [1,3]. The MAL was shown to contain a molten globule-like structure and cause apoptosis-induction in a variety of transformed and immature cells, but not in healthy differentiated cells [1,2,4]. The affected cells were carcinomas of lung, kidney, throat, bladder, colon, ovaries and the prostate, glioblastomas in the brain, melanomas, and leukemias. In contrast, native monomeric α-LA isolated from human whey showed no apoptosis-inducing activity.

The difference in cytotoxic activity between the native-state α-LA and α-LA in MAL was not a consequence of post-translational modifications, but rather a conformational change altering the tertiary structure while leaving the secondary structure intact, resulting in the partial unfolding and molten globule-like conformation of the protein in MAL [3].

The link between the altered folding of α-LA and apoptosis induction was subsequently identified by deliberate *in vitro* conversion of apo-state α-LA deprived of Ca^{2+}, to an apoptosis-inducing complex on an anion-exchange chromatography column previously exposed to human milk casein [5]. Additional studies showed the requirement of a lipid cofactor in the stabilization of the partially unfolded state of α-LA, by the finding that α-LA in the complex was preserved in a molten globule-like conformation even at physiological pH and in the presence of Ca^{2+} at 25-37°C [2]. The lipid cofactor was identified as oleic acid (C18:1 cis Δ9) by chemical extraction of casein-conditioned column matrices and GC-MS analysis [5,6]. The name MAL was subsequently changed to HAMLET. A similar complex with bovine α-lactalbumin has been called BAMLET.

2. Structure of α-LA

Human α-LA is an acidic globular protein composed of 123 amino acids with a molecular mass of 14.2 kDa and homologous with the lysozyme protein family. It is secreted by the epithelia cells of the mammary gland during lactation, and this is the only tissue which expresses the protein [7-9]. The crystal structure of human and bovine α-LA has been resolved by X-ray crystallography [10,11]. The protein consists of two domains, a large one composed mainly of α-helical structures and a smaller β-sheet domain. The overall three dimensional structure is stabilized by a Ca^{2+} binding loop [12,13] (Figure 1).

The large domain contains four α-helixes corresponding to amino acids 5-11 (A, dark blue), 23-34 (B, light blue), 86-98 (C, yellow) and 105-109 (D, orange) and three short 3_{10}-helical domains corresponding to amino acids 12-16, 101-104 and 115-119. The smaller β-sheet domain contains a triple-stranded antiparallel β-sheet (amino acids 40-50, green) and a short 3_{10}-helical domain (amino acids 76-82) [14]. The two domains are connected through a cysteine bridge (amino acids 73,91) creating the Ca^{2+} binding loop region. Additionally, two cysteine bridges are located in the α-helical structures (amino acids 6,120 and 28,111) and finally one bridge in the β-sheet domain (amino acids 61,77), aiding in the stabilization of the native conformation of α-LA [10,12]. The binding of Ca^{2+} to α-LA is required for correct formation of the disulfide bonds during protein folding [15].

3. Calcium-binding properties of α-LA

The native human α-LA is a metalloprotein containing a high-affinity primary Ca^{2+} binding site and Ca^{2+}-binding is required for the stabilization and structural integrity of the native fold [13,16-19]. The strong Ca^{2+} binding site, with an apparent association constant of 3×10^{8} M^{-1} at 20°C, is located in a loop between the two domains, and is coordinated by the side chain carboxylates of Asp82, Asp87, Asp88 and carbonyl oxygens of Lys79 and Asp84 [10,20,21]. In Figure 1 only the side chains Asp82, Asp87 and Asp88 are shown. Furthermore, two water molecules also participate in the coordination of Ca^{2+} at the site, creating a distorted pentagonal bipyramidal structure [10-13]. A secondary surface-located and low-affinity Ca^{2+} binding site was discovered by X-ray crystallography, which only binds Ca^{2+} at high concentrations. However, this binding site is believed to play no structural role in human α-LA [22].

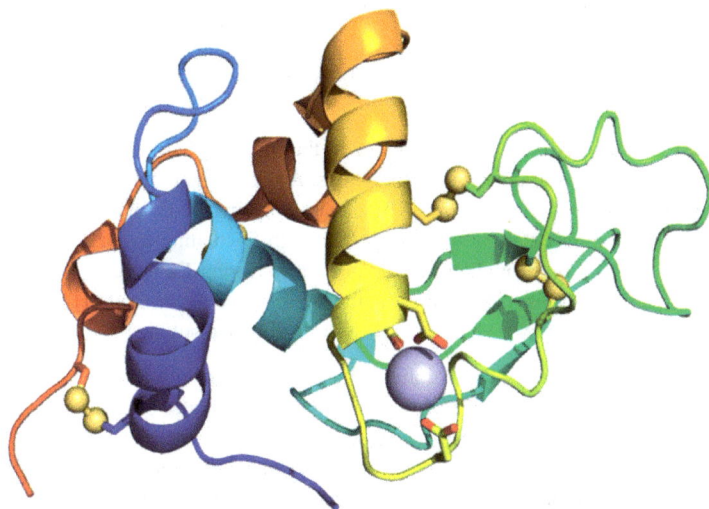

Figure 1. Model of human α-lactalbumin drawn by PyMOL Molecular Graphics System, Version 1.5.0.1 Schrödinger, LLC using the coordinates from [10]. The Ca^{2+} is shown as a ball in purple, helix A in dark blue, helix B in light blue, helix C in yellow, helix D in orange, while the β-sheet is in green. The sulfur atoms in the four disulfide bonds are indicated by small yellow balls.

Other divalent cations, such as Mn^{2+}, Mg^{2+}, Na^{+} and K^{+} can also bind α-LA, competing with Ca^{2+} for the same binding site [13,23,24]. A distinct zinc binding site different from the calcium binding site has also been localized [25]. However, binding of these cations causes only minor structural changes and does not play a major structural role in α-LA, compared to the binding of Ca^{2+} [13].

The binding of Ca^{2+} to α-LA through re-normalization of the solvent Ca^{2+}, temperature or pH conditions, and the release of Ca^{2+} through low pH, EDTA or heat treatment, have been shown in both circular dichroism spectroscopy and fluorescence spectroscopy data analysis, to cause structural and functional changes mainly in the tertiary structure, while leaving a native-like secondary structure [13,18,20,24,26]. The native folded Ca^{2+}-bound α-LA has higher structural stability against increased temperatures, pressures or denaturant concentrations [13,24,27-30].

4. Biological function of α-LA

Alpha-lactalbumin folding occurs in the lumen of the endoplasmatic reticulum and the protein is subsequently transported to the membrane surface of the Golgi apparatus, where it interacts with galactosyltransferase creating the lactose synthase complex [13,31]. In the absence of α-lactalbumin, the enzyme is a non-specific galactosyltransferase involved in transferring galactosyl groups from UDP-galactose to a range of different substrates [13,31]. Alpha-lactalbumin functions as a substrate modifier, increasing specificity and affinity of the galactosyltransferase for glucose, when catalyzing the final step of lactose production in a lactating mammary gland [7,32].

5. Folding of α-LA

A noteworthy property of α-LA is the ability of the protein to adopt relatively stable partially unfolded intermediates under various conditions. Accordingly, it has been intensely used as a protein folding model in many studies. The classical molten globule of α-LA was defined as the acid denatured compact state of α-LA at pH 2.0 with fluctuating tertiary structure [13,33,34]. However, α-LA can adopt molten globule-like states during Ca^{2+}-free (apo-state) conditions, during high temperatures at 90°C and in the presence of denaturants [14,17,35,36]. Similar conformational states can also be formed by reduction of the disulfide bonds in α-LA [37,38].

The molten globule-like conformational states of α-LA are not significantly populated at physiological conditions [26]. In addition, stable unfolded states of α-LA only persist at low pH or in the absence of metal ions, since the protein instantly returns to the native conformation, if solution conditions are re-normalized [19]. In the molten globule conformation, α-LA has a native-like secondary structure, a less well-defined tertiary structure and a larger stokes radius compared to the native protein [14,34,39].

The apo-state of α-LA, a molten globular-like conformation, is required for the formation of HAMLET. This conformation has high instability and sensitivity towards the ionic strength of the solution, compared to the native state of the protein [19]. However, the apo-state α-LA is more hydrophobic than the native state of the protein, rendering it prone to fatty acid binding [2]. The apo-state of α-LA is thus stabilized by binding to specific fatty acids in the HAMLET complex [6]. The proposed requirement of a special molten globular-like conformation of the proteinaceous component in HAMLET-like complexes has been challenged, as bovine β-lactoglobulin in complex with OA had cytotoxic activity exceeding that of HAMLET without retaining such a structure [40].

6. Fatty acid binding profile of α-LA

Alpha-lactalbumin possesses several fatty acid binding sites, and has been shown to bind stearic acid, palmitic acid and oleic acid spin-labelled fatty acid analogs [6,41]. The release of the Ca^{2+} ion and subsequent unfolding of the protein triggers α-LA to achieve a new fatty acid binding profile, with a higher affinity for cis-unsaturated fatty acids than for the corresponding trans conformations.

The interaction between apo-state α-LA and distinctive fatty acids has been studied, with the observation that certain fatty acids interact with apo-state α-LA in a stereo-specific manner [6]. Is has been found, that saturated C18 fatty acids, unsaturated C18:1 *trans* fatty acids and fatty acids with longer or shorter carbon chains only form insignificant amounts of complexes with apo-state α-LA [6]. All unsaturated *cis* fatty acids are able to form stable complexes with apo-state α-LA, with oleic acid showing the highest conversion efficiency [6]. It was found, however, that only oleic acid (C18:1 *cis* Δ9 – in HAMLET) and vaccenic acid (C18:1 *cis* Δ11) in complex with apo-state α-LA showed biological activity in apoptosis assay, suggesting that unsaturated C18:1 *cis* fatty acids have the correct stereo-specificity to bind apo-state α-LA in the formation and stabilize the HAMLET complex [6]. Another study has found that vaccenic acid, palmitoleic acid linoleic acid had similar activity as oleic acid while elaidic acid and stearic acid shoved a weaker activity [42]. The fatty acid profile of α-LA thus seems to be rather broad and not as specific as initially anticipated.

Based on the three-dimensional structures of native and apo-state α-LA, two tentative hydrophobic regions and possible fatty acid binding sites within α-LA have been proposed. One is suggested in the interface between the two sub-domains including the C- and D-helix and the β-sheet domain [6,43,44]. The other is anticipated to be formed by residues within the A, B and 3_{10}-helices.

The binding of oleic acid to bovine apo-state α-LA has previously been shown to induce changes in the secondary structure of the protein, resulting in a typical molten globule state of the protein [45]. Far-UV CD of oleic acid binding to human apo-state α-LA showed similar results, with an increase in α-helical structure, largely independent of the temperature conditions [30].

Upon Ca^{2+} release structural changes in α-LA also occur in the cleft between the two domains, with the slight expansion of the Ca^{2+}-binding loop tilting the 3_{10} helix toward the C helix, causing disruption of the aromatic cluster composed of Trp60, Trp104, Phe53 and Tyr103 in the interface between the two domains [11]. X-ray crystallography and NMR spectroscopy studies have revealed that apo-state bovine α-LA contains a largely intact α-helical domain with native side chain packing and an unstructured β-sheet domain [11,46]. Similar findings have been made for the molten globule-like conformation of human apo-state α-LA [47]. Therefore, it has been proposed that oleic acid binds in the area between the two domains, thus stabilizing the molten globule-like state of HAMLET and allowing Ca^{2+} binding without any effect on the tumoricidal activity [2,6]. Nevertheless, it is still unclear exactly how oleic acid binds to apo-state α-LA. Although the apo-state of α-LA is negatively charged, it may be hydrophobic interactions mediating the binding of the oleic acid, considering the increased hydrophobicity of apo-state α-LA compared to native α-LA [11,46,48,49]. Furthermore, even though the apo-state is negatively charged, it may also possess positively charged residues that are able to bind to the negatively charged polar head-groups of the oleic acids through electrostatic interactions [48,49]. Accordingly, it has been suggested that hydrophobic amino acids in the interface between the two domains may bind the fatty acid tail, while positively charged amino acids such as Arg70, Lys94 and Lys99 are plausible coordination residues for the fatty acid head group [2].

In contrast to these above-mentioned suggestions, one of the same studies showed that several proteolytically generated fragments of α-LA were able to incorporate oleic acid, implying that there is not only one single fatty acid binding site in HAMLET [48]. The exact mechanism of fatty acid binding to α-LA remains uncertain.

7. Structure of HAMLET

HAMLET is defined as a conversion complex between apo-state α-LA and a C18:1 *cis* fatty acid co-factor, which has the ability to induce apoptosis selectively in tumor cells [6]. HAMLET has been described as the first discovered example of a protein that exhibits a well-defined function in the native structure, but additionally acquires a potentially beneficial function following partial unfolding [1,5,38].

Studies of the HAMLET structure by tryptophan fluorescence spectroscopy, near-UV CD spectroscopy and far-UV CD spectroscopy have revealed that the complex retains a stable, partially unfolded, molten globule-like conformation even at physiological conditions, in contrast to partially unfolded α-LA, which reverts to the native conformation, if solution conditions are re-normalized [3,5,14,19,29,45]. In this molten globule-like conformation, α-LA in the HAMLET complex showed retention of the secondary structure, near-complete loss of the tertiary structure, a larger stokes radius, and increased exposure of hydrophobic surfaces, compared to native α-LA [3,5,14,34,39,50].

8. Characteristics of HAMLET

The formation, stabilization and cytotoxic activity of the HAMLET complex have been shown to require both partial unfolding of α-LA and the binding of a fatty acid co-factor [5,6,19,51]. The partial unfolding of α-LA, e.g. through metal ion depletion, results in destabilization of the β-sheet domain, while leaving the α-helix domain largely unchanged, paving the way for fatty acid binding and subsequent HAMLET formation [19,50]. In accordance with this, it has been shown that the tumoricidal activity of HAMLET is largely independent of the β-sheet domain and C-terminal portion of human α-LA, since the isolated α-helix domain of α-LA was able to form a cytotoxic complex with OA [52].

Additional changes in the α-LA structure can induce the formation of HAMLET or HAMLET analogs. It has been proposed that the entire 123-residue sequence of α-LA is not required for cytotoxic activity, as fragments of α-LA in a complex with OA were able to induce apoptosis in Jurkat tumor cells [48]. Furthermore, it has been shown that a recombinant variant of human α-LA, where all cysteine residues were substituted with alanine residues, had the ability to form cytotoxic HAMLET complex analogs [38]. In addition, an α-LA mutant where Asp87 was shifted to an alanine with no Ca^{2+} binding activity was also able to form cytotoxic complexes with oleic acids, implying that a functional Ca^{2+} site is not required for the conversion of α-LA to the active complex or to cause cell death [19]. This being the case despite the fact that HAMLET has a high Ca^{2+}-affinity, with a Ca^{2+} association constant of 5.3×10^6 M^{-1} at physiological salt conditions [19]. However, the structural changes induced upon Ca^{2+} binding have no effect on the cytotoxic activity of the complex [19,30].

Interestingly, human α-LA is not the only variant of the protein able to form cytotoxic complexes with OA. Alpha-lactalbumin of bovine, equine, caprine, and porcine origin also have the ability to form complexes with OA, showing HAMLET-like cytotoxicity, while natural HAMLET formation in acid precipitates of casein was unique to human milk [9]. Comparison of highly purified human and bovine α-LA's ability to form complexes with OA showed that the two proteins behaved very similarly and that the cytotoxic activity was comparable [53].

The revelations that neither native-state, apo-state nor otherwise partially unfolded α-LA alone shows significant cytotoxic activity, in addition to the above-mentioned findings, have caused researchers to focus on the possibility that the cytotoxic component of HAMLET may be the fatty acid and not the α-LA protein [3,5,19,50]. As a consequence, it has been suggested that α-LA is merely a carrier or synergist of the tumor-killing activity of the associated oleic acid [48,49]. In agreement with this, studies have found that the cytotoxicity of OA alone is very similar to that of BAMLET, the bovine ortholog of HAMLET, and that of HAMLET-like complexes composed of bovine β-lactoglobulin and pike parvalbumin in complex with OA [40,54]. The cytotoxic activity of the β-lactoglobulin and paralbumin complexes even exceeded that of HAMLET, indicating that the proteinaceous component of the complexes is of less importance than the OA component in relation to the cytotoxic activity [40]. The potent toxicity of free oleic acid against Jurkat and HL-60 tumor cells has been shown, supporting the proposal that the fatty acid could be the cytotoxic component of the HAMLET complex [54,55]. However, in contrast with this proposal, α-LA has previously been found to be cytotoxic in the absence of oleic acid [56-60].

9. Stoichiometry of the protein and fatty acid in HAMLET and HAMLET-like complexes

The stoichiometry of protein and fatty acid in HAMLET or HAMLET-like complexes remains a topic of debate. Initial gas chromatography and mass spectrometry data suggested that the average number of oleic acid molecules bound in the HAMLET complex was 0.9, with some batch variation [6]. The assumption of a 1:1 protein/fatty acid ratio was also reported in the preparation of BAMLET by chromatography [61]. By gas chromatography and mass spectrometry a later study estimated an α-LA to OA ratio of 1:5.4 in a HAMLET complex prepared by the OA-preconditioned anion exchange chromatography method [38]. An α-LA to OA ratio of 1:8.2 in HAMLET determined by LC-MS has also been reported [52]. A ratio of 1:10-1:13 was found for BAMLET and HAMLET complexes formed by a new alternative preparation method, using two consecutive DEAE-Sepharose ion exchange columns for the purification of α-LA and subsequent complex formation, but high variation in stoichiometry was common within different batches [53].

The average number of oleic acid molecules bound per α-LA in LA/OA complexes has been spectrofluorimetrically estimated to be 2.9 for LA/OA formed at 17°C and 9 for LA/OA formed at 45°C [30]. Another study has reported that the bovine LA/OA complex, prepared by direct mixing of the constituents, consists of 4-5 protein molecules with 68-85 bound molecules of OA, thus indicating that every α-LA binds on average 17 OA molecules [49].

A possible explanation for the variation in the observed α-LA to OA ratios might be that the different preparation techniques and experimental conditions (e.g. changes in temperature or altered α-LA to OA ratios) could have a major impact on the observed stoichiometry. In addition, it has been proposed that some of the OA might be present in an unbound form resulting in a higher OA to α-LA ratio [30].

10. Methods of HAMLET or HAMLET-analog preparation *in vitro*

Several different methods have been utilized for the *in vitro* preparation of HAMLET or its orthologs from α-LA and OA. These include OA-conditioned anion exchange chromatography using either a DEAE-Trisacryl M in the conventional method [3,5,6,19] or a DEAE-Sepharose column in an alternative method [53]. Direct mixing under heat denaturation [30,36,52], heat denaturation of α-LA and OA-conditioned anion exchange chromatography [62], direct mixing in solution at room temperature with alternated pH conditions [45,48,49,63], direct mixing followed by anion exchange chromatography [61] and alkaline conditions [64] have also been used.

The conventional method of HAMLET preparation is OA-preconditioned anion exchange chromatography [5]. This method is initiated with the purification of native-state α-LA by hydrophobic interaction chromatography and subsequent EDTA-treatment, resulting in the loss of Ca^{2+} and partial unfolding of the protein. The apo-state α-LA is subjected to a DEAE-Trisacryl M ion exchange chromatographic column previously exposed to human milk or pre-loaded with fatty acids. The apo-state α-LA interacts with the OA pre-loaded column, forming the HAMLET complex, which elutes at a salt concentration of 1M. The eluted complex is desalted by dialysis and lyophilized to obtain HAMLET dry powder.

Regardless of the preparation method used in the different experiments, it has been shown that the cytotoxicity of the resulting complexes was similar to that of conventionally prepared HAMLET or BAMLET [30,36,48,49,54,61-65].

11. Cellular trafficking of HAMLET

In order to clarify the subcellular localization of HAMLET and the interactions of the complex with different tumor cell compartments, researchers have used biotinylated [5,66], I^{125} radioactive labeled [66,67] and Alexa Fluor 568-labeled HAMLET complexes in confocal microscopy and subcellular fractionation experiments [67-69]. Site-specific labeling of HAMLET and HAMLET-like complexes with aminooxy-Alexa Fluor 488 or biotin molecules has also been used [52].

It has previously been described that α-LA interacts with cell membranes and lipid bilayers, as well as fatty acids [41,70-74]. In addition, it has been observed that α-LA is able to rapidly insert itself into the lipid bilayer at pH 2 [75]. However, native human α-LA is only inefficiently bound to and internalized by tumor cells, and does not translocate to the nucleus, nor influence tumor cell viability [3,5,68,69,76]. Even partial unfolding of human α-LA has been shown to be insufficient for increased protein internalization into tumor cells [38]. In addition, human α-LA and oleic acid alone were unable to alter membrane structures, further indicating the necessity of interaction between unfolded α-LA and an associated fatty acid for efficient internalization of the HAMLET complex in tumor cells [76].

Even though a number of studies have focused on investigating the mechanism of HAMLET uptake by tumor cells, it still remains unclear. HAMLET has been shown to interact with the cell surface of tumor cells, be internalized and subsequently accumulate in the nucleus [3,5,67,68]. In the case of healthy cells, HAMLET interacted with the cell membrane and was internalized, but did not accumulate in the nuclei of the cells [4,67].

A study of the interaction between HAMLET and natural or artificially generated, negatively charged, model lipid membranes has revealed that the interaction results in loss of cell membrane integrity, leakage of vesicular contents and morphological changes of the membranes [76]. Interestingly, HAMLET disturbs the integrity of the tested membranes, even under physiological conditions, as opposed to native-state α-LA that only disrupts liposomes at acidic pH [77]. Similarly, at physiological conditions HAMLET readily distorts lipid monolayers at low concentrations, while forming pore-like oligomeric structures resembling annular oligomers at higher concentrations [78]. The ability to form oligomers is suggested to be a property of the α-LA polypeptide chain, enhanced by the fatty acid component, which might be important for the cytotoxic activity of HAMLET [78]. It has been suggested that the degree to which HAMLET and other LA-OA-complexes are able to disturb the membrane integrity, depends on the lipid composition and physical characteristics of the membrane [74,76].

HAMLET binds to the cell surface of intact tumor cells in a patchy distribution, indicating the possible targeting (e.g. receptor-binding) of the complex to specific membrane regions [76]. Another study found that HAMLET and an α-LA/OA complex formed at 17°C, interacts with and alters trans-membrane integral currents of artificial vesicles and natural plasma membranes of the green alga *C. coralline* [65]. The binding of the OA-complexes caused suppression of Ca^{2+} current and Ca^{2+}-activated Cl^- current, as well as increased nonspecific K^+ leakage, indicating nonselective permeability of the depolarized plasma membrane [65]. Furthermore, all the OA-bound states of α-LA had a higher affinity for interaction with the membranes compared to α-LA alone, indicating a Ca^{2+}-independent association. Similar results were obtained regarding the bactericidal activity and membrane depolarization effect of HAMLET, bovine β-lactoglobulin and pike parvalbumin in complex with OA on *S. pneumococci* bacteria [40,64]. Accordingly, it has been proposed that once the complex has adsorbed to the plasma membrane, it is most likely internalized by nonspecific pinocytosis, while other mechanisms of endocytosis, such as receptor-mediated endocytosis have not been ruled out [65].

Translocation of HAMLET from the membrane to the cytoplasm has been followed by confocal microscopy, showing the formation of intracellular aggregates in both normal cells and in tumor cells, although accumulation of HAMLET was higher in tumor cells [4]. The mechanism of HAMLET redistribution from the cytoplasm to different organelles of the cell, especially the nucleus, remains to be clarified. Initial studies found that the nuclear uptake of α-LA/OA occurred through the nuclear pore complex in a Ca^{2+}-independent manner, and that the accumulation of α-LA in the nucleus resulted in Ca^{2+}-dependent DNA fragmentation in the tumor cells [66]. Later, it was suggested that the interaction between HAMLET and ribosomal proteins could initiate nuclear translocation and targeting of the complex [2].

12. Tumor cell death mechanisms of HAMLET

A key unresolved issue regarding HAMLET is by what biological mechanisms the complex induces tumor cell death. A broad variety of transformed and immature tumor cells, but not healthy differentiated cells, have been found to be affected by the complex and even antibiotic-resistant strains of *S. pneumoniae* bacteria seem to be prone to the cytotoxic activity, showing apoptosis-like morphological and mechanistic changes [4,79,80]. HAMLET is believed to have multiple intracellular targets and has metaphorically been called a "Lernaean Hydra", a mythological creature that uses numerous heads to attack enemies [51]. It has been shown that HAMLET induces at least three major responses, namely an apoptotic pathway, an autophagic pathway and chromatin structure disorder [1,67,81]. All the suggested tumor cell death mechanisms of HAMLET or HAMLET-like complexes are listed in Table 1.

Besides the three major responses, it has been shown that BAMLET activates a caspase-independent lysosomal cell death pathway mainly in tumor cells, causing lysosomal membrane permeabilization possibly contributing to BAMLET-induced cell death [61]. In a study of HAMLET binding to α-actinin, it has been indicated that the apoptosis-promoting p38 pathway is the top-scoring pathway in HAMLET-induced cell death, with p38 inhibition delaying death of HAMLET-treated tumor cells [82].

An elevated expression level of c-Myc, an oncogene that has the ability to bind a significant fraction of all known gene promoters, has been shown to increase HAMLET-sensitivity of various tumor cells, suggesting that the level of c-Myc expression could be a direct determinant of HAMLET susceptibility [83]. Furthermore, a reduction in the extracellular glucose levels or degree of glycolysis, predominantly the aerobic glycolysis of tumor cells known as the Warburg Effect, of A549 lung carcinoma cells resulted in an enhanced HAMLET-sensitivity [83]. HAMLET was shown to bind and reduce the activity of the glycolytic enzyme hexokinase 1 in the A549 carcinoma cells, which could partly explain the inhibitory effect of HAMLET on glycolysis [83].

HAMLET has been shown to interact directly with 20S proteasome subunits *in vitro* as well as *in vivo*, resulting in structural modifications and partial inhibition of 20S proteasome activity, possibly leading to accumulation of incorrectly folded proteins contributing to cell death [69]. Moreover, HAMLET was found by N-terminal sequencing and MALDI-TOF-MS to bind intact ribosomes, as well as individual ribosomal proteins L4, L6, L15, L13a, L30,

L35a, S12 and L21 [2]. The plausible effects of the interaction are protein translational blocking and ribosome-initiated nuclear targeting of HAMLET [2].

Responses	Effects
Major responses:	
Apoptotic pathway [2,54,84,87,90,91]	MAL co-localizes with mitochondria and causes release of cytochrome c from the IMM to the cytosol
	MAL causes loss of mitochondrial membrane potential ($\Delta\Psi_m$) and abnormal MPT
	HAMLET causes p53-independent apoptotic or apoptotic-like cell death in tumor cells
	BAMLET-induced cell death varies according to cell type
Autophagic pathway [81,91]	HAMLET induces an macroautophagic response in tumor cells
Chomatin structure disorder [67,90,100]	HAMLET accumulates in the nucleus of tumor cells
	HAMLET binds to histones, independent of the histone tail, resulting in chromatin structure disorder in tumor cells
	HAMLET causes caspase-dependent and caspase-independent chromatin condensation and acts in synergy with histone deacetylase inhibitors
Other responses:	
Anti-adhesion [1,82,102]	MAL and HAMLET facilitates tumor cell detachment
c-Myc oncogene status [83]	Elevated expression of c-Myc oncogene sensitizes tumor cells to HAMLET-induced tumor cell death
Inhibition of 20S proteasome activity [69]	HAMLET binds to 20S proteasome subunits causing structural modifications and partial activity inhibition
Inhibition of glycolysis [83]	HAMLET binds to and reduces the activity of HK1
Lysosomal pathway [61]	BAMLET activates a caspase-independent lysosomal pathway in tumor cells leading to lysosomal membrane permeabilization
p38 pathway [82]	Indicated to be the top-scoring cell death pathway of HAMLET
Ribosome interactions [2]	HAMLET binds to intact ribosomes and individual ribosome proteins

Table 1. The tumor cell death mechanisms of HAMLET and HAMLET-like complexes.

For elaboration of the specific tumor cells prone to the mentioned effect, the reader is encouraged to see the references. HK1, hexokinase 1; IMM, inner mitochondrial membrane; MAL, multimeric α-lactalbumin; MPT, mitochondrial permeability transition.

13. HAMLET and apoptosis

Several studies have focused on clarifying the precise role of apoptosis in the HAMLET-induced cell death of tumor cells. Initially, it was shown by the use of an anti-CD95 antibody that the CD95 receptor-mediated apoptotic pathway had no effect on the apoptosis-inducing activity of MAL [84]. It was also found that MAL co-localized with mitochondria and caused release of cytochrome c from the inner mitochondrial membrane space to the cytosol, resulting in activation of caspase-3-like enzymes and, to a lesser degree, caspase-6-like enzymes of the caspase cascade [84]. The activation of cytosolic caspases by apogenetic factors released from the inner mitochondrial membrane (e.g. cytochrome c), and subsequent Apaf-1, procaspase-9 and dATP association, resulting in formation of the apoptosome complex, are normal events of the classical apoptotic pathway [85-87]. It was later revealed that MAL induced a loss of the mitochondrial membrane potential ($\Delta\Psi_m$) and abnormal mitochondrial permeability transition (MPT) in isolated mitochondria through opening of the MPT pore, resulting in Ca^{2+}-dependent release of cytochrome c to the cytosol [87].

HAMLET-treated Jurkat leukemia cells and A549 lung carcinoma tumor cells exhibited classical apoptotic changes, as well as apoptosis-like changes, such as proapoptotic caspase activation, phosphatidyl serine externalization, DNA fragmentation, apoptotic body formation and compacted chromatin condensation [88-90]. However, classical apoptosis was not the cause of death, as a pan-caspase inhibitor zVAD-fmk was unable to rescue HAMLET-treated cells from dying [84,90]. This was confirmed by the finding that HAMLET-induced cell death is independent of the anti-apoptotic Bcl-2 and Bcl-xL proteins in Jurkat leukemia, K562 promyelocytic leukaemia and FL5.12 murine prolymphocytic tumor cells, as over-expression of Bcl-2 failed to prevent cell death [90]. In addition, the tumoricidal activity of HAMLET is independent of the tumor suppressor protein p53, as there was no difference in HAMLET susceptibility of tumor cells with wild-type, deleted or mutated p53 gene [2,90,91]. Analyzing the effect of BAMLET on Jurkat and THP1 cells by flow cytometry indicated that the death of Jurkat cells looked more apoptotic than the death of THP1 cells which were more necrotic, showing that the actual mechanism of cell death apparently varies between different types of cells [53].

14. HAMLET and macroautophagy

Autophagy has been a topic of comprehensive debate concerning its function as an alternative caspase-independent type II programmed cell death pathway, in addition to serving as a survival mechanism during cellular stresses [92-95]. Macroautophagy, the only autophagic response included in type II programmed cell death, is an adaptive stress response observed in cells exposed to cellular stresses (e.g. starvation), which triggers the recycle of organelles and long-lived proteins as a nutritional source utilized to prolong cell survival [91,92, 96].

HAMLET induces a macroautophagic response in treated tumor cells, with the appearance of cytoplasmic vacuoles and double-membrane enclosed vesicles [81,91]. Furthermore, HAMLET was shown to alter the staining-pattern of LC3-GFP-transfected cells from uniform (LC3-I) to granular (LC3-II), reflecting the translocation of LC3 to autophagosomes during macroautophagy [81]. HAMLET also induced LC3-II accumulation, detected by a Western Blot, when lysosomal degradation was inhibited, which is a clear indicative of macroautophagy [81].

The inhibition of macroautophagy in HAMLET-treated tumor cells by RNA interference of Beclin-1 synthesis resulted in reduced cell death and inhibition of the increase in granular LC3-GFP staining, suggesting that macroautophagy is an important response pathway in HAMLET-induced cell death [81,91]. In accordance with this finding, it was shown that mTOR, an inhibitor of macroautophagy, is inactivated in tumor cells in response to HAMLET [81,97]. An important note is that the mitochondrial damage observed in the different tumor cells of the mentioned experiments also has the potential to trigger macroautophagy in the cells [81,98,99].

15. HAMLET cell nuclei interactions

A striking feature of HAMLET is the ability to move through the cytoplasm and accumulate in the nucleus of tumor cells, as initially elucidated by the study of the active human milk fraction [3,66]. Healthy, differentiated cells did not accumulate biotinylated MAL or Alexa-labeled HAMLET in the nucleus, although uptake in the cytoplasm was observed [66,67].

Through combination of *in vitro* and *in vivo* experiments, HAMLET was found to co-localize with histones and bind strongly to histone H3, as well as histones H4 and H2B to a lesser degree, resulting in perturbation of the chromatin structure and chromatin assembly in tumor cells, possibly impairing transcription, replication and recombination [67,100]. The binding of HAMLET to the histones was independent of the histone tail, and the binding impairs histone deposition on DNA.

In a different study it has also been shown that monomeric Ca^{2+}-loaded α-LA and apo-state α-LA alone can bind histone H3 *in vitro*, possibly through electrostatic interactions with several α-LA molecules bound per histone protein [101]. However, the finding should have no major implications on HAMLET studies, as native monomeric α-LA is unable to reach the cell nuclei of intact tumor cells [3,5,31].

HAMLET has been shown to act in synergy with histone deacetylase inhibitors, as pre-treatment of Jurkat tumor cells with these resulted in enhanced lethal effect of HAMLET and an increased hyperacetylation response [100]. Caspase-independent DNA damage and DNA fragmentation was observed after treatment, suggesting that apoptosis was not the key pathway involved in the combined effect [100]. In addition, HAMLET caused chromatin condensation, as HAMLET-treatment of stably transfected HeLa cells resulted in a decrease of nuclear size. The ability of HAMLET to induce both caspase-dependent and caspase-independent chromatin condensation pattern was found in another study, indicating that the response to HAMLET involves both classical apoptosis and caspase-independent cell death pathways [90].

16. Anti-adhesive properties of HAMLET

The ability of HAMLET to facilitate tumor cell detachment has been described both in early and recent literature. Initially, MAL was found to possess anti-adhesive properties against bacterial attachment to alveolar type II lung carcinoma cells [1]. Subsequently, an *in vivo* experiment studying the therapeutic effects of HAMLET on bladder cancer showed that HAMLET triggers massive shedding of dead tumor cells into the urine, further indicating the anti-adhesive properties of HAMLET towards tumor cells [102]. Later, *in vitro* studies revealed that HAMLET binds to α-actinin-4, in addition to interacting with α-actinin-1, and causes detachment of A549 lung carcinoma cells [82]. Furthermore, a reduction in β1 integrin staining, as well as in FAK and ERK1/2 phosphorylation was observed, suggesting that HAMLET-treatment induced disruption of integrin-dependent cell adhesion signaling.

17. Therapeutic applications of HAMLET

A major reason for studying the cytotoxic activity of HAMLET is the apparent specificity towards cancer cells. Although this specificity of cytotoxic HAMLET-induced cell death remains unclear, there have been continuous indications in many *in vitro* studies that HAMLET could have a therapeutic potential. As a consequence, researchers started conducting *in vivo* studies of the possible applications of HAMLET [68,102-105]. The proposed therapeutic applications of HAMLET and the observed effects are listed in Table 2. One problem in using HAMLET as a potential cancer agent is its interaction with albumin, which through binding of free fatty acids could neutralize the cytotoxic activity of HAMLET or BAMLET [53,87,102]. As albumin is present in nearly all physiologic fluids, it will in many cases be difficult for HAMLET to reach the target cancer cells in an active form.

The observed effects mentioned are *in vivo* unless otherwise noted. GBM: glioblastoma multiforme.

18. HAMLET as a skin papilloma treatment

Papillomas are characterized as premalignant lesions of the skin and mucosal surfaces, caused by human papillomavirus transformation of keratinocytes [4,104]. The therapeutic treatment options are ineffective and include cryotherapy, immuno modulators, curettage, cautery, salicylic acid, CO_2 laser, antimitotic agents, or photo-dynamic theraphy [106,107].

Skin papillomas were the first model selected for the examination of the *in vivo* effects of HAMLET [104]. Patients resistant to conventional treatments were included in this randomized, placebo-controlled, double-blinded study. HAMLET was shown to reduce the papilloma volume in 100% (20/20) of the patients and 96% of their lesions (88/92), compared to 15% in the placebo group patients (3/20) and 20% (15/74) of their lesions. By studying the long-term effects of HAMLET-treatment against skin papillomas, it was found that all lesions had completely resolved in 83% of the patients after 2 years. Furthermore, there were no observed differences between immunocompetent and immunosuppressed patients, indicating the potential of HAMLET-treatment for immunosuppressed skin papilloma patients, instead of the currently used invasive methods [104].

Applications	Effects
Bladder cancer treatment [2,102,105]	Induces bladder carcinoma cell death (*in vitro*) Increases daily shedding of dead tumor cells into the urine Reduces tumor size or area and changes tumor character Delays progression of bladder cancer
Gliblastoma (multiforme) treatment in a rat model [68]	Induces apoptosis in GBM biopsy spheroids (*in vitro*) Reduces intracranial tumor volume Delays onset of pressure symptoms
Skin papilloma treatment [104]	Reduces the volume of papillomas Leads to long-term resolution of lesions Similar effects of treatment in immunosuppressed patients as in immunocompetent patients

Table 2. Possible therapeutic applications of HAMLET.

19. HAMLET as a bladder cancer treatment

The development of bladder cancers is common and remains a massive worldwide challenge, despite continuous advances in the therapeutic options available. The prevalence of bladder cancer is 1 in 4000 people, accounting for about 5% of all cancers, making it the fourth most common malignancy in the United States and the fifth most common in Europe [108]. The current treatments of bladder cancer typically involve intravesical instillation of antitumor agents such as *Bacillus* Calmette-Guerin, thiotepa, epirubicin and mitomycin C [109].

Early *in vitro* studies found that HAMLET induced rapid cell death in numerous cell lines, including bladder carcinoma cells [2]. The direct *in vivo* effect of HAMLET on bladder cancers was initially shown, when 9 male superficial bladder cancer patients received intravesical HAMLET-treatment during the week before scheduled transurethral surgery [102]. HAMLET stimulated a rapid increase in the daily shedding of dead tumor cells into the urine in 8 of 9 patients and an apoptotic response was detected by the TUNEL DNA fragmentation assay in 6 of 9 patients. In addition, a reduction in tumor size or a change in tumor character was observed at surgery in 8 of 9 patients.

The therapeutic potential of HAMLET was further studied in a mouse bladder carcinoma model, with MB49 carcinoma cells installed via a catheter into the bladder of anesthetized mice, followed by five intravesical installations of HAMLET [105]. The treatment resulted in significantly decreased tumor areas and delayed the progression of bladder cancer compared to controls. Furthermore, whole body imaging of Alexa Fluor 568-labeled HAMLET revealed that the uptake and retention of the complex were tumor tissue-specific.

20. HAMLET in treatment of glioblastoma

Gliomas are a heterogeneous group of intracranial neoplasms originating from neuroglial cells, which accounts for over 60% of all primary brain tumors and have an unfavorable prognosis [110-113]. Glioblastoma multiforme (GBM), grade IV on the current WHO grading-scale for tumors in the central nervous system, is the most malignant of the gliomas, showing a mean survival time of less than a year [113,114]. The current treatment is only palliative, involving surgery, radiotherapy and chemotherapy [115].

HAMLET has been shown to induce apoptosis in GBM biopsy spheroids *in vitro*, as detected by the TUNEL DNA fragmentation assay [68]. Similarly, *in vivo* studies of the effect of HAMLET-treatment on GBM's, established by xenotransplantation of human glioblastoma biopsy spheroids into nude rat brains, showed that HAMLET reduced the intracranial tumor volume and delayed the onset of pressure symptoms in tumor-xenografted rats [68,116].

21. HAMLET and involution of the mammary gland

The development of the mammary gland through puberty, pregnancy and lactation has been intensively studied and much is known about the factors responsible for regulating the processes. When suckling stops, the gland goes into involution, and despite much research little is known regarding the factors involved in this remodeling of the gland.

Milk synthesis occurs in mammary epithelial organized in small hollow "balls" called alveoli. It is known that the accumulation of milk in the alveoli results in down regulation of protein synthesis as well as reduction in the number of secretory cells by apoptosis [117]. It is also believed that local factors secreted into the milk are responsible for the effect, but molecular details of such compounds are lacking. A protein called FIL (Feedback Inhibitor of Lactation) from milk has been isolated and partly characterized [118], but at present its existence is unconfirmed.

Some sea animals have a special lactation cycle due to long periods of foraging trips away from the offspring. Remarkably, despite the absence of suckling for many days no apoptosis occurs in the mammary gland of e.g. Cape fur seals, and this phenotype has been correlated with lack of α-LA in their milk [119]. The possibility that α-LA either alone or in a complex with fatty acids is involved in the regulation of involution has been suggested as apoptotic cell death in the mammary gland was enhanced by the introducing of HAMLET in the gland of lactating mice [103].

22. Conclusion

The milk protein alpha-lactalbumin (α-LA) is one of the most studied proteins with respect to structure and function and has been a model for investigating intermediate folding conformations generally called molten globule states. It has a well-described function as part of the enzyme lactose synthetase, where α-LA is one component of the two-polypeptide enzyme. The other part is a non-specific galactosyl transferase, which in complex with α-LA becomes highly specific for forming lactose from UDP-galactose and glucose.

In the late nineties an entirely different activity of α-LA was discovered. When α-LA is in a partly unfolded conformation it can form a complex with oleic acid (OA) where the complex shows cell-killing activity specific for cancer cells. This partly unfolded conformation of α-LA can be obtained by removing a tightly bound calcium ion, leaving the protein in the apo-state capable of interacting directly with OA and thereby forming the cytotoxic complex. The complex between human α-LA and OA has been called HAMLET (Human Alpha-lactalbumin Made LEthal to Tumor cells) and similar complexes can be formed with α-LA from other species.

In cell culture systems HAMLET shows cytotoxic activity in μM concentrations but the activity varies greatly between cell types. The exact mechanism of cell-killing is not clear, but HAMLET has been shown to interact with numerous cell organelles including the nucleus, lysozymes, mitochondria, proteasomes and ribosomes. Apoptosis, autophagy and chromatin structure disorder are three different cytotoxic pathways described when HAMLET is added to cell cultures. It is therefore likely that different cells will show different death mechanisms dependent on the experimental conditions.

When the cytotoxic activity of α-LA/OA complexes are measured in a dose-response manner, it is characteristic that very steep curves are obtained where as little as a four-fold dilution can result in no activity compared to 100% cell death in the undiluted solution. This underlines the importance of defining the experimental conditions exactly when measuring cytotoxic activity as small changes in concentration might lead to contradictory conclusions. This aspect is especially important when the activity of HAMLET is compared between healthy differentiated normal cells and tumor cells. According to numerous reports HAMLET kills tumor cells as well as undifferentiated cells but leave normal cells untouched. This has recently been questioned as it was found that both normal white blood cells and fully developed erythrocytes were highly sensitive to the cell-killing activity.

The potential of HAMLET as a therapeutic agent has been investigated in three different models. First, skin papillomas were treated in a double-blinded study showing positive results compared to the controls, secondly, patients with bladder cancer indicated reduction in tumor size when the bladder was flushed with a solution of HAMLET, and finally, in a glioblastoma rat model HAMLET treatment resulted in a decrease of the intracranial tumor volume. This suggests that tumor cells in some cases might be more sensitive to HAMLET when compared with the healthy cells from which the tumors originate.

It has been theorized that HAMLET could be naturally formed in the digestive system of breast-fed children, due to low pH conditions of the stomach, serving the function as a natural scavenger and selective cytotoxic killer of cancerous cells in early infancy. The development of normal epithelia cells in the digestive system could also be affected. Obviously, many dairy products contains α-LA in fairly high concentrations and numerous production techniques, such as acid pH and heat treatment, facilitate the formation of partly unfolded α-LA making the protein ready to bind fatty acids and other similar compounds. To what degree such complexes in fact are formed in dairy products remains to be seen.

Author details

Marcel Jøhnke and Torben E. Petersen*
Department of Molecular Biology and Genetics, Aarhus University, Aarhus, Denmark

23. References

[1] Håkansson A, Zhivotovsky B, Orrenius S, Sabharwal H, Svanborg C (1995) Apoptosis induced by a human milk protein. Proc. natl. acad. sci. USA 92: 8064-8068.

[2] Svanborg C, Agerstam H, Aronson A, Bjerkvig R, Düringer C, Fischer W, Gustafsson L, Hallgren O, Leijonhuvud I, Linse S, Mossberg AK, Nilsson H, Pettersson J, Svensson M (2003) HAMLET kills tumor cells by an apoptosis-like mechanism--cellular, molecular, and therapeutic aspects. Adv. cancer res. 88: 1-29.

[3] Svensson M, Sabharwal H, Håkansson A, Mossberg AK, Lipniunas P, Leffler H, Svanborg C, Linse S (1999) Molecular characterization of α-lactalbumin folding variants that induce apoptosis in tumor cells. J. biol. chem. 274: 6388–6396.

[4] Gustafsson L, Hallgren O, Mossberg AK, Pettersson J, Fischer W, Aronsson A, Svanborg C (2005) HAMLET kills tumor cells by apoptosis: structure, cellular mechanisms, and therapy. J. nutr. 135: 1299-1303.

[5] Svensson M, Håkansson A, Mossberg AK, Linse S, Svanborg C (2000) Conversion of α-lactalbumin to a protein inducing apoptosis. Proc. natl. acad. sci. USA. 97: 4221-4226.

[6] Svensson M, Mossberg AK, Pettersson J, Linse S, Svanborg C (2003) Lipids as cofactors in protein folding: stereo-specific lipid-protein interactions are required to form HAMLET (human α-lactalbumin made lethal to tumor cells). Protein sci. 12: 2805-2814.

[7] Hill RL and Brew K (1975) Lactose synthetase. Adv. enzymol. relat. areas mol. biol. 43: 411-490.

[8] Stinnakre MG, Vilotte JL, Soulier S, Mercier JC (1994) Creation and phenotypic analysis of α-lactalbumin-deficient mice. Proc. natl. acad. sci. USA 91: 6544-6548.

[9] Pettersson J, Mossberg AK, Svanborg C (2006) α-Lactalbumin species variation, HAMLET formation, and tumor cell death. Biochem. biophys. res. commun. 345: 260-270.

[10] Acharya KR, Ren J, Stuart DI, Phillips DC, Fenna RE (1991) Crystal structure of human α-lactalbumin at 1.7 Å resolution. J. mol. biol. 221: 571-581.

* Corresponding Author

[11] Chrysina ED, Brew K, Acharya KR (2000) Crystal structures of apo- and holo-bovine α-lactalbumin at 2.2-Å resolution reveal an effect of calcium on inter-lobe interactions. J. biol. chem. 275: 37021-37029.

[12] Iyer LK and Qasba PK (1999) Molecular dynamics simulation of α-lactalbumin and calcium binding c-type lysozyme. Protein eng. 12: 129-139.

[13] Permyakov EA and Berliner LJ (2000) α-lactalbumin: structure and function. FEBS lett. 473: 269–274.

[14] Kuwajima K (1996) The molten globule state of α-lactalbumin. FASEB j. 10: 102–109.

[15] Ewbank JJ and Creighton TE (1993) Pathway of disulfide-coupled unfolding and refolding of bovine alpha-lactalbumin. Biochemistry 32: 3677-3693.

[16] Hiraoka Y, Segawa T, Kuwajima K, Sugai S, Murai N (1980) α-Lactalbumin: a calcium metalloprotein. Biochem. biophys. res. commun. 95: 1098–1104.

[17] Wu LC, Schulman BA, Peng ZY, Kim PS (1996) Disulfide determinants of calcium-induced packing in α-lactalbumin. Biochemistry 35: 859-563.

[18] Anderson PJ, Brooks CL, Berliner LJ (1997) Functional identification of calcium binding residues in bovine α-lactalbumin. Biochemistry 36: 11648-11654.

[19] Svensson M, Fast J, Mossberg AK, Düringer C, Gustafsson L, Hallgren O, Brooks CL, Berliner L, Linse S, Svanborg C (2003). α-Lactalbumin unfolding is not sufficient to cause apoptosis, but is required for the conversion to HAMLET (human α-lactalbumin made lethal to tumor cells). Protein sci. 12: 2794-2804.

[20] Permyakov EA, Yarmolenko VV, Kalinichenko LP, Morozova LA, Burstein EA (1981) Calcium binding to alpha-lactalbumin: structural rearrangement and association constant evaluation by means of intrinsic protein fluorescence changes. Biochem. biophys. res. commun. 100: 191-197.

[21] Kronman MJ, Sinha SK, Brew K (1981) Characteristics of the binding of Ca^{2+} and other divalent metal ions to bovine alpha-lactalbumin. J. biol. chem. 256: 8582-8587.

[22] Chandra N, Brew K, Acharya KR (1998) Structural evidence for the presence of a secondary calcium binding site in human α-lactalbumin. Biochemistry 37: 4767-4772.

[23] Permyakov EA, Kalinihenko LP, Morozova LA, Yarmolenko VV, Burstein EA (1981) α-Lactalbumin binds magnesium ions: study by means of intrinsic fluorescence technique. Biochem. biophys. res. commun. 102: 1-7.

[24] Permyakov EA, Morozova LA, Burstein EA (1985) Cation binding effects on the pH, thermal and urea denaturation transitions in alpha-lactalbumin. Biophys. chem. 21: 21-31.

[25] Ren J, Stuart DI, Acharya KR (1993) α-lactalbumin possesses a distinct zinc binding site. J. biol. chem. 268: 19292-19298.

[26] Fast J, Mossberg AK, Nilsson H, Svanborg C, Akke M, Linse S (2005) Compact oleic acid in HAMLET. FEBS lett. 579: 6095-100.

[27] Griko YV, Freire E, Privalov P L (1994) Energetics of the alpha-lactalbumin states: a calorimetric and statistical thermodynamic study. Biochemistry 33: 1889-1899.

[28] Dzwolak W, Kato M, Shimizu A, Taniguchi Y (1999) Fourier-transform infrared spectroscopy of the pressure-induced changes in the structure of the bovine α-lactalbumin: the stabilizing role of the calcium ion. Biochim. biophys. acta 1433: 45-55.

[29] Fast J, Mossberg AK, Svanborg C, Linse S (2005) Stability of HAMLET – a kinetically trapped α-lactalbumin oleic acid complex. Protein sci. 14: 329-340.

[30] Knyazeva EL, Grishchenko VM, Fadeev RS, Akatov VS, Permyakov SE, Permyakov EA (2008) Who is Mr. HAMLET? Interaction of human α-lactalbumin with monomeric oleic acid. Biochemistry 47: 13127–13137.

[31] Mossberg AK, Mok K H, Morozova-Roche LA, Svanborg C (2010) Structure and function of human α-lactalbumin made lethal to tumor cells (HAMLET)-type complexes. FEBS j. 277: 4614-4625.

[32] Brew K and Grobler J (1992) α-Lactalbumin. Adv. dairy chem. 1 (Fox, P., ed): 191-229. Elsevier, London.

[33] Finkelstein AV and Ptitsyn OB (1977) Theory of protein molecule self-organization. I. Thermodynamic parameters of local secondary structures in the unfolded protein chain. Biopolymers 16: 469-495.

[34] Dolgikh D, Gilmanshin R, Brazhnikov E, Bychkova V, Semisotnov G, Venyaminov S, Ptitsyn O (1981) Alpha-Lactalbumin: compact state with fluctuating tertiary structure? FEBS lett. 136: 311-315.

[35] Pfeil W (1987) Is thermally denatured protein unfolded? The example of α-lactalbumin. Biochim. biophys. acta 911: 114-116.

[36] Kamijima T, Ohmura A, Sato T, Akimoto K, Itabashi M, Mizuguchi M, Kamiya M, Kikukawa T, Aizawa T, Takahashi M, Kawano K, Demura M (2008) Heat-treatment method for producing fatty acid-bound alpha-lactalbumin that induces tumor cell death. Biochem. biophys. res. commun. 376: 211-214.

[37] Redfield C, Schulman BA, Milhollen MA, Kim PS, Dobson CM (1999) α-Lactalbumin forms a compact molten globule in the absence of disulfide bonds. Nat. struct. biol. 6: 948-952.

[38] Pettersson-Kastberg J, Mossberg AK, Trulsson M, Yong YJ, Min S, Lim Y, O'Brien JE, Svanborg C, Mok KH (2009) α-Lactalbumin, engineered to be nonnative and inactive, kills tumor cells when in complex with oleic acid: a new biological function resulting from partial unfolding. J. mol. biol. 394: 994-1010.

[39] Schulman BA, Redfield C, Peng Z, Dobson CN, Kim PS (1995) Different subdomains are most protected from hydrogen exchange in the molten globule and native states of human α-lactalbumin. J. mol. biol. 253: 651-657.

[40] Permyakov SE, Knyazeva EL, Khasanova LM, Fadeev RS, Zhadan AP, Roche-Hakansson H, Hakonsson AP, Akatov VS, Permyakov EA (2012) Oleic acid is a key cytotoxic component of HAMLET-like complexes. Biol. chem. 393: 85-92.

[41] Cawthern KM, Naryan M, Chaudhuri D, Permyakov EA, Berliner LJ (1997) Interactions of α-lactalbumin with fatty acids and spin label analogs. J. biol. chem. 272: 30812-30816.

[42] Brinkman CR (2011) Cytotoxic activity of a bovine alpha-lactalbumin:oleic acid complex. PhD dissertation, Faculty of Science, Aarhus University.

[43] Wu LC and Kim PS (1998) A specific hydrophobic core in the α-lactalbumin molten globule. J. mol. biol. 280: 175-182.

[44] Saito M (1999) Molecular dynamics model structures for the molten globule state of α-lactalbumin: aromatic residue clusters I and II. Protein eng. 12: 1097-1104.

[45] Polverino de Laureto P, Frare E, Gottardo R, Fontana A (2002) Molten globule of bovine α-lactalbumin at neutral pH indued by heat, trifluoroethanol, and oleic acid: a comparative analysis by circular dichroism spectroscopy and limited proteolysis. Proteins 49: 385-397.

[46] Wijesinha-Bettoni R, Dobson CM, Redfield C (2001) Comparison of the structural and dynamical properties of holo and apo bovine α-lactalbumin by NMR spectroscopy. J. mol. biol. 307: 885-898.

[47] Paci E, Smith LJ, Dobson CM, Karplus M (2001) Exploration of partially unfolded states of human α-lactalbumin by molecular dynamics simulation. J. mol. biol. 306: 329-347.

[48] Tolin S, De Franceschi G, Spolaore B, Frare E, Canton M, Polverino de Laureto P, Fontana A (2010) The oleic acid complexes of proteolytic fragments of α-lactalbumin display apoptotic activity. FEBS j. 277: 163-73.

[49] Spolaore B, Pinato O, Canton M, Zambonin M, Polverino de Laureto P, Fontana A (2010) α-Lactalbumin forms with oleic acid a high molecular weight complex displaying cytotoxic activity. Biochemistry 49: 8658-67.

[50] Casbarra A, Birolo L, Infusini G, Dal Piaz F, Svensson M, Pucci P, Svanborg C, Marino G (2004) Conformational analysis of HAMLET, the folding variant of human α-lactalbumin associated with apoptosis. Protein sci. 13: 1322-1330.

[51] Mok KH, Pettersson J, Orrenius S, Svanborg C (2007) HAMLET, protein folding, and tumor cell death. Biochem. biophys. res. commun. 354: 1-7.

[52] Mercer N, Ramakrishnan B, Boeggeman E, Qasba PK (2011) Application of site-specific labeling to study HAMLET, a tumorricidal complex of α-lactalbumin and oleic acid. PLoS ONE 6: e26093.

[53] Brinkmann CR, Thiel S, Larsen MK, Petersen TE, Jensenius JC, Heegaard CW (2011) Preparation and comparison of cytotoxic complexes formed between oleic acid and either bovine or human α-lactalbumin. J. dairy sci. 94: 2159-2170.

[54] Brinkmann CR, Heegaard CW, Petersen TE, Jensenius JC, Thiel S (2011) The toxicity of bovine α-lactalbumin made lethal to tumor cells is highly dependent on oleic acid and induces killing in cancer cell lines and non-cancer derived primary cells. FEBS j. 278: 1955-1967.

[55] Cury-Boaventura MF, Pompéia C, Curi R (2004) Comparative toxicity of oleic acid and linoleic acid on Jurkat cells. Clin. nutr. 23: 721-732.

[56] Thompson MP, Farrell HM, Mohanam S, Liu S, Kidwell WR, Bansal MP, Cook RG, Medina D, Kotts CE, Bano M (1992) Identification of human-milk α-lactalbumin as a cell-growth inhibitor. Protoplasma 167: 134-144.

[57] Sternhagen LG and Allen JC (2001) Growth rates of a human colon adenocarcinoma cell line are regulated by the milk protein alpha-lactalbumin. Adv. exp. med. biol. 501: 115-120.

[58] Xu M, Sugiura Y, Nagaoka S, Kanamaru Y (2005) IEC-6 intestinal celle death induced by bovine milk α-lactalbumin. Biosci. biotechnol. biochem. 69: 1082-1089.

[59] Xu M, Sugiura Y, Nagaoka S, Kanamaru Y (2005) Involvement of SDS-stable higher Mr forms of bovine normal milk α-lactalbumin in inducing intestinal IEC-6 cell death. Biosci. biotechnol. biochem. 69: 1189-1192.

[60] Lin IC, Su SL, Kuo CD (2008) Induction of cell death in RAW 264.7 cells by alpha-lactalbumin. Food chem. toxicol. 46: 842-853.

[61] Rammer P, Groth-Pedersen L, Kirkegaard T, Daugaard M, Rytter A, Szyniarowski P, Høyer-Hansen M, Povlsen LK, Nylandsted J, Larsen JE, Jäättelä M (2010) BAMLET activates a lysosomal cell death program in cancer cells. Mol. cancer ther. 9: 24-32.

[62] Lišková K, Kelly AL, O'Brien N, Brodkorb A (2010) Effect of denaturation of alpha-lactalbumin on the formation of BAMLET (bovine α-lactalbumin made lethal to tumor cells). J. agric. food chem. 58: 4421–4427.

[63] Zhang M, Yang F Jr, Yang F, Chen J, Zheng CY, Liang Y (2009) Cytotoxic aggregates of α-lactalbumin induced by unsaturated fatty acid induce apoptosis in tumor cells. Chem biol interact. 180: 131-42.

[64] Permyakov SE, Knyazeva EL, Leonteva MV, Fadeev RS, Chekanov AV, Zhadan AP, Hakansson AP, Akatov VS, Permyakov EA (2011) A novel method for preparation of HAMLET-like protein complexes. Biochimie 93: 1495-1501.

[65] Zherelova OM, Kataev AA, Grishchenko VM, Knyazeva EL, Permyakov SE, Permyakov EA (2009) Interaction of antitumor α-lactalbumin-oleic acid complexes with artificial and natural membranes. J. bioenerg. biomembr. 41: 229-237.

[66] Håkansson A, Andréasson J, Zhivotovsky B, Karpman D, Orrenius S, Svanborg C (1999) Multimeric α-lactalbumin from human milk induces apoptosis through a direct effect on cell nuclei. Exp. cell res. 246: 451-460.

[67] Düringer C, Hamiche A, Gustafsson L, Kimura H, Svanborg C (2003) HAMLET interacts with histones and chromatin in tumor cell nuclei. J. biol. chem. 278: 42131-42135.

[68] Fischer W, Gustafsson L, Mossberg AK, Gronli J, Mork S, Bjerkvig R, Svanborg C (2004) Human α-lactalbumin made lethal to tumor cells (HAMLET) kills human glioblastoma cells in brain xenografts by an apoptosis-like mechanism and prolongs survival. Cancer res. 64: 2105-2112.

[69] Gustafsson L, Aits S, Onnerfjord P, Trulsson M, Storm P, Svanborg C (2009) Changes in proteasome structure and function caused by HAMLET in tumor cells. PLoS ONE 4: e5229.

[70] Bañuelos S and Muga A (1995) Binding of molten globule-like conformations to lipid bilayers. Structure of native and partially unfolded α-lactalbumin bound to model membranes. J. biol. chem. 270: 29910-29915.

[71] Bañuelos S and Muga A (1996) Structural requirements for the association of native a partially folded conformations of α-lactalbumin with model membranes. Biochemistry 35: 3892-3898.

[72] Grishcenko VM, Kalinichenko LP, Deikus GY, Veprintsev DB, Cawthern KM, Berliner LJ, Permyakov EA (1996) Interactions of α-lactalbumin with lipid vesicles studied by tryptophan fluorescence. Biochem. mol. biol. int. 38: 453-466.

[73] Halskau O, Froystein NA, Muga A, Martinez A (2002) The membrane-bound conformation of α-lactalbumin studied by NMR-monitored 1H exchange. J. mol. biol. 321: 99-110.

[74] Agasoester AV, Halskau O, Fuglebakk E, Froeystein NA, Muga A, Holmsen H, Martinez A (2003) The interaction of peripheral proteins and membranes studied with α-lactalbumin and phospholipid bilayers of various compositions. J. biol. chem. 278: 21790-21797.

[75] Cawthern KM, Permyakov E, Berliner LJ (1996) Membrane-bound states of alpha-lactalbumin: implications for the protein stability and conformation. Protein sci. 5: 1394-1405.

[76] Mossberg AK, Puchades M, Halskau O, Baumann A, Lanekoff I, Chao Y, Martinez A, Svanborg C, Karlsson R (2010) HAMLET interacts with lipid membranes and perturbs their structure and integrity. PLoS One 5: e9384.

[77] Rodland I, Halskau O, Martinez A, Holmsen H (2005) Alpha-Lactalbumin binding and membrane integrity - effect of charge and degree of unsaturation of glycerophospholipids. Biochim. biophys. acta 1717: 11-20.

[78] Baumann A, Gjerde AU, Ying M, Svanborg C, Holmsen H, Glomm WR, Martinez A, Halskau Ø (2012) HAMLET forms annular oligomers when deposited with phospholipids monolayers. J. mol. biol. doi:10.1016/j.jmb.2012.02.006.

[79] Håkansson A, Svensson M, Mossberg AK, Sabharwal H, Linse S, Lazou I, Lonnerdal B, Svanborg C (2000) A folding variant of α-lactalbumin with bactericidal activity against Streptococcus pneumoniae. Mol. microbiol. 35: 589-600.

[80] Hakansson AP, Roche-Hakansson H, Mossberg AK, Svanborg C (2011) Apoptosis-like death in bacteria induced by HAMLET, a human milk lipid-protein complex. PLoS ONE 6: e17717.

[81] Aits S, Gustafsson L, Hallgren O, Brest P, Gustafsson M, Trulsson M, Mossberg AK, Simon HU, Mograbi B, Svanborg C (2009) HAMLET (human α-lactalbumin made lethal to tumor cells) triggers autophagic tumor cell death. Int. j. cancer 124: 1008-1019.

[82] Trulsson M, Yu H, Gisselsson L, Chao Y, Urbano A, Aits S, Mossberg AK, Svanborg C (2011) HAMLET binding to α-actinin facilitates tumor cell detachment. PLoS ONE 6: e17179.

[83] Storm P, Aits S, Puthia MK, Urbano A, Northen T, Powers S, Bowen B, Chao Y, Reindl W, Lee DY, Sullivan NL, Zhang J, Trulsson M, Yang H, Watson JD, Svanborg C (2011)

Conserved features of cancer cells define their sensitivity to HAMLET-induced death; c-Myc and glycolysis. Oncogene 30: 4765-4779.

[84] Köhler C, Håkansson A, Svanborg C, Orrenius S, Zhivotovsky B (1999) Protease activation in apoptosis induced by MAL. Exp. cell res. 249: 260-268.

[85] Cryns VL and Yuan JY (1998) Proteases to die for. Genes dev. 12: 1551-1570.

[86] Zou H, Li Y, Liu X, Wang X (1999) An APAF-1•cytochrome c multimeric complex is a functional apoptosome that activates procaspase-9. J. biol. chem. 274: 11549-11556.

[87] Köhler C, Gogvadze V, Håkansson A, Svanborg C, Orrenius S, Zhivotovsky B (2001) A folding variant of human α-lactalbumin induces mitochondrial permeability transition in isolated mitochondria. Eur j. biochem. 268: 186-191.

[88] Kerr JF, Wyllie AH, Currie AR (1972) Apoptosis: a basic biological phenomenon with wide-ranging implications in tissue kinetics. Br. j. cancer 26: 239-257.

[89] Ellis HM and Horvitz HR (1986) Genetic control of programmed cell death in the nematode C. elegans. Cell 44: 817-829.

[90] Hallgren O, Gustafsson L, Irjala H, Selivanova G, Orrenius S, Svanborg C (2006) HAMLET triggers apoptosis but tumor cell death is independent of caspases, Bcl-2 and p53. Apoptosis 11: 221-233.

[91] Hallgren O, Aits S, Brest P, Gustafsson L, Mossberg AK, Wullt B, Svanborg C (2008) Apoptosis and tumor cell death in response to HAMLET (human α-lactalbumin made lethal to tumor cells). Adv. exp. med. biol. 606: 217-240.

[92] Levine B and Klionsky DJ (2004) Development of self-digestion: molecular mechanisms and biological functions of autophagy. Developmental cell 6: 463-477.

[93] Yorimitsu T and Klionsky DJ (2005) Autophagy: molecular machinery for self-eating. Cell death differ. 12 (Suppl. 2): 1542-1552.

[94] Gonzalez-Polo RA, Boya P, Pauleau AL, Jalil A, Larochette N, Souquere S, Eskelinen EL, Pierron G, Saftig P, Kroemer G (2005) The apoptosis/autophagy paradox: autophagic vacuolization before apoptotic death. J. cell sci. 118: 3091-3102.

[95] Høyer-Hansen M and Jäättelä M (2008) Autophagy: an emerging target for cancer therapy. Autophagy 4: 574-580.

[96] Schweichel JU and Merker HJ (1973) The morphology of various types of cell death in prenatal tissues. Teratology 7: 253-266.

[97] Meijer AJ, and Codogno P (2006) Signalling and autophagy in health, aging and disease. Mol. aspects med. 27: 411-425.

[98] Elmore SP, Qian T, Grissom SF, Lemasters JJ (2001) The mitochondrial permeability transition initiates autophagy in rat hepatocytes. FASEB j. 15: 2286-2287.

[99] Kundu M and Thompson CB (2005). Macroautophagy versus mitochondrial autophagy: a question of fate? Cell death differ. 12: 1484-1489.

[100] Brest P, Gustafsson M, Mossberg AK, Gustafsson L, Düringer C, Hamiche A, Svanborg C. (2007) Histone deacetylase inhibitors promote the tumoricidal effect of HAMLET. Cancer res. 67: 11327-34.

[101] Permyakov SE, Pershikova IV, Khokhlova TI, Uversky VN, Permyakov EA (2004) No need to be HAMLET or BAMLET to interact with histones: binding of monomeric α-lactalbumin to histones and basic poly-amino acids. Biochemistry 43: 5575-5582.

[102] Mossberg AK, Wullt B, Gustafsson L, Mansson W, Ljunggren E, Svanborg C (2007) Bladder cancers respond to intravesical instillation of HAMLET (human α-lactalbumin made lethal to tumor cells). Int. j. cancer, 121: 1352-1359.

[103] Baltzer A, Svanborg C, Jaggi R (2004) Apoptotic cell death in the lactating mammary gland is enhanced by a folding variant of α-lactalbumin. Cell mol. life sci. 61: 1221-1228.

[104] Gustafsson L, Leijonhufvud I, Aronsson A, Mossberg AK, Svanborg C (2004) Treatment of skin papillomas with topical α-lactalbumin-oleic acid. N. engl. j. med. 350: 2663–2672.

[105] Mossberg AK, Hou Y, Svensson M, Holmqvist B, Svanborg C (2010) HAMLET treatment delays bladder cancer development. J. urol. 183: 1590–1597.

[106] Leman JA and Benton EC (2000) Verrucas: Guidelines for management. Am. j. clin. dermatol. 1: 143-149.

[107] Gibbs S, Harvey I, Sterling JC, Stark R (2003) Local treatments for cutaneous warts. BMJ 325: 461.1.

[108] Sengupta N, Siddiqui E, Mumtaz FH (2004) Cancers of the bladder. J. r. soc. promot. health 124: 228.

[109] Schenkman E and Lamm DL (2004) Superficial bladder cancer therapy. Sci. world j. 4: 387-399.

[110] Russel DS and Rubinstein LJ (1989) Pathology of tumors of the nervous system. 5th ed. Williams and Wilkins, Baltimore, MD.

[111] Daumas-Duport C, Scheithauer B, O'Fallon J, Kelly P (1988) Grading of astrocytomas: a simple and reproducible method. Cancer 62: 2152-2165.

[112] Kim TS, Halliday AL, Hedley-Whyte ET, Convery K (1991) Correlates of survival and the Daumas-Diport grading system for astrocytomas. J. neurosurg. 74: 27-37.

[113] Gundersen S, Lote K, Hannisdal E (1996) Prognostic factors for glioblastoma multiforme-development of a prognostic index. Acta oncol. 35 (Suppl. 8): 123-127.

[114] Louis DN, Ohgaki H, Wiestler OD, Cavenee WK, Burger PC, Jouvet A, Scheithauer BW, Kleihues P (2007) The 2007 WHO classification of tumours of the central nervous system. Acta neuropathol. 114: 97–109.

[115] Grossman SA, and Bastor JF (2004) Current management of glioblastoma multiforms. Semin. oncol. 31: 649-659.

[116] Engebraaten O, Hjortland GO, Hirschberg H, Fodstad O (1999) Growth of precultured human glioma specimens in nude rat brain. J. neurosurg. 90: 125-132.

[117] Li M, Liu X, Robinson G, Bar-Peled U, Wagner K-U, Young WS, Hennighausen L, Furth PA (1997) Mammary-derived signals activate programmed cell death during the first stage of mammary gland involution. Proc. natl. acad. sci. USA 94: 3425-3430.

[118] Wilde CJ, Addey CVP, Boddy LM, Peaker M (1995) Autocrine regulation of milk secretion by a protein in milk. Biochem. j. 305: 51-58.

[119] Sharp JA, Lefèvre C, Nicholas KR (2008) Lack of functional alpha-lactalbumin prevents involution in Cape fur seals and identifies the protein as an apoptotic milk factor in mammary gland involution. BMC biology 6: 48.

Processing on Milk Proteins

Complexation Between Casein Micelles and Whey Protein by Indirect UHT- Processing of Milk: Influence of Surface Hydrophobicity and Dye-Binding Characteristics of Micelles in Relationship with Their Physico-Functionality

Bärbel Lieske

Additional information is available at the end of the chapter

1. Introduction

Milk is nearly always heat treated and the aim of heating was to make the milk safe for human health by killing microorganisms and inactivating the enzymes. Now there has been consistent interest in more novel and sophisticated strategies to improve the rheological properties and stability of acid gels by a more tailored complexation between casein and whey protein *via* SH/SS-exchange. The most frequented approach explored has been (ultra)high-pressure processing. Pressure treatment alone decreases the casein micelle size; however, this effect was less marked when heat and pressure treatments were combined and gels formed on acidification revealed a range of firmness, yield stresses and yield strains which were not to related to whey protein denaturation by oneself (1).Today, the understanding of the effects of high pressure seems quite advanced, e.g.; that the casein micelle properties were changed and that the denaturation of whey protein affected the gel assembly process in cheese and yoghurt (2). Investigations in the effect of high pressure (in the range of 50 to 350 MPa) on constituents of the colloidal phase of milk revealed that non-denatured β-lactoglobulin is left reaching maximum value of 62% in the range of 100 to 350 MPa (7). Other studies addressed the effects of pressure, pH, and temperature on the casein micelle dissociation in skim milk during high-pressure treatment with measurements of light scattering. A relationship between the barostability of casein micelles and pH was established, whereas no effect of temperature was ascertainable, and a mechanism for high pressure-induced disruption of micelle integrity is suggested in which the state of calcium plays a crucial role in the micelle dissociation process (8). A study in high pressure effects on the structure of casein micelles in milk by cryo-transmission electron microscopy showed that in the range 150 to 300 MPa a large number of small micelles coexisted with a fraction of large micelles, which appeared perfectly spherical with smooth and well-defined surfaces (6).

The manufacture of acid gels from ultra-high pressure homogenization-treated milk in combination with a thermal treatment (90 °C/5min) resulted in a firmer gel quality to be used for improving the rheological properties and stability of yoghurt and, thus decreasing the need for additives (5). This expectation proved to be right as shown in two further experimental studies dealing with acid coagulation properties of set and stirred yoghurt (3, 4).

Thermo-sonification has proved to be an alternative technique to extent the shelf life of whole milk (10) and the combination of preheating (45°C) with thermosonification (24 kHz for 10 min) allowed the preparation of yoghurts with rheological properties superior (9) to those of control yoghurts produced from conventionally heated milk (90°C for 10min) (11). The effects of preheating milk (80°C for 10min) for preparing acid milk gels were studied with reference to the preheating pH by dynamic oscillation measurements resulting in firmer gel qualities at a higher pH (>6,65), probably due to greater involvement of S-S interactions (12). A short-time of preheating in terms of procedure is also an element of ultra-high temperature processing (UHT) of milk which is well-tried in the dairy practice today. Indirect UHT processing was used to prepare dairy desserts for studying the influence of varying concentrations of additives on the dessert rheology. As a result, a more extensive whey protein denaturation and subsequent complexation with casein micelles is believed to contribute to the improvement of rheological properties of the UHT desserts (13).

Own research into the effects on indirect UHT processing are in agreement with this and confirmed that perfect complexation may result only by the indirect procedure (14). Moreover, perfect complexation and inherent physico-functionality of the protein were to set not only by the technology alone but also by the current colloid-chemical status of the casein micelles in milk. In herd milk, this status may vary considerably in one period of lactation and inherent negative consequences are seen too late in the finished product.

This research was aimed at the investigation into the influence of a varying colloidal status of casein micelles on the resulting micellar complexation with whey protein on indirect UHT processing of skim milk. For this an own colloid-chemical approach is used to explain the structural changes of casein micelles with given examples of four herd milk samples and one dairy-based product for the purpose of confirmation

2. Materials and methods

2.1. Raw milk and thermal treatments

Bulked herd milk was collected from one farm in Schleswig-Holstein for a period of time of 8 month. Both separation of milk and thermal treatment were carried out using certified equipment. The skimmed milk was first regenerative (principle of heat recovery) preheated to 70-85 °C and then heated indirectly (126 °C for 4 s) and after that cooled down to 5 °C.

2.2. Spectrophotometric analysis of ß-lactoglobulin (ß-Lg)

For quantification of β-Lg (15) from the pH 4,6 -filtrate a spectrophotometric method was applied using a molar absorption coefficient of 0.950 for 0, 1 % (w/w) β-Lg at 280 nm. All spectrophotometric results are means of triplicate determinations repeated twice.

2.3. Surface hydrophobicity of proteins in milk

The development of this method is dated back to 1990/91 and was refined further until 2004. The initial interest was to know the reason of interference of the non-ionic detergent Tween 80 in the Bio-Rad Protein Assay with a variety of proteins. The detergent binding turned out to be a measure of hydrophobic surface area of integral proteins and thus constituting the basis for an assay procedure to determine the hydrophobic potential on the surface of proteins. At first, the methodical research was turned to analyze only isolated proteins (16). A methodical description is given below.

2.3.1. Reagent and sample preparation

Detergent reagent: 0,25 % Tween ("commercial-grade") is prepared in destilled and/or deionized water, stored at room temperature for max. 3 days.

Dye reagent: The dye reagent is purchased as a five-fold concentrate, which must be diluted and filtered through a large-pored filter prior further use.

Sample preparation: Proteins are diluted to about 0,05-0,10 % using *aqua dest.* or 0,05M phosphate buffer. As a rule, the colour intensity of 50 µL diluted sample developed with 2,5 mL dye reagent should not exceed A 595nm, 1cm = 0,500 vs. *aqua dest.* or buffer.

Additional items required:

- Spectrophotometer: allowing measurements at 595 nm.
- Polystyrene cuvettes: 10 mm path length as semi-micro cuvettes.
- Test tubes: polystyrene test tubes (13 x 64 mm) each fitted with a mixing spatula (Boeringer, Mannheim, Germany)
- Dispenser and microliterpipets are necessary for precise dispensing the dye reagent resp. for adding the sample.
- Rack: Test tube rack to store the test tubes containing samples and blanks.

2.3.2. Assay procedure

The hydrophobicity of proteins and casein micelles is calculated from two different protein assays, which are developed at the same time. Triplicates for each single measurement are necessary.

1a) 50 µL protein (A sample) is placed on the bottom of a dry and clean test tube whereas 50 µL *aqua dest.* or buffer are used as blank (A blank).

1b) 50 µL 0,25 % Tween 80 is placed on the bottom of a dry and clean test tube and in addition to 50 µL protein is placed onto the droplet of Tween 80 (B sample) whereas *aqua dest.* or buffer and Tween 80 are prepared as blank (B blank).

2) Only the detergent-containing tubes (1b) are shaken for 10 min. (avoid foaming) to complete detergent binding at temperatures between 18-22 °C.

3) Add 2,5 mL diluted dye reagent to each test tube, insert mixing spatulas by use of a pincette to avoid any skin contact. Move the spatula several times up and down without foaming.

4) Allow standing for 12 min. to develop the colour. The colour intensity of each tube is measured at 595 nm vs. *aqua dest.* Avoid any warming of cuvettes by a prolonged standing inside the cuvette department of the photometer.

5) Calculation

Protein Hydrophobicity (PH) is defined as following (Nakai and Li-Chan)   (17)

PH = (nonpolar residues)/ (nonpolar residues) – (polar residues)

Using this definition on detergent binding according the proposed method, PH is calculated as

$$\mathrm{SHP}\ (\%)\ = \frac{\left(A_{sample} - A_{blank}\right) - \left(B_{sample} - B_{blank}\right) * 100}{\left(A_{sample} - A_{blank}\right)}$$

Analyzing the casein in its natural colloidal status (18) this method was modified to protect the micelle structure from an to early dissociation on dilution with water for a short time. Among some other things, the purity of water is one crucial point for the colloid-chemical analysis requiring a special conditioning procedure in some laboratories and need to do some research into the causes of this. The water quality plays also an important role in the dilution of the detergent for analysis; for this "protein-grade" Tween 80 is to be used. In original packing the detergent is diluted 10 % (v/v) in distilled, deionized and filtrated water and applicable for a limited period of time after opening. At present, all Tween-preparations on the market are plant-based and replaced the preferred high-grade detergent products of animal origin.

The estimation of micellar SHP is carried out pH-depended between pH 6,0 and 7,5. For this, skimmed milk samples were held at 4-5 °C during pH adjustments at pH 6,0, 6,2, 6,4 and 6,6 ca. 8 gL^{-1} Glucono-delta-lacton (GDL) or NaOH (4 gL^{-1}) was used for adjustments at pH 6,8, 7,0, 7,2 and 7,4, respectively.

The eight samples were sealed and stored at 4-6 °C overnight. Before analysis the samples are equilibrated in a water bath; first, at 36 °C for 30 min while stirring and then 70 min rocking (120 rpm) at room temperature (19-21 °C). The procedure of equilibration is based on measurements of micellar turbidity. The SHP is analyzed from a watery dispersion (100-200 µL milk in 25 mL highly pure water) at 21 ± 1 °C. The measurements of samples were carried out in triplicate and six fold in case of the protein-free standards (blanks). The experiments were repeated if the spread of readings do not agree with those permitted for the Bio-Rad- method.

3. Results and discussion

3.1. The influence of UHT-processing at 126 °C/ 4s on some major components in three different skim milk samples

Effects of indirect UHT-processing on some major components in the three skim milk samples are summarized in Tab.1.

No.	Milk Sample	pH	Protein %	β-Lg gL⁻¹	SHP %
1a	Skim milk ,	6,82	3,52	3,34	74,2
1b	indirect UHT (126°C/4s)	6,77	3,51	1,37	82,7
2a	Skim milk,	6,77	3,51	3,34	65,4
2b	indirect UHT (126°C/4s)	6,77	3,49	1,35	66,3
3a	Skim milk,	6,72	3,47	3,23	70,2
3b	indirect UHT (126°C/4s)	6,72	3,45	1,29	75,5

Table 1. Composition of three bulked raw milk samples before and on indirect UHT-processing at 126 °C for 4 s

The pH-value of samples No. 1 and 2 is between 6,77-6,82 pointing to an increased oxidative potential in both milk samples and thus to natural changes of the SH/SS- status of the proteins (19).

The high pH determined in sample No. 1a compared to No. 1b is explained by fact that this UHT- processing happened at the day of milk production but two days later to meet the demands of sample preparation in the colloid-chemical analysis (see 2.3). Within this period of time the pH of No. 1 increased in parallel with the decline of the inherent oxidative potential. In this course of events casein serves as the major radical scavenger in milk, whereas β-Lactoglobulin (β-Lg) and α-Lactalbumin (α-La) seem to be far less active (20). In agreement with it, the own results of β-Lg analysis in Tab.1 show a nearly constant rate of denaturation in all heat-processed samples and are probably more linked up with definite conditions on heat- processing.

From this point of view, samples No. 3 and No. 3b seem to be without criticism. In this instance the SHP gained about 5,3 % on UHT-processing. It is a definite indication that casein and whey protein became associated (reversible and/or irreversible) in this sample. Moreover, complexation is also verifiable for the other samples, the increase of SHP amounts to 8,5 % and 0,9 % for No. 1b and 2b, respectively. In both milk samples the participation of casein in complexation is varying. The underlying changes of the micelle status in milk are investigated and discussed below.

3.2. Research into demands on the colloid-chemical status of natural casein micelles relative to their suitability as a physico-functional food ingredient

3.2.1. The influence of oxidative stress on the colloidal-chemical status of casein and on results of thermal complexation by indirect UHT

In Figs. 1-3, the colloid-chemical profiles of three skimmed bulk milk samples (Figs.1a-3a) are compared with their profiles resulting from UHT-processing at 126 °C/ 4s (Figs.1b-3b). In the profiles two colloid-chemical criteria are compared vs. the pH-value, i.e., (i) the overall surface hydrophobicity of micelles SHP) and (ii) the corresponding absorbance of the protein-dye complex (O.D. at 525nm) developed with the Bio-Rad protein assay. Data of absorbance are of use for a better understanding of the natural pH-depended structural changes of micelles in milk.

Here the profiles of SHP and O.D. vs. pH are applied to visualize the influence of the colloidal micellar status in raw milk on the complex formation between casein and whey protein by indirect UHT-processing.

First, the profile of skimmed raw milk depicted in Fig.1a will be viewed as this milk is (i) is suitable to visualize a nearly perfect thermal modification (Fig.1b) and thus (ii) suited to explain the interpretation of an ordinary colloid-chemical profile of milk, very briefly.

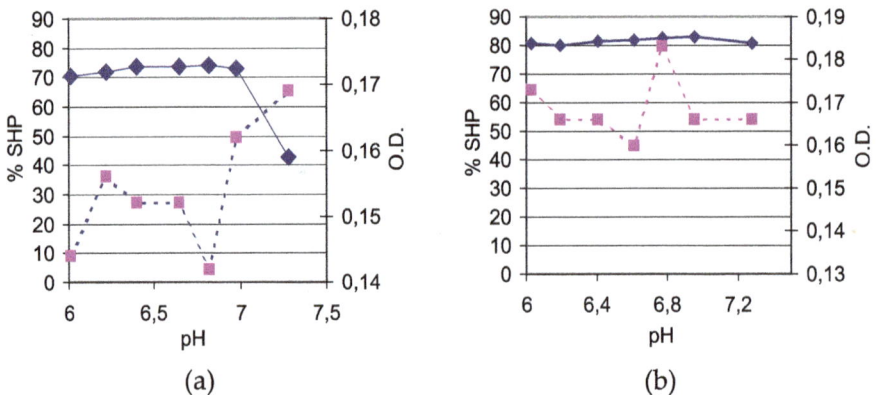

(a)   (b)

Figure 1. a: Effect pH on micellar surface hydrophobicity and dye-binding characteristics of skimmed raw milk. Micellar surface hydrophobicity (SHP; ♦—♦) and dye-binding (O.D.; ■---■) were determined from pH 6 to 7.5; orig. pH 6,82. N = 3.
b: Effect pH on micellar surface hydrophobicity and dye-binding characteristics of milk treated by indirect UHT-processing (126 °C/4s). Micellar surface hydrophobicity (SHP; ♦—♦) and dye-binding (O.D.; ■---■) were determined from pH 6 to 7.5; orig. pH 6,77. N = 3.

At natural pH of milk, casein micelles are very voluminous and exhibit both, a great surface area (O.D.) and a high hydrophobic potential (SHP) of the protein, too. A very hydrophilic surface layer, developed by the C-terminal glycosylated residues of κ-casein, is protecting the hydrophobic core of micelles to get collapsed in contact with other micelles. On lowering the pH, the colloidal calcium phosphate becomes increasingly dissolved. This is paralleled to a gradual decrease in micellar surface area and thus to a more dense protein structure. In the profile of milk (Fig.1a) two minima of dye binding (O.D.) are detectable, one at natural pH

and the other at pH 6,0 and both are pointing to a denser micelle structure. At pH 6,0 the total dissolution of colloidal calcium is completed whereas the dense micelle structure at the natural pH stems from a partial oxidation of micellar κ-casein at the surface *via* SH/ SS-exchange; a result of immunological reactions in the bovine mammary gland (see 3.1). At pH ≥ 6,82 the SHP decreases considerably due to the beginning of micellar disintegration with a stronger binding of calcium to individual caseins. It is paralleled with the dissociation of whole κ-casein from the micellar surface at pH 6,9. With advanced disintegration (pH > 6,9) the dye binding increases further showing intersection with the graph of SHP (Fig. 1a) This intersection is a common feature of the colloidal profiles of raw milk.

Next, the colloid-chemical profile of the heat-processed milk in Fig.1b will be viewed. On this treatment the micellar SHP is determined on very high level (≥80 %) irrespective of the pH-value. The course of O.D. indicates a maximum at natural pH whereas next to the natural pH there is a slight increase towards pH 6,0 and constant readings towards the neutral point. This means that the micellar colloidal status became static and thus without the known dynamic equilibriums between micelles in the colloidal milieu of milk, such as the equilibrium between dissolved and colloidal calcium and phosphate. It looks very much as if most of the colloidal calcium remained associated with the heat-modified micelles and this might explain the distinct white colour of this milk. The colloidal changes have positive effects on the rheological properties in both directions, in sour milk products as well in dietetic foods giving priority to a more neutral pH-value.

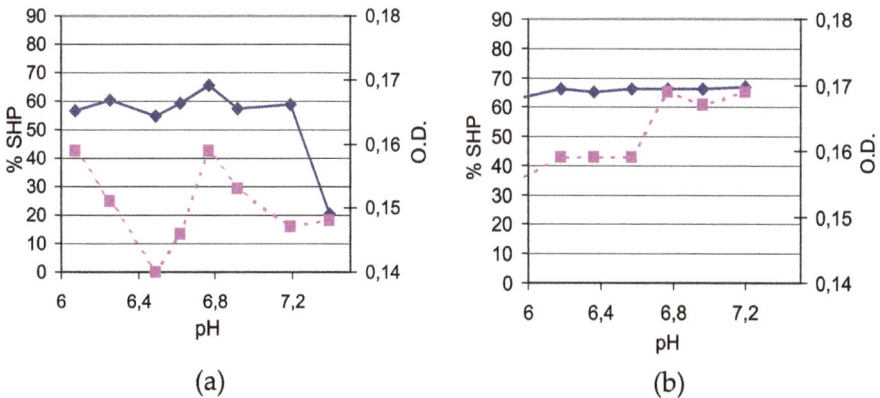

(a)   (b)

Figure 2. a: Effect pH on micellar surface hydrophobicity and dye-binding characteristics of skimmed raw milk representing elevated oxidative stress resulting from increased incidence of herd mastitis. Micellar surface hydrophobicity (SHP; ♦—♦) and dye-binding (O.D.; ■---■) were determined from pH 6 to 7.5; orig. pH 6,77. N = 3.

b: Effect pH on micellar surface hydrophobicity and dye-binding characteristics of milk treated by indirect UHT-processing (126 °C/4s) representing elevated oxidative stress resulting from increased incidence of herd mastitis. Micellar surface hydrophobicity (SHP; ♦—♦) and dye-binding (O.D.; ■---■) were determined from pH 6 to 7.5; orig. pH 6,77. N = 3.

In the following the profiles of samples No. 2a and 2b will be viewed. The herd milk was sampled at the time of increased udder infections. The somatic cell count (SCC) and colony forming units (CFU) are determined at $5,81·10^5$ per mL and $5,1·10^3$ per mL, respectively, and inherent effects on the bulked raw milk are seen in the graph of SHP in Fig. 2a. The natural pH was determined at pH 6,77. At this point a partial dissociation of κ-casein is seen, however, complete dissociation of κ-casein from the micellar surface as well as intersection of the two graphs occurred at pH 7,4 pointing to an exceptionally high oxidative potential in this milk sample. Here, the dissociation of κ-casein in two stages is put down to the fact that the herd milk was collected in one milk tank comprising both, first-grade raw milk and milk taken from infected mammary glands. In practice, all casein micelles are contributed in reducing the oxidative potential in milk on udder infections. This oxidation is the reason for poor/non renneting milk with insufficient gel matrix formation (19). In agreement with this, the micellar defects of the milk sample in Fig. 2a led to a partial dissipation of the gel matrix of Mozzarella Cheese on prolonged storage. The interpretation of micellar self-oxidation in the profiles SHP/O.D. vs. pH was very difficult (21) compared to the confirmative proof of mykotoxins or antibiotics (next chapter). Here, the presence of mykotoxins in milk was tested by a rapid immunoaffinity-based method for determination by fluorometry and antibiotics by use of the qualitative BR Test "AS".

Usually, the course of micellar dissociation is associated with a significant increase in protein-dye binding due to the occurrence of a vast number of submicellar particles. In the profile of raw milk, in Fig, 2a, the value of O.D. at pH 7,4 is even below that determined at pH 6,0. From this it can be drawn that the dissociation of casein micelles results in significant larger protein particles than in high-quality milk. The inherent change of the SH/SS-status blockades the free sulfhydryl groups in some places at the surface and within the individual micelles and thus reducing their chance to be fully involved in complexation with whey proteins by UHT *via* SH/SS-exchange as seen in Fig.2b. On UHT- processing at 126 °C/4s a constant SHP is determined between pH 6,0 and pH 7,2 amounting to ca. 66 %. It is paralleled with a continuous increase of O.D. indicating to a less stable complexation due to the scavenger function of free SH-groups of casein discussed before.

3.2.2. Influence of protein-bound antibiotics on the colloid-chemical status of casein and on inherent thermal complexation of whole milk protein

The udder health problems on herd level mentioned above were regulated by a medical treatment with antibiotics (AB). The milk samples used for the colloid-chemical analysis were collected some days after the legitimated waiting period. At that time the "PR" Test reacted still positively on AB's. The results obtained are summarized in Figs.3a and 3b.

The profile of the raw milk (Fig. 3a) shows maximum hydrophobicity at natural pH (pH 6,72) and thereafter the SHP decreases steadily coming up to total dissociation of micelles at pH 7,4. The overall graph of O.D. is on low level indicating to a denser micelle structure. The peak of dye binding at pH 7,2 has been turned out to be a distinguishing feature of protein-bound AB's in raw milk and appeared never in the profiles of high-grade milk samples.

Another group of foreign substances in milk (e.g. toxins of chemical and microbial origin) are occurring protein-bounded as well and are recognizable by their typical signs in the colloidal

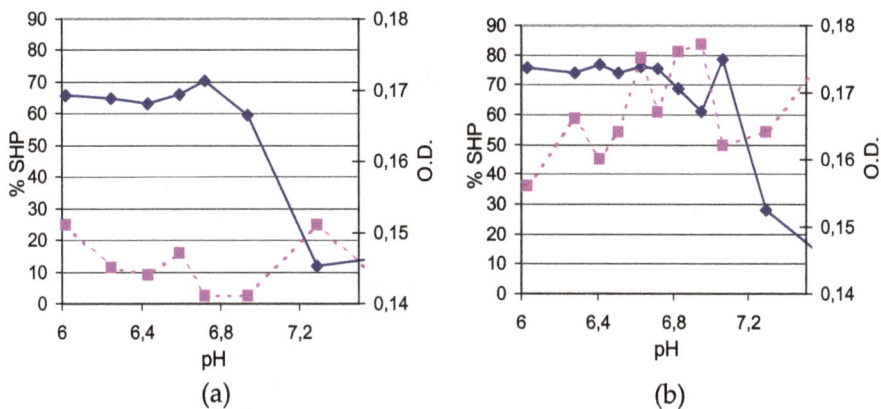

(a)   (b)

Figure 3. a: Effect pH on micellar surface hydrophobicity and dye-binding characteristics of skimmed raw milk containing protein-bound antibiotics. Micellar surface hydrophobicity (SHP; ◆—◆) and dye-binding (O.D.; ■---■) were determined from pH 6 to 7.5; orig. pH 6,72. N = 3.
b: Effect pH on micellar surface hydrophobicity and dye-binding characteristics of milk treated by indirect UHT-processing (126 °C/4s) containing protein-bound antibiotics. Micellar surface hydrophobicity (SHP; ◆—◆) and dye-binding (O.D.; ■---■) were determined from pH 6 to 7.5; orig. pH 6,77. N = 3, repeated twice

profiles, except for mykotoxins. These were taken up by the micellar surface and thus evident at each pH of analysis. Mykotoxins are competing increasingly with the non-ionic detergent for the places at the casein micelle surface to dock with on gentle warming, e.g.; on a prolonged standing inside the cuvette department of the photometer. In this case the readings of absorbance must happen right after the cuvette is placed into the photometer.

Next, the graphs of SHP and O.D. resulting from UHT-processing will be viewed (Fig.3b). Between pH 6,0 and pH 6,8 a high SHP is determined and a further peak of SHP occurred at pH 6,95 paralleled with a somewhat lower value of O.D. Generally, the graph of O.D shows a great variation *vs.* pH-value. The high readings of O.D. are indicating to a labile structure of the complexed protein on account of protein-bound AB's embedded. For this reason the complexed protein begins to dissociate at an early stage at pH 6,8 and inherent dissociation of the protein-bound AB's happens in the peak of SHP at pH 6,95. The bounding of AB`s to casein in Fig. 3b is noticeable increased compared to the raw milk (Fig. 3a).It might explain that protein-bound AB's in heated milk samples are underestimated owing to their lacking analytical availability. This event gives rise to a common misinterpretation believing that AB's in milk are degradable on heating (22).

Own investigations of evidencing protein-bound AB's immediately resulted in a qualitative rapid method derived from HPLC working with the Source 15 PHE PE 4,6/100 for separating the individual casein fractions by Hydrophobic Interaction Chromatography (HIC)(21). Whenever a casein sample with protein-bound AB's is dissolved and prepared for injection onto the column a very stable turbidity is perceptible. This turbidity has been used to indicate protein-bound AB's in milk sample. A brief description of the methodical principle is given now: an approximate protein concentration of 7 mg/mL is dissolved in 4 M guanidine thiocyanate (GdmSCN) prepared in 0,1 M phosphate buffer, pH 7,2 (PBS) and then homogenized using the Ultra-Turrax TP 18/10 (Janke& Kunkel KG, IKA Werk Staufen; FRG). The completely dispersed protein sample is diluted to 10 mL with PBS and vortex-mixed. The turbidity is measured at 320-330 nm. The readings are compared with AB-free milk analyzed in parallel.

In theory, the raw milk sample in Fig. 3a meets important requirements of a thermal complexation of whole milk protein; in particular, there is no oxidative potential owing to the immune system in the mammary gland is turned off as long as AB's are left (23).

The increased SHP occurring between pH 6,0 and pH 6,8 (Fig. 3b) bears out this assumptions, however, the structural stability indicated left to the original pH of milk is wearing off near to the neutral point owing to protein-bound AB's in this sample.

This will be substantiated with another profile shown in Fig. 4 assigned to a liquid foodstuff (pH 7,0) produced from concentrated skimmed milk. The packaged foodstuff showed emulsion-problems on short storage and was recalled. It should be added that the qualitative BR Test "AS" turned out negatively whereas the qualitative "Turbidity"-test reacted positively on AB's.

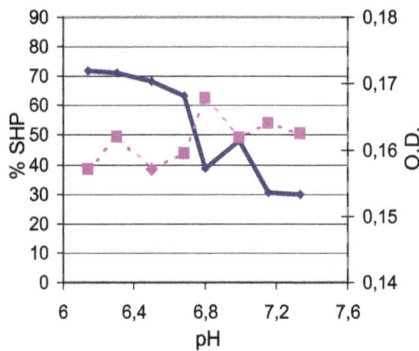

Figure 4. Effect pH on micellar surface hydrophobicity and dye-binding characteristics of milk protein as dietetic constituent in "ready-to-meal" product representing elevated emulsifying problems due to protein-bound antibiotics. Micellar surface hydrophobicity ((SHP; ♦—♦) and dye-binding (O.D.; ■---■) were determined from pH 6 to 7.5. N = 6.

4. Conclusions

Milk protein of high physico-functionality resulting from indirect heat processing at 126 °C/4s makes definite demands on the colloid-chemical status of casein micelles in raw milk. The ideal complexation between casein and whey protein in milk required that all superficial sulfhydryl-groups of the micelles were free and available to take part in SH/SS-exchange between the milk proteins. In practice, a less demanding standard is sufficient for normal commercial use. The analysis of the colloid-chemical status of casein micelles in milk can give an answer whether a thermal modification is to make good economic sense or not.

With the colloid-chemical research two reasons for an inadequate complex formation on indirect UHT-processing were turned out and, both were to trace back to bacterial infections of the mammary gland; i.e., *(i)* the self-oxidation of micellar sulfhydryl groups in milk whereby the casein is acting as the main scavenger in reducing the natural oxidative potential in milk and *(ii)* the presence of protein-bound antibiotics used to control the pathogens of mastitis.

Effects attributable to oxidation *(i)* are negligible as far as it remained on low level. On higher oxidative stress it requires a thermal processing very soon after the milk production is finished or alternatively, the fresh raw milk is bulked separately according to quality for avoiding a further spreading of the oxidative activities in whole bulked milk. The prospects for improving the physico-functionality of casein *via* complexation are steadily decreasing on advancing oxidation. The effects of micellar self-oxidation are paralleled with an increased pH of milk what is just right to be used as criterion to make a distinction between varying raw milk qualities.

The presence of protein-bound antibiotics in milk *(ii)* is a serious problem in various respects. Here the complexation of raw milk containing protein-bound antibiotics is considered. Under these circumstances the immune reactions of the mammary gland is inactive and thus all micellar sulfhydryl-groups were available to get involved in intermolecular complexation and in addition to also the pH of milk agrees with the theory (24). The fact is however that the molecular-structural organization inside the micelles is messed up and thus only a labile complexation between casein micelles and the whey protein can be formed on indirect UHT-processing.

Author details

Bärbel Lieske
Germany

5. References

[1] Anema, S.G. Heat and/or high-pressure treatment of skim milk: changes to the casein micelle size, whey proteins and the acid gelation properties of the milk. *International Journal of Dairy Technology* (2008) 61 (3) 245-252

[2] Kelly, A.L.; Kothary, K.I.; Voigt, D.D.; Huppertz, T. Improving technical and functional properties of milk by high-pressure processing. In *Dairy-derived ingredients: food and neutraceutical uses (Edited by Corredig, M.J.* Cambridge, UK; Woodhead Publishing Ltd (2009) 417-441 ISBN 978-1-84569-465-4

[3] Kielczewska, K.; Krok, A.; Czernewicz, M.; Kopeć, M. Effect of high pressure on constituents of the colloidal phase of milk. *Milchwissenschaft* (2009) 64 (4) 358-360

[4] Orlien, V.; Boserup, L.; Olsen, K. Casein micelle dissociation in skim milk during high-pressure treatment: effects of pressure, pH, and temperature. *Journal of Dairy Science* (2010) 93 (1) 12-18

[5] Knudsen, J.C.; Skibstedt, L.H. High pressure effects on the structure of casein micelles in milk as studied by cryotransmission electron microscopy. *Food Chemistry* (2010) 119 (1) 202-208

[6] Hernandez, A.; Harte, F.M. Manufacture of acid gels from skim milk using high-pressure homogenization. *Journal of Dairy Science* (2008) 91 (10) 3761-3767

[7]   Serra, M.; Trujillo, A.J.; Jaramillo, P.D., Guamis, B.; Ferragut, V. Ultra-high pressure homogenization-induced changes in skim milk: impact on acid coagulation properties. *Journal of Dairy Research* (2008) 75 (1) 69-75

[8]   Serra, M.; Trujillo, A.J.; Guamis, B.; Ferragut, V. Evaluation of physical properties during storage of set and stirred yoghurts made from ultra-high pressure homogenization –treated milk. *Food Hydrocolloids* (2009) 23 (1) 82-91

[9]   Bermúdez-Aguirre, D.; Mawson, R.; Versteeg, K.; Barbosa-Canovas, G.V. Composition properties, physicochemical characteristics and shelf life of whole milk after thermal and thermo-sonication treatments. *Journal of Food Quality* (2009) 32 (3) 283-302

[10] Riener, J.; Noci, F.; Cronin, D.A.; Morgan, D.J.; Cyng, J.G. A comparison of selected quality characteristics of yoghurt prepared from thermo-sonicated and conventionally heated milks. *Food Chemistry* (2010) 119 (3) 1108-1113

[11] Cayot, P.; Fairise, J.F.; Colas, B.; Lorient, D.; Brulé, G. Improvement of rheological properties of firm acid gels by skim milk heating is conserved after stirring. *Journal of Dairy Research* (2003) 70 (4) 424-431

[12] Lakemond, C.M.M.; Vliet, T. van Acid skim milk gels: the gelation process as affected by preheating pH. *International Dairy Journal* (2008) 18 (5) 574-584

[13] Depypere, F.; Verbeken, D.; Torres, J.D. Dettwinck, K. Rheological properties of dairy desserts prepared in an indirect UHT pilot plant. *Journal of Food Engineering* (2009) 91 (1) 140-145

[14] Lieske, B. The effects of direct and indirect ultra-high heat treatment of skim milk on the status of complexation between casein micelles and whey protein. *Milchwissenschaft* (2011) 66 (3) 254-257

[15] Lieske, B., Konrad, G., Faber, W. A new spectrophotometric assay for native β-Lactoglobulin in raw and processed bovine milk. *International Dairy Journal* (1997) 7 805-812

[16] Lieske, B., Konrad, G. A new approach to estimate surface hydrophobicity of proteins. *Milchwissenschaft* (1994) 49 663-666

[17] Nakai, S, Li-Chang, E. Procedures for measuring protein hydrophobicity *in: Hydrophobic interaction in food systems.* (ed. CRC Press, Boca Raton 1988) p.180-184

[18] Lieske, B. Effects of aging of raw milk on some structural properties of native casein micelles. *Milchwissenschaft* (1998) 53 562-565

[19] Lieske, B., Valbuena, R. Variation in the colloid-chemical status of casein micelles with influence on the chymosin coagulation properties of raw milk samples. *Milchwissenschaft* (2008) 63 (3) 247-250

[20] Clausen, M.R., Skipsted, L.H., Stagsted, J. Characterization of major radical scavenger species in bovine milk through size exclusion chromatography and functional assays. *Journal of Agricultural and Food Chemistry* (2009) 57 (7) 2912-2919

[21] Lieske, B., Valbuena, R. Hydrophobic Interaction High Liquid Chromatography using the Source 15 PHE 4,6/100 for separation and determination of major bovine casein fractions. *Milchwissenschaft* (2008) 63 (2) 178-181

[22] Zorraquino, M.A., Fernandez, N., Molina, M.P. Heat inactivation of β-lactam antibiotics in milk. *Journal of Food Protection* (2008) 71 (6) 1193-1198

[23] Lieske, B., Jantz, A., Finke, B. An improved analytic approach for the determination of bovine serum albumin in milk. *Lait* (2005) 85 237-248

[24] Walstra, P., Jenness, R. Outline of milk composition and structure. In *Dairy Chemistry and physics. Edited by John Wiley & Sons;* New York-Chicester-Brisbane-Toronto-Singapore (1984) page 8

Chapter 6

Maillard Reaction in Milk – Effect of Heat Treatment

Tomoko Shimamura and Hiroyuki Ukeda

Additional information is available at the end of the chapter

1. Introduction

Milk is usually subjected to heat treatment to ensure microbiological safety before retail and consumption. There are three types of heat treatment; (1) low temperature long time (LTLT) pasteurization, (2) high temperature short time (HTST) pasteurization, and (3) ultra-high temperature (UHT) treatment. In all types of heat treatment, the Maillard reaction occurs in milk.

The Maillard reaction (nonenzymatic glycation) is a chemical reaction between amino group and carbonyl group; it is the extremely complex reaction that usually takes place during food processing or storage. In the case of milk, lactose reacts with the free amino acid side chains of milk proteins (mainly ε-amino group of lysine residue) to proceed to early, intermediate, and advanced stages of Maillard reaction and forms enormous kinds of Maillard reaction products. The reactions of lactose and milk proteins have been frequently investigated and the formations of various Maillard reaction products in milk during heat treatment have been demonstrated [1]. In the general Maillard reaction, firstly an Amadori product is generated, and it progresses to the 3-deoxyosone or 1-deoxyosone route depending on the reaction pH. In the case of the Maillard reaction of disaccharides such as lactose, there is a third reaction route. It is the 4-deoxyosone route. A main carbohydrate in milk is lactose. Thus, the Maillard reaction in milk progresses via the above described three routes. Finally, the Maillard reaction results in the formation of melanoidins (browning compounds).

2. Effect of the Maillard reaction on milk proteins

The Maillard reaction shows various effects on milk proteins such as bioavailability, solubility, forming property, emulsifying property, and heating stability [1-4]. In addition, the formation of flavor compounds and browning compounds is caused as the consequences of the Maillard reaction between lactose and milk proteins [1, 5].

As for the effect of the Maillard reaction on the bioavailability of milk proteins, various studies were performed. Generally, in the Maillard reaction in milk, lactose mainly reacts with ε-amino group of lysine residue of milk proteins. Thus, the lysine loss by the Maillard reaction increases with a severity of heat treatment. The modified lysine cannot be available as a nutrient any more. For example, steam injection process (direct heating) generated 3.6% (120°C for 400 sec) and 6.8% (130°C for 290 sec) of the blocked lysine in whole milk. The indirect heating at 115°C for 10 to 40 min increased the modified lysine from 11.0 to 13.0% [6]. In addition, it was revealed that the lysine residues in skim milk powder were more susceptible to heating than those in skim milk [7].

Le et al. [3] recently suggested that the Maillard reaction was responsible for the solubility loss in milk protein concentrate powder. It was also reported that the glycated β-lactoglobulin was more stable at acidic pH and more stable against heating. The glycation of β-lactoglobulin, moreover, could improve its forming and emulsifying properties [4]. These results suggested the usefulness of the Maillard reaction for enabling milk proteins to have different properties.

3. Monitoring of the Maillard reaction of milk using XTT assay

The Maillard reaction has a lot of effects on the function of milk proteins and sensory property of milk and dairy products as described above. Particularly, in the manufacturing of milk, the excess progress of Maillard reaction and the formation of melanoidins are undesired, because a commercial value of milk is drastically decreased by them. Therefore, the detection of the Maillard reaction products is important for the quality control of milk. So far, several heat-induced markers have been proposed to control and check the heat treatment given to milk and dairy products. For example, furosine, hydroxymethylfurfural (HMF), and lactulose concentrations have been recognized to be the most promising indicators, since these concentrations increase with the heat treatment [1]. In Japan, protein reducing substance value (PRS) obtained by a ferricyanide assay is also widely-used conventional indicator [8]. It is based on the detection of reducing substances such as sulfhydryl group which are generated by heating in the fraction of acid-precipitated milk protein. These method, however, are generally time-consuming and complicated.

We proposed an assay method for determining the ability of milk to reduce 3'-[1-[(phenylamino)-carbonyl]-3,4-tetrazolium]-bis(4-methoxy-6-nitro)benzensulfonic   acid hydrate (XTT: Figure 1) as a method of evaluating the extent of the Maillard reaction [8-11]. The tetrazolium salt XTT is reduced to water-soluble formazan which is suitable for the spectrophotometric measurement. Taking account into an economical view and a rapidity of assay, we tried to develop a microplate assay. The assay conditions were as follows: XTT concentration, 0.5 mM; reaction pH, 7.0; menadione concentration, saturation level (ca. 0.55 mM); reaction temperature, room temperature; volume of XTT solution, 60 μL; volume of sample solution, 40 μL; detection wavelength, 492 nm; reference wavelength, 600 nm;

reaction time, 20 min. The procedure of the XTT assay was as follows: the XTT solution was added into each well in a microplate. Afterward, the sample solution was added to the well. After mixing on a microplate shaker at 500 rpm for 15 sec, a difference in the absorbance between 492 and 600 nm was read on a microplate reader as the absorbance at 0 min. After 20 min, the absorbance difference was read again. An increase in the absorbance for 20 min was recorded as the ability of the sample to reduce XTT (XTT reducibility).

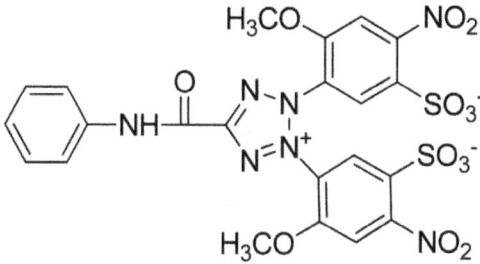

Figure 1. Structure of tetrazolium salt XTT.

Using the XTT assay described above, the XTT reducibility of milk was examined (Figure 2). In this test, an LTLT milk (Milk A: 65°C for 30 min) and two kinds of UHT milk (Milk B: 130°C for 2 sec, Milk C: 140°C for 3 sec) were used. When the milk was mixed with XTT solution, intense orange color which was derived from the XTT formazan was recognized. As shown in Figure 2, Milk C showed the highest XTT reducibility and the order of XTT reducibility was Milk C, B, and A. At the same time, the HMF content in Milk B and C was also determined as the conventional indicator of the Maillard reaction, because it was reported that the HMF content clearly increased with the severity of the heating treatment of milk [12, 13]. As a result, the content of HMF in Milk C (12.6 μM) was higher than that of Milk B (8.95 μM). From these results, it was revealed that the XTT assay could estimate the degree of thermal stress delivered to the milk as well as the HMF value.

In addition, the changes in the XTT reducibility during storage of UHT milks (Milk B and C) were investigated. For purpose of comparison, the PRS of Milk B and C were also examined by the ferricyanide assay [8]. The UHT milks were stored at 4°C and 37°C for about 4 weeks (Figure 3, 4). The XTT reducibility of Milk B and C gradually decreased depending on the storage period and the rate of decrease in the XTT reducibility was clearly larger at higher storage temperature (Figure 3). This result strongly suggested that, when the XTT assay is applied to the milk heated under a given condition, the result can serve to estimate not only the heating condition but also the storage period after heat treatment if the storage temperature is known or if storage period is known.

On the other hand, the PRS of Milk B and C were almost constant at all storage conditions (Figure 4). This result showed that the ferricyanide-reducing substances might be stable unlike the XTT-reducing substance. In addition, we found that the HMF value of Milk C did not significantly change during its storage for 60 days at 5°C [11]. It was in accordance with the result of Fink and Kessler about HMF [13]. They also reported that the lactulose concentration of UHT milk was constant throughout the storage period for 70 days at room temperature [13]. It could be concluded, therefore, that the XTT assay is applicable for the estimation of storage conditions which was impossible by the conventional method.

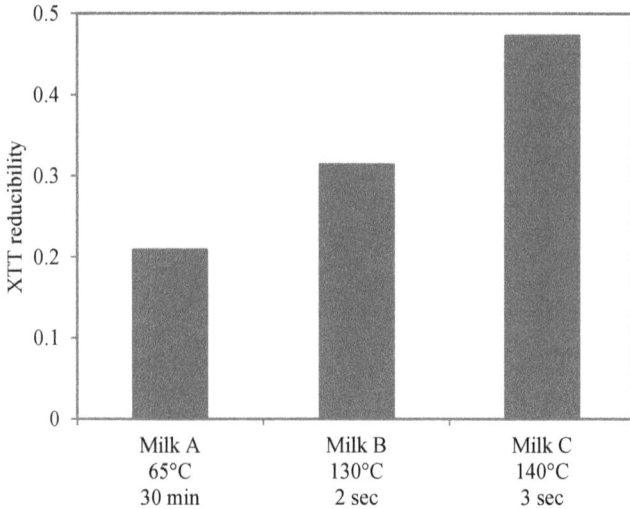

Figure 2. Comparison of XTT reducibility of LTLT milk (Milk A) and UHT milks (Milk B and C).

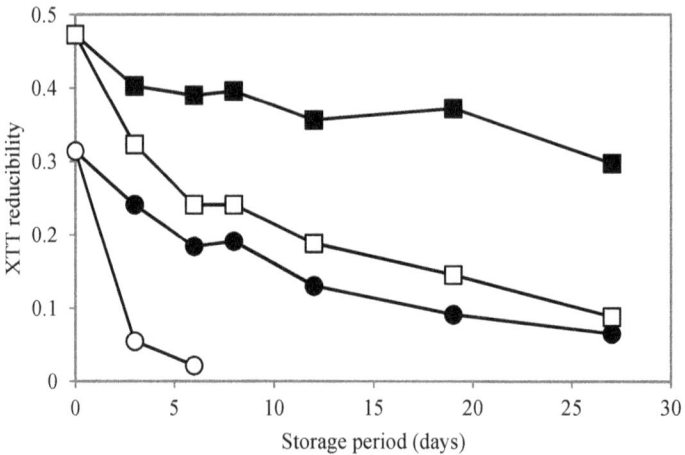

Milks were stored at 4°C (●, Milk B; ■, Milk C) and 37°C (○, Milk B; □, Milk C).

Figure 3. Changes in XTT reducibility during storage.

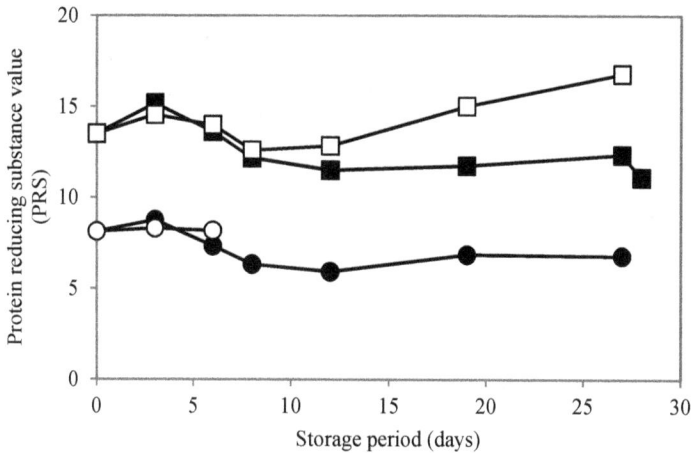

Milks were stored at 4°C (●, Milk B; ■, Milk C) and 37°C (○, Milk B; □, Milk C).

Figure 4. Changes in PRS during storage.

4. Demonstration of the presence of aminoreductone formed during the Maillard reaction in milk

4.1. In the model system of lactose and butylamine [9]

In order to clarify the XTT-reducing substance that is formed during the Maillard reaction in milk, we firstly used a model system consisting of lactose and butylamine, and then performed the spectrophotometric analysis of the heated model solution. The model solution of lactose-butylamine heated at 80-100°C for 0-30 min showed a characteristic UV absorption maximum at 320 nm. During the heating at 80-100°C for 30 min, the changes in the absorbance at 320 nm (Figure 5) and the XTT reducibility (Figure 6) were investigated. As a result, both indices increased gradually in accordance with the rise in temperature and heating time. This result indicated that the compound with the absorption maximum at 320 nm was formed by heating. Moreover, the behavior of increase in absorbance at 320 nm was similar to that of the XTT reducibility. Since the similar trend was recognized in the time course of both indices, the XTT reducibility was plotted against the absorbance at 320 nm. In consequence, a significant relationship between them was recognized with a correlation coefficient of 0.967 (n = 19, $p < 0.001$). From these results it was found that the compound with the absorption maximum at 320 nm might be responsible for the reduction of XTT.

The ^{13}C- and ^{1}H-NMR analyses of the compound with the absorption maximum at 320 nm which was extracted from lactose-butylamine model solution heated at 100°C for 15 min were performed. The signals of the ^{13}C- and ^{1}H-NMR could be assigned the compound as the aminoreductone, 1-(butylamino)-1,2-dehydro-1,4-dideoxy 3-hexulose (Figure 7). This compound was reported as the Maillard reaction product formed in the 4-deoxyosone route.

It was also reported as the characteristic compound in the Maillard reaction of disaccharides [14]. In addition, we demonstrate a linear relationship between the XTT reducibility and the amount of aminoreductone which was determined more specifically by HPLC [10]. These results strongly indicated that the aminoreductone formed during Maillard reaction of lactose was mainly responsible for the reduction of XTT.

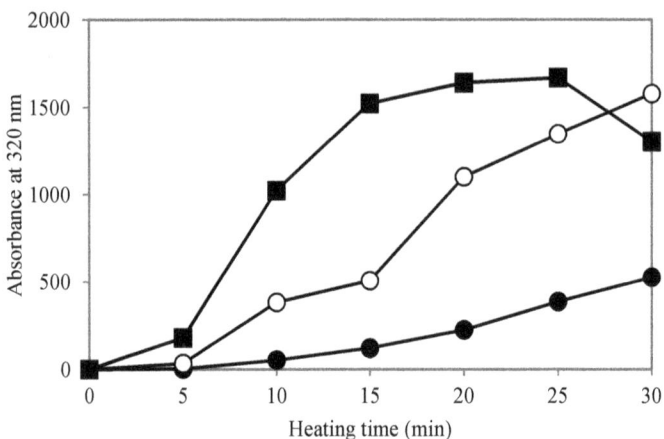

Lactose (262 mM) and butylamine (1.16 M) in 1.28 M phosphate buffer (pH 7.0) were heated at 80°C (●), 90°C (○), and 100°C (■).

Figure 5. Effect of heating temperature and time on the absorbance at 320 nm.

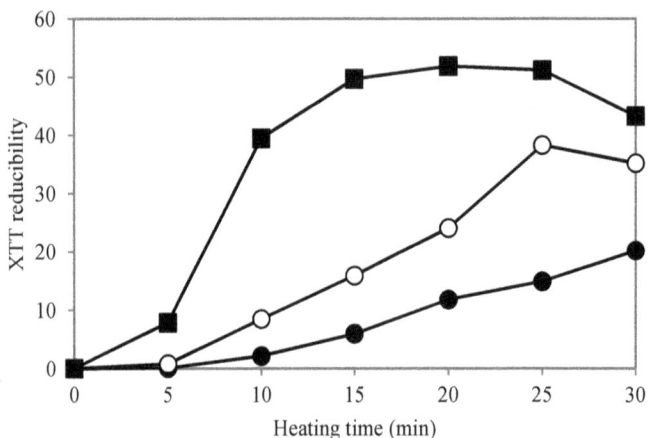

Lactose (262 mM) and butylamine (1.16 M) in 1.28 M phosphate buffer (pH 7.0) were heated at 80°C (●), 90°C (○), and 100°C (■).

Figure 6. Effect of heating temperature and time on the XTT reducibility.

$$HC-NHR$$
$$\overset{||}{C}-OH$$
$$\overset{|}{C}=O$$
$$\overset{|}{C}H_2$$
$$\overset{|}{C}HOH$$
$$\overset{|}{C}H_2OH$$

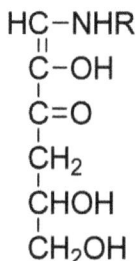

Figure 7. Structure of aminoreductone generated by the Maillard reaction of lactose and butylamine. (R = butyl group)

From these results, we presumed that the aminoreductone is formed by the Maillard reaction between lactose and ε-amino groups of milk proteins, and then it is responsible for the reduction of XTT. However, at that time, there was no report to prove the presence of aminoreductone in milk. Thus, we tried to demonstrate it using model system consisting of lactose and milk proteins and UHT milk.

4.2. In the model system of lactose and milk proteins

As a model system of milk, the solution consisting of lactose (4.6%) and casein (2.6%), α-lactalbumin (0.12%), or β-lactoglobulin (0.32%) was used and heated at 130°C for 15 min. After heating, the characteristic absorption maximum or shoulder at 320 nm was recognized. In addition, the changes of the absorbance at 320 nm (Figure 8) and the XTT reducibility (Figure 9) were investigated. In all model systems, the increases in the absorbance at 320 nm and the XTT reducibility depended on the heating time. Because similar tendencies were observed between two indices in all model systems, correlations were examined. Consequently, there were significant linearities as follows: casein ($r = 0.993$, $n = 6$, $p < 0.001$), α-lactalbumin ($r = 0.996$, $n = 6$, $p < 0.001$), and β-lactoglobulin ($r = 0.975$, $n = 6$, $p < 0.001$). From these results, it was suggested that aminoreductone is generated in the Maillard reaction between lactose and milk proteins and it is responsible for the reduction of XTT.

4.3. In milk [15]

As described above, a possibility of the formation of aminoreductone on the milk proteins during the Maillard reaction with lactose was clearly shown. However, direct demonstration of the presence of aminoreductone in milk had not been accomplished because of a difficulty in isolation of an intact aminoreductone from milk proteins. For instance, aminoreductone is labile and hence not suitable for enzyme hydrolysis and multiple extraction steps. To achieve the practical application of the XTT assay in food industries including dairy products, it was essential to demonstrate the presence of aminoreductone in milk. Because of this background, we attempted to isolate aminoreductone from milk proteins using 2,4-dinitrophenylhydrazine (DNP), a common labeling reagent for the carbonyl group, and Cu^{2+} [16]. A mechanism of derivatization of aminoreductone in milk is shown in Figure 10. In this derivatization step, Cu^{2+} plays as an oxidizing agent against aminoreductone, and the oxidized aminoreductone (OAR) has two or three carbonyl groups. Finally, it was assumed that two or three carbonyl groups in OAR are derivatized by DNP (OAR-DNP).

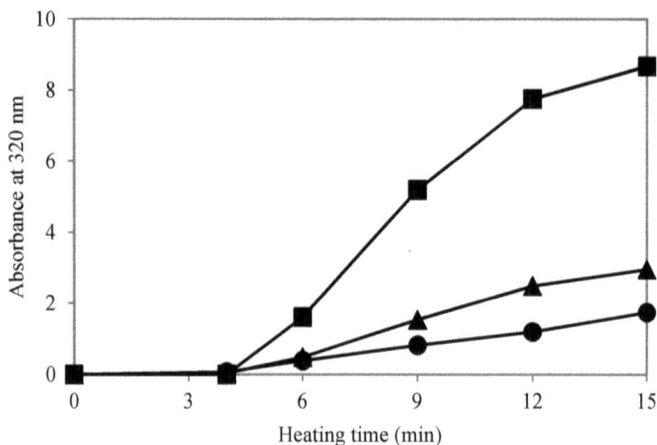

Lactose (4.6%) and casein (■: 2.6%), α-lactalbumin (•: 0.12%), or β-lactoglobulin (▲: 0.32%) in 20 mM phosphate buffer (pH 6.7) were heated at 130°C.

Figure 8. Effect of heating time on the absorbance at 320 nm.

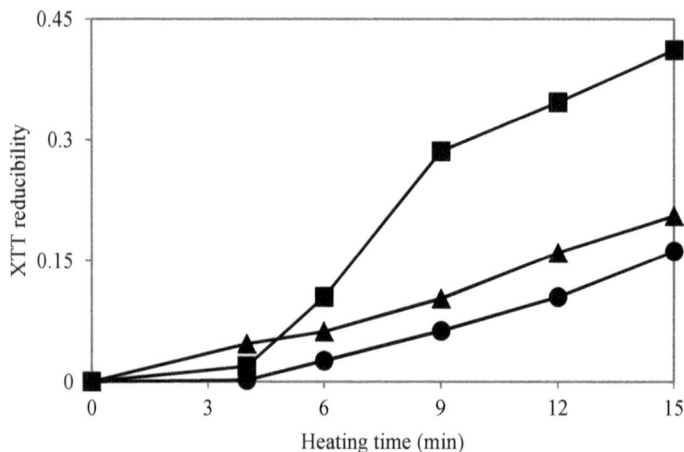

Lactose (4.6%) and casein (■: 2.6%), α-lactalbumin (•: 0.12%), or β-lactoglobulin (▲: 0.32%) in 20 mM phosphate buffer (pH 6.7) were heated at 130°C.

Figure 9. Effect of heating time on the XTT reducibility.

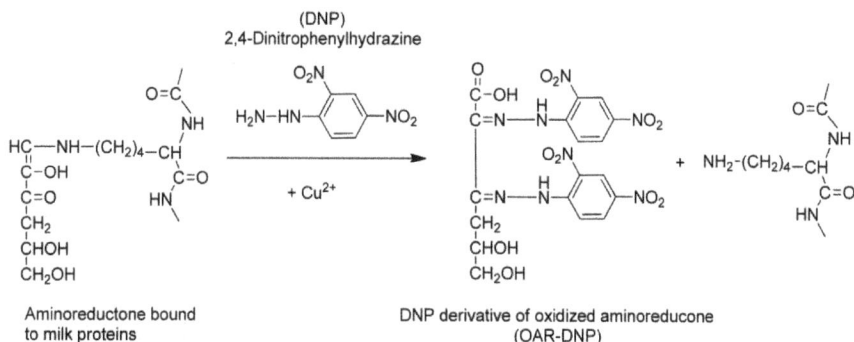

Figure 10. Derivatization mechanism of aminoreductone in milk using DNP and Cu^{2+}.

The derivatization using DNP and Cu^{2+} was applied to aminoreductone in UHT milk (140°C, 3 sec). In this study, the UHT milk was reheated at 130°C for 15 min in order to increase the content of aminoreductone, because the original content of aminoreductone in the commercially available UHT milk was not so high. As a result, the reheating of UHT milk could increase the content of DNP derivative by 40 times. The DNP derivative which was thought to be corresponding to OAR-DNP in the reheated UHT milk was purified by preparative normal-phase HPLC and preparative reversed-phase HPLC. Finally, the purified compound (4.2 mg) was obtained from 980 mL of UHT milk and analyzed by ^{13}C- and ^1H-NMR. The NMR signals of the DNP derivative from UHT milk could be assigned to the structure of OAR-DNP shown in Figure 10. In addition, the NMR signals of DNP derivative from UHT milk were nearly the same as those of the OAR-DNP from lactose-butylamine model system. These results demonstrated that aminoreductone was formed by the Maillard reaction on the milk proteins and present in milk.

Therefore, considering the above, the principle of the present XTT assay can be concluded as follows (Figure 11): (1) Lactose and ε-amino groups of lysine residue in milk proteins react non-enzymatically to form the Amadori product by the heating process. (2) Aminoreductone structure is formed on the milk proteins after elimination of galactose moiety from lactose through 4-deoxyosone pathway. (3) Aminoreductone is oxidized by XTT, whereas XTT is simultaneously reduced to the corresponding water-soluble formazan.

It would be thought that the above mentioned steps (1) and (2) progresses depending on the time and temperature of heating process in milk production, so the XTT assay can differentiate the extent of heat treatment of milk. Based on the study using model system of lactose and butylamine, the relationship between aminoreductone concentration and XTT reducibility was examined. As a result, there was a good linearity was recognized ($r = 0.98$) and a regression equation was $y = 0.606\ x + 0.046$, in which x and y represented the concentration of aminoreductone (mM) and the XTT reducibility [17]. Based on this equation, the concentration of aminoreductone in UHT milk could be estimated as 0.44 mM.

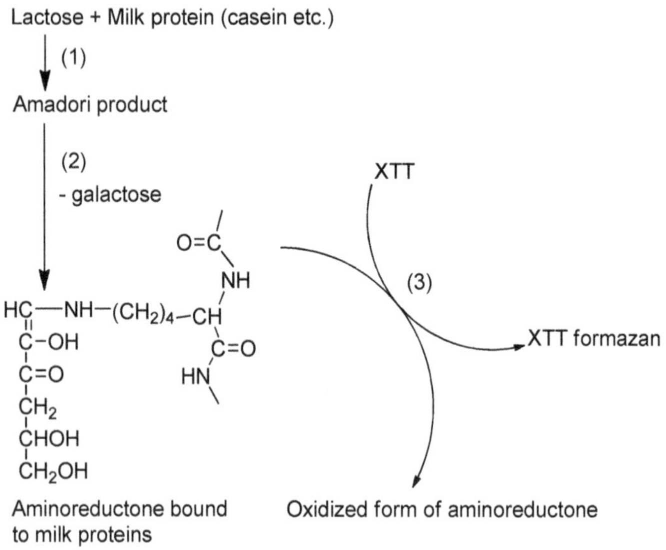

Figure 11. Principle of XTT assay.

As described above, during the course of elucidation of the XTT-reducing substance, Cu^{2+} was used as the oxidizing agent (Figure 10), because we empirically knew that it was easily decomposed by the addition of Cu^{2+} [18]. In fact, the aminoreductone extracted from the heated model solution of lactose and butylamine was rapidly oxidized by the addition of Cu^{2+}, and simultaneously the XTT reducibility was also lost. On the other hand, in our previous work, it was revealed that the commercially available UHT milk contains Cu^{2+} at the concentration of 30 µg/L [18]. Thus, it was easily presumed that the aminoreductone formed by the heating process was gradually oxidized by endogenous Cu^{2+} in milk during storage period. This was the reason why that the XTT reducibility decreased depending on the storage period (Figure 3). The detailed investigation about the relationship between the aminoreductone concentration, Cu^{2+} concentration, and storage stability of milk is now in progress.

5. Functionality of aminoreductone

The functionalities of aminoreductone have attracted interest and, so far, some studies were performed. Trang et al. [17] reported a protective effect of aminoreductone against riboflavin (vitamin B2) photolysis. It is well known that the milk is important source of riboflavin and its content is 1.5 mg/L. It is stable to heat and oxidation, but is rapidly photo-degraded. In experimental condition at 7000 lux light intensity, the riboflavin (1.5 mg/L) was almost completely degraded for 150 min. On the other hand, the addition of aminoreductone (0.22 mM: half concentration in UHT milk) could extend the half-life period of riboflavin. The protective effect of aminoreductone against riboflavin photolysis was higher than that of ascorbic acid which was famous antioxidant. In addition, the antioxidative activity of aminoreductone was reported [19]. From these results, it was suggested that the aminoreductone formed by the Maillard reaction would contribute to keep the nutritional value and sensory quality of milk.

Furthermore, aminoreductone showed antimicrobial activity against *Helicobacter pylori* (*H. pylori*). *In vitro* it effectively inhibited the growth of 24 kinds of *H. pylori* strains including antibiotic-resistant strains and had a bactericidal activity [20, 21]. The Killing ability was observed even in acidic condition. In addition, aminoreductone also had the antimicrobial activity against methicillin-resistant *Staphylococcus aureus* (MRSA) [21, 22]. These results indicated that foods containing aminoreductone, such as milk and dairy products, have a potential health benefits in medical practice.

6. Conclusion

In this chapter, the demonstration of the presence of aminoreductone formed by the Maillard reaction in milk and the specific assay method for aminoreductone were focused and introduced. Since the novel functionality of aminoreductone have come out one after another, the information of aminoreductone obtained by the XTT assay would gain importance in the quality control in milk and dairy products.

Author details

Tomoko Shimamura* and Hiroyuki Ukeda
Faculty of Agriculture, Kochi University, Nankoku, Japan

7. References

[1] van Boekel M. A. J. S. (1998) Effect of heating on Maillard reactions in milk. Food Chem. 62: 403–414.

[2] Evangelisti F., Calcagno C., Nardi S., Zumin P. (1999) Deterioration of protein fraction by Maillard reaction in dietetic milks. J. Dairy Res. 66: 237-243.

[3] Le T. T., Bhandari B., Deeth H. C. (2011) Chemical and physical changes in milk protein concentrate (MPC80) powder during storage. J. Agric. Food Chem. 59: 5465-5473.

[4] Augustin M. A., Udabage P. (2007) Influence of processing on functionality of milk and dairy proteins. In: Taylor S. L., editor. Advances in Food and Nutrition Research. Amsterdam: Elsevier. pp. 1-38.

[5] Freedman M. (1996) Food browning and its prevention: an overview. J. Agric. Food Chem. 44: 631-653.

[6] Finot P. A., Deutsch R., Bujard E. (1981) The extent of the Maillard reaction during the processing of milk. Prog. Fd. Nutr. Sci. 5: 345-355.

[7] Henle T., Walter H., Klostermeyer H. (1991) Evaluation of the extent of the early Maillard-reaction in milk products by direct measurement of the Amadori-product lactuloselysine. Z. Lebensm. Unters. Forsch. 193: 199-122.

---

* Corresponding Author

[8] Shimamura T., Ukeda H., Sawamura M. (2001) Quality evaluation of milk using XTT assay –comparison with ferricyanide assay. Nippon Shokuhin Kagaku Kogaku Kaishi (in Japanese) 48: 840-843.

[9] Shimamura T., Ukeda H., Sawamura M. (2000) Reduction of tetrazolium salt XTT by aminoreductone formed during the Maillard reaction of lactose. J. Agric. Food Chem. 48: 6227-6229.

[10] Shimamura T., Ukeda H., Sawamura M. (2004) Relationship between the XTT reducibility and aminoreductone formed during the Maillard reaction of lactose: the detection of aminoreductone by HPLC. Food Sci. Technol. Res. 10: 6-9.

[11] Ukeda H., Goto Y., Sawamura M., Kusunose H., Kamikado H., Kamei T. (1995) Reduction of tetrazolium salt XTT with UHT-treated milk: its relationship with the extent of heat treatment and storage conditions. Food Sci. Technol. Int. 1: 52-57.

[12] Kessler H. G., Fink R. (1986) Changes in heated milk and stored milk an interpretation by reaction kinetics. J. Food Sci. 51: 1105-1111.

[13] Fink R., Kessler H. G. (1988) Comparison of methods for distinguishing UHT treatment and sterilization of milk. Milchwissenschaft 43: 275-280.

[14] Pischetsrieder M., Schoetter C., Severin T. (1998) Formation of an aminoreductone during the Maillard reaction of lactose and N^α-acetyllysine or proteins. J. Agric. Food. Chem. 46: 928-931.

[15] Shimamura T., Kurogi Y., Katsuno S., Kashiwagi T., Ukeda H. (2011) Demonstration of the presence of aminoreductone formed during the Maillard reaction in milk. Food Chem. 129: 1088-1092.

[16] Shinohara K., Fukumoto Y., Tseng Y.-K., Inoue Y., Omura H. (1974) Action of amino reductones on nucleic acids. Nippon Nogei Kagaku Kaishi (in Japanese) 48: 499-506.

[17] Trang V. T., Kurogi Y., Katsuno S., Shimamura T., Ukeda H. (2008) Protective effect of on the photo-degradation of riboflavin. Int. Dairy J. 18: 344-348.

[18] Ukeda H., Shimamura T., Hosokawa T., Goto Y., Sawamura M. (1998) Monitoring of the Maillard reaction based on the reduction of tetrazolium salt XTT. Food Sci. Technol. Int. Tokyo 4: 258-263.

[19] Pischetsrieder M., Rinaldi F., Gross U., Severin T. (1998) Assessment of the antioxidative and prooxidative activities of two aminoreductones formed during the Maillard reaction: effects on the oxidation of β-carotene, N^α-acetylhistidine, and cis-alkenes. J. Agric. Food Chem. 46: 2945-2950.

[20] Trang V. T., Takeuchi H., Kudo H., Aoki A., Katsuno S., Shimamura T., Sugiura T., Ukeda H. (2009) Antimicrobial activity of aminoreductone against Helicobacter pylori. J. Agric. Food Chem. 57: 11343-11348.

[21] Takeuchi H., Shimamura T., Kudo H., Ukeda H., Sugiura T. (2012) Antimicrobial activity of natural products and food components. Current Research in Agriculture and Food Chemistry 1: 1-12.

[22] Trang V. T., Takeuchi H., Kudo H., Katsuno S., Shimamura T., Kashiwagi T., Son V. H., Sugiura T., Ukeda H. (2011) In vitro antimicrobial activity of aminoreductone against the pathogenic bacteria methicillin-resistant Staphylococcus aureus (MRSA). J. Agric. Food Chem. 59: 8953-8960.

Milk Protein Allergy

Chapter 7

Protein Profile Characterization of Donkey Milk

Paolo Polidori and Silvia Vincenzetti

Additional information is available at the end of the chapter

1. Introduction

Milk is a biological fluid designed to contain all nutritional requirements of a specific mammalian newborn; therefore, the composition of milk differs by the needs of the neonate of different species. Although much research has been devoted to milk composition in the domestic horse, donkey's milk has recently aroused scientific interest, above all among paediatric allergologists and nutritionists. Clinical studies have demonstrated that donkey milk may be considered a good replacer for dairy cow's milk in feeding children with severe Ig-E mediated cow's milk protein allergy, when human milk can not be given [1]. For these patients, donkey milk is not only useful [2], but also safer compared with milk obtained by other mammalian species [3], due to the high similarity with human milk, especially considering protein fractions content [4, 5].

2. Milk proteins

In 1838, a Swedish scientist called Jacob Berzelius suggested the term "protein" after the greek word "proteios", which means "primary" or "of the first rank"; since then, many scientific discoveries have been made about these large molecules indispensable for the good functioning of the body's cells, tissues and organs. In the past, it was believed that the function of proteins was restricted to tissue-formation, while carbohydrates and lipids provided energy to the body. Another out-of-date concept was the belief that dietary proteins were completely hydrolyzed in the gastro-intestinal tract and only free amino acids could be adsorbed from the gut. The new concept is that macro- and microelements (such as vitamins and minerals) may interact to perform different functions in the body [6]; amino acids and peptides formed in the digestion of natural proteins are adsorbed and incorporated (anabolism) into various tissues and organs as body proteins.

Milk protein is a very heterogeneous group of molecules and, for ease description, could be classified into five main categories: caseins, whey proteins, milk fat globule proteins, enzymes and other miscellaneous minor proteins [7]. Milk proteins appear to be an exciting link between nutrition, dietetics and therapy; today, consumers can expect more than just nutrition from intake of milk proteins. In fact, milk contains a variety of bioactive compounds with special properties associated with the development, growth and survival of infants beyond those provided by nutrition alone [8]. The major antimicrobial proteins in milk are immunoglobulins, lactoferrin, lactoperoxidase and lysozyme [9]. Immunoglobulins (IgG, IgM and secretory IgA) act by a specific mode of action involving antigen-antibody reactions. The other three proteins are non-specific protective factors, and their antimicrobial mechanisms of action differ from each other.

Lactoferrin, also called lactotransferrin, is an iron binding protein present in milk, saliva, tears and mucus secretions, its bacteriostatic effects are usually attributed to lactoferrin's ability to bind environmental iron ions. Lactoferrin inhibits the growth of many kinds of Gram-negative and Gram-positive bacteria as well as some species of fungi and yeast [9]. Lactoperoxidase, a heme-containing protein, catalyzes an oxidation reaction involving hydrogen peroxide (H_2O_2) and functions as a component of host defense system. The lactoperoxidase system is known to be effective to preserve raw milk without refrigeration and it is used in the production of many dairy products [10].

Lysozyme exerts its antimicrobial activity by the hydrolysis of glycosidic bonds of mucopolisaccarides in bacterial cell walls [11]. Lysozyme, together with other peptides including immunoglobulins, lactoferrin and lactoperoxidase, is active in the infant's digestive tract in order to reduce the incidence of gastro-intestinal infections [12].

3. Cow milk allergy

Food allergy is the clinical syndrome resulting from sensitization of an individual to dietary proteins or other food allergens present in the intestinal lumen [13]. Food allergy is much more common among children than adults, and is more common among younger children than older children [14]. Cow Milk Allergy (CMA) is a frequent disease in infants, but its etiologic mechanisms are not clear [15]. Clinical symptomology for patients allergic to bovine milk proteins include: rhinitis, diarrhea, vomiting, asthma, anaphylaxis, urticaria, eczema, chronic catarrh, migraine, colitis and epigastric distress. Cow milk allergy is clinically an abnormal immunological reaction to cow milk proteins, which may be due to the interaction between one or more milk proteins and one or more immune mechanisms, and resulting in immediate IgE-mediated reactions [16]. On the other side, reactions not involving the immune system are defined as cow milk protein intolerance.

Cow milk is one of the most common food allergies in children, occurring in between 0.3 and 7.5% of the infant population [17]. The clinical diagnosis of milk allergy differs widely due to the multiplicity of symptoms. Cow milk contains more than 20 proteins (allergens) that can cause allergic reactions [18]. The main proteins are casein and whey protein; casein is fractionated into α-, β- and k-casein, whey proteins include α-lactalbumin, β-lactoglobulin, bovine serum albumin (BSA) and immunoglobulin (Igs). In addition to those, several minor proteins are also present in cow milk. Most studies revealed that casein and β-lactoglobulin are the main allergens in cow milk. The type of immune response after intrusion of foreign proteins is extremely variable, depending on the animal species, the age of the host, the quality and the quantity of antigens absorbed, the location of the absorption, the pathophysiological state and the genetic background [19].

For human beings cow's milk represents the most common feeding during the infant weaning, but also the first allergen in life. The European Academy of Allergy and Clinical Immunology distinguishes allergy from intolerance [20]. Allergy is an adverse reaction to food with an involvement of the immune system; intolerance is an adverse reaction to food that does not involve the immune system, does not reply to a precise and single fault and shows different symptoms. In many countries cow's milk is the most important food allergen in babies and children [21]. Adverse reactions to cow's milk were found in 2% of babies during the first year of life: 30% of cases at the first month, 60% before the third and 96% within the twelfth [22, 23]. Symptoms can even appear during the breast-feeding because newborn reacts against a small amount of cow milk proteins present in maternal milk [24]. Children followed for the first 3 years of life, 56% of cases had recovered from cow's milk allergy at 1-year age, 77% at 2 years and 87% at 3 years age [25] (Host and Halken, 1990). However allergy can persist for all life.

Considering the possible use of alternative milk sources for human in cases of cow's milk allergy, the use of goat's milk should be avoided because of the high risk of cross-reactivity , while mare's and donkey's milks, used in popular practice for allergic children, are valid alternative protein sources when appropriately evaluated from the hygienic point of view [26].

4. Donkey milk

The donkey (*Equus asinus*) is a member of the horse family and its progenitor was the small gray donkey of northern Africa (*Equus africanus*) domesticated around 4000 BC on the shores of the Mediterranean Sea. It worked together with humans for centuries; the most common role was for transport. It still remains an important work animal in the poorer Regions.

Compared with ruminant's milk, donkey milk has been studied less in the past, but in the last years research interest and capital investment in donkey milk have increased because its composition is similar to that of human milk (see Table 1).

The protein composition is significantly different from cow's milk: the total content is lower (13-28 mg/ml) and quite similar to that of human and mare milk: this condition avoids an excessive renal load of solute [28]. The main difference is the proportion of whey proteins: they are 35-50% of the nitrogen fraction while they represent only 20% in cow's milk [29]. Comparing donkey's and mares milk, the casein to whey protein ratio in mares milk is 0.2:1 immediately post-partum, and changes to 1.2:1 during the first week of lactation [30].

	Donkey	Human
pH	7.0-7.2	7.0-7.5
Protein (g/100g)	1.5-1.8	0.9-1.7
Fat (g/100 g)	0.3-1.8	3.5-4.0
Lactose (g/100 g)	5.8-7.4	6.3-7.0
Ash (g/100 g)	0.3-0.5	0.2-0.3
Total Solids (g/100 g)	8.8-11.7	11.7-12.9
Caseins (g/100 g)	0.64-1.03	0.32-0.42
Whey Proteins (g/100 g)	0.49-0.80	0.68-0.83

Table 1. Comparison of chemical composition and physical properties of donkey and human milk [27] (copyright permission obtained).

The donkey's three major whey proteins are α-lactalbumin, β-lactoglobulin and lysozyme. Donkey's milk α-lactalbumin has two isoforms with different isoeletric point [31]. Recently, it has been shown that α-lactalbumin presents antiviral, antitumor, and anti-stress properties. In particular in human breast milk it was shown that the α-lactalbumin forms a complex with oleic acid called HAMLET (Human Alpha-lactalbumin Made Lethal to Tumor cells) that proved to be able to induce tumour-selective apoptosis. This complex may be considered as a potential therapeutic agent against various tumour cells [32]. Furthermore it was shown that α-lactalbumin possesses anti-inflammatory activity exerted by the inhibition of cyclooxygenase-2 (COX-2) and phospholipase A2 [33].

One of the main allergens in children is β-lactoglobulin that is the major whey protein in cow milk [17] while it is absent in human milk [34]. In donkey milk the content of β-lactoglobulin is approximately 40% of the whey proteins equal to the level in mare milk and lower than that in cow milk [34]. This condition may be related to the hypoallergenic characteristic of donkey milk [1, 12, 35]. The mechanism for tolerance may be related to the specific levels of the major allergenic components in the milk. Donkey's milk has three genetic variants for β-lactoglobulin: one presents three amino acid substitutions while the others have two amino acid exchanges [29]. Donkey milk β-lactoglobulin is a monomer whereas this protein is a dimer in ruminant's milk. β-lactoglobulin is a protein of the lipocalin family and has high affinity for a wide range of compounds opening the way to various suppositions about its function. In fact it has been shown that this protein is involved in hydrophobic ligand transport and uptake, enzyme regulation, and the neonatal acquisition of passive immunity, other authors demonstrated that β-lactoglobulin forms complexes with folic acid suggesting that these complexes could be used as an effective carrier of folic acid in functional foods [36].

The high content of lysozyme may be responsible for the low bacterial concentration in donkey milk [4, 5]. Donkey milk lysozyme presents two isoforms that differ in three amino acid substitutions at position 48, 52 and 61 [29]. The concentration of lysozyme in human milk increases strongly after the second month of lactation, suggesting that this enzyme plays an important role in fighting infections in breast-fed infants during the late lactation [30].

The lactose content (7%) of donkey milk is similar to that of human milk and is much higher if compared with cows milk. The high content is responsible for the good palatability and facilitates the intestinal absorption of calcium that is essential for infant's bone mineralization. Donkey milk shows a lower fat content compared to human milk, presenting for this reason a reduced energetic value [37]. The large number of fatty acids present in the lipid fraction of milk makes it one of the most complex naturally occurring fats. Saturated fatty acids are the most represented class in donkey's milk compared to monounsaturated and polyunsaturated fatty acids, even though a wide variability can be observed in the data available in literature, most likely related to dietary and/or body condition differences [4]. Within a well balanced and integrated diet, donkey's milk is a good source of essential fatty acids; this category of fatty acids are very important in the diet of patients with Cow Milk Allergy (CMA), especially if affected by multiple food allergy. These subjects are in fact at risk of developing a deficiency in essential fatty acids and particularly in PUFA n – 3, which are absolutely necessary for adequate growth, neurological development and cardiovascular health [38]. Donkey's milk shows an high content of both linoleic (C18:2) and linolenic (C18:3) acids, respectively 9.0 g/100 g and 5.1 g/100 g of total fatty acids, when compared with ruminant species milks, in which the contents of the above mentioned Polyunsaturated Fatty Acids are always lower. To increase the total fat content in donkey milk, clinical studies [17, 35] suggested to enrich donkey milk with medium-chain triglycerides, in order to obtain a final fat content similar to human milk.

The mineral composition is very close to that of human milk except for the highest level of calcium and phosphorus but the Ca-P ratio is similar. The milk produced in the first month of lactation, when it is the only nutritional source for the foal, contained the highest levels of mineral elements that may be related to the fast growth stage of the foal. Afterwards, the mineral supply in milk decreases considerably.

Basically, donkey milk has nutritional properties that make it more similar to human milk than another mammalian one. Therefore it could be used not only as a breast milk substitute for allergic children but also as a new dietetic food for human consumption. This chapter would be a further contribution to increase the characterization of donkey milk, evaluating the nutritional qualities of donkey milk using different proteomic approaches.

5. Proteomic approaches

The term proteome was introduced to describe "all proteins expressed by a genome or tissue [39]. Another definition of proteome, but similar, is: "a set of all expressed proteins in a cell, tissue or organism at a certain point in time" [40]. After genomics and transcriptomics, proteomics is considered the next and more articulate step in the study of biological systems. In fact, while the genome is more or less constant, any proteins may exist in multiple forms that vary from cell to cell and from time to time. In fact the proteome analysis reveals translational, post-translational modification, regulatory and degradation processes that affect protein structure, localization, function, and turnover. Proteomics is the study of multi-protein systems and their roles as part of a larger system or network. Therefore the context of proteomics is system biology, rather than structural biology, since it is a tool used to characterize the behaviour of the system rather than the behaviour of any single component.

Milk proteins have been studied in depth for well over 50 years and lot of studies were performed in order to analyse the various milk protein components, in various milk from different mammals. However, many questions concerning milk protein expression, structure and protein modifications remain still not completely covered such as some details of protein modifications due to disease and processing. The milk proteome is extremely complex because of the presence of post-translational modifications, alternative splicing and different genetic variants. The post-translational modifications are: glycosylation, phosphorylation, disulphide bonds formation and proteolysis, they create a large number of different protein variants from a single gene product. In milk the molecular composition of proteins is very important since it influences the functional properties of milk proteins such as solubility, clotting aptitude, thermal denaturation and the nutritional properties of the milk. Usually, the analytical procedures used for structural analysis of milk proteins are based on chromatographic techniques (ionic exchange, reversed phase chromatography, size-exclusion chromatography) followed by one-dimensional electrophoretic techniques (PAGE, Urea-PAGE and SDS-PAGE), or more efficiently on bi-dimensional electrophoresis technique (2-DE).

In our work we approached the study of donkey milk protein profile by different techniques for protein separation: initially they were based on chromatographic techniques followed by sodium dodecyl sulphate polyacrylamide gel electrophoresis SDS-PAGE [5]. Successively, the milk was analyzed through two-dimensional electrophoresis (2-DE) followed by N-terminal sequencing, in order to give a more detailed panoramic view of the proteins that are present in donkey milk [41].

5.1. Chromatographic approaches followed by SDS-PAGE

Donkey's milk casein fraction was characterized by different chromatographic approaches using an Äkta Purifier HPLC system: ion-exchange chromatography, and reversed-phase. After chromatography, each protein was subjected to SDS-PAGE. The purified caseins were identified by N-terminal sequencing [5]. By cation-exchange chromatography (Mono S HR 5/5 column, GE Healthcare,1.0 ml bed volume), performed at pH 5 and 7, followed by 15% SDS-PAGE (Mini Protean III apparatus, Bio-Rad) it was possible to separate 9 peaks that were identified as β-caseins (sequence: REKEELNVSS) and α_{s1}-caseins (sequence: RPKLPHRQPE), having different molecular weight, as shown in figure 1 and table 2.

Reversed-phase chromatography on HPLC (RP-HPLC) followed by 15% SDS-PAGE and N-terminal analysis was performed on the skimmed donkey's milk giving as a result three main peaks (K, L, M) identified as lysozyme (sequence, KVFSKXELA), α-lactalbumin, (sequence, KQFTKXELSQVLXSM), and β-lactoglobulin (sequence TNIPQTMQ), respectively

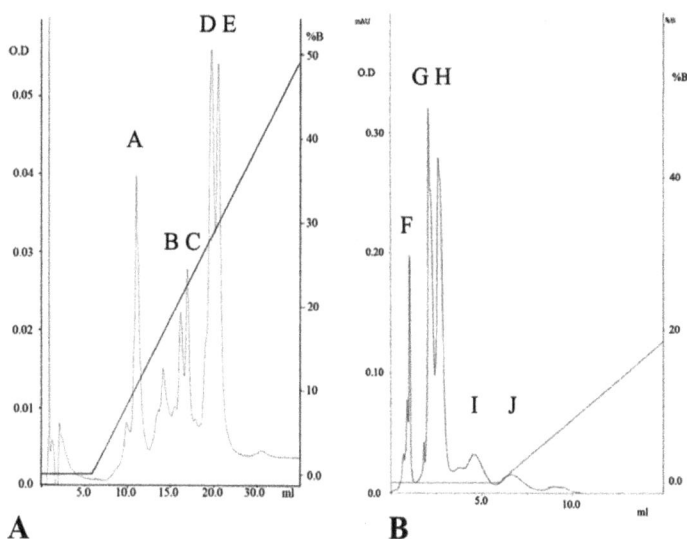

Figure 1. Cationic-exchange chromatography on HPLC (Mono S HR 5/5) analysis on whole casein performed at pH 5.5 (A) and at pH 7.0 (B) [5] (copyright permission obtained).

Peak/Chromatography	Protein	kDa	N-terminal sequence
A: cationic exchange, pH 5.5	B-casein	35.40	REKEELNVS
C, D, E: cationic exchange, pH 5.5	αs_1-casein	33.00	RPKLPHRQPE
	αs_1-casein	30.70	RPKLPHRQPE
F: cationic exchange, pH 7.0	B-casein	33.30	REKEELNVS
G, H, I, J: cationic exchange, pH 7.0	αs_1-casein	31.30	RPKLPHRQPE
	αs_1-casein	29.40	RPKLPHRQPE

(copyright permission obtained).

Table 2. Donkey's milk caseins identified by cationic exchange chromatography [5].

(Figure 2A and table 3). RP-HPLC was also performed on the donkey's milk casein fraction after their precipitation from skimmed milk at pH 4.6. Five peaks were recovered (N-R) each of them submitted to 13% SDS-PAGE and N-terminal analysis and the results, showed in figure 2B and in table 3, indicated mainly the presence of αS1-caseins and β-caseins. Furthermore, the β-casein sequence of peak R (REKEALNV) showed an E→A substitution in the fifth aminoacid [5].

This study revealed the presence of β-caseins (sequence: REKEELNVSS) and αs_1-caseins (sequence: RPKLPHRQPE), which presented marked homology with αs_1- and β-caseins from mare's milk [42], while the presence of other types of caseins, such as αs_2-, γ- and k- were not determined in donkey milk. This result show another high similarity between donkey and human milk: in fact, the presence of αs_2-caseins in human milk has not been demonstrated [34].

Peak/RP-HPLC	Protein	kDa	N-terminal sequence
K	Lysozyme	14.60	KVFSKXELA
L	α-lactalbumin	14.12	KQFTKXELSQVLXSM
M	β-lactoglobulin	22.40	TNIPQTMQ
P	αs1-casein	33.30	RPKLPHQPE
Q	β-casein	37.50	REKEELNVS
R	β-casein	37.50	REKEALNVS

Table 3. Donkey's milk protein fraction identified by reversed phase chromatography in HPLC (see also figure 2 A and B). [5] (copyright permission obtained).

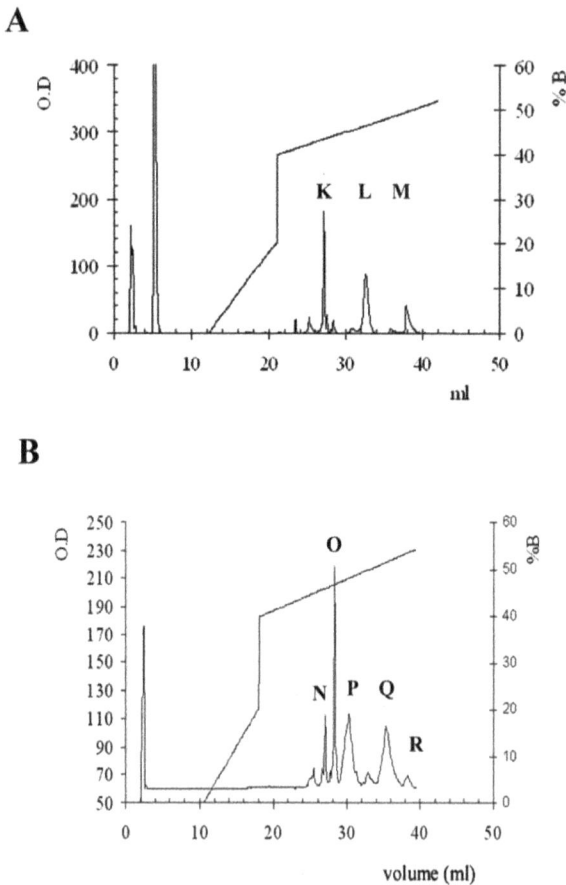

Figure 2. (A) Reversed-phase HPLC of: (A) skimmed donkey's milk and (B) casein fraction [5] (copyright permission obtained).

5.2. Quantitative determination of lysozyme, β-lactoglobulin, α-lactalbumin and lactoferrin.

Thanks to RP-HPLC analysis, it was possible also to calculate the lysozyme, β-lactoglublin and α-lactalbumin concentrations (in mg/ml) at different stages of lactation (60, 90, 120, 160 and 190 days after parturition), the results are shown in Table 4.

Days after parturition	Lysozyme (mg/ml)	β-lactoglobulin	α-lactalbumin
60	1.34	Not determined	0.81
90	0.94	4.13	1.97
120	1.03	3.60	1.87
160	0.82	3.69	1.74
190	0.76	3.60	1.63

Table 4. Quantitative determination of lysozyme, β-lactoglobulin, α-lactalbumin in different stages of lactation [5] (copyright permission obtained).

The amount of lysozyme in donkey's milk varied considerably during the different stages of lactation, with a mean value of 1.0 mg/ml, and proved to be higher with respect tothat in bovine (traces), human (0.12 mg/ml) and goat'smilk (traces), whereas, it was very close to mare's milk (0.79 mg/ml) [43]. The mean β-lactoglobulin content in donkey's milk (3.75 mg/ml) was very close to that of bovine milk (3.3 mg/ml) and pony mare's milk (3.0 mg/ml), whereas in human milk the β-lactoglobulin is absent [34]. The α-lactalbumin content increased in the three months after parturition till the value of 1.8 mg/ml, close to the α-lactalbumin content in human milk (1.6 mg/ml) but lowest compared to the pony mare's α-lactalbumin content (3.3 mg/ml) [5].

Lactoferrin was purified by a cationic exchange chromatography (Mono S HR5/5 column) and its identity was confirmed by N-terminal sequencing and by western blot analysis using anti-lactoferrin antibodies (see figure 3A) [41]. The quantitative determination of donkey's milk lactoferrin (Figure 3B) gave a result of 0.080 ±0.0035 g/L, similar to that found in mare (0.1 g/L), cow (0.02-0.2 g/L), goat (0.06-0.40 g/L), and sheep milk (0.135 g/L), but lower when compared with the lactoferrin content in human milk, in which values are usually in the range 1.0-6.0 g/L [44, 45].

Lactoferrin is an iron-binding protein that displays many biological functions: regulation of iron homeostasis, cellular growth, anti-microbial and anti-viral functions, and protection against cancer development and metastasis [46]. Lactoferrin exerts its antibacterial effect by two different mechanisms involving two separate domains of the protein. In the first one, the antimicrobial effect is due to the high iron binding affinity of the protein that deprives some iron-requiring bacteria of iron and consequently inhibits their growth [46, 47]. The second antimicrobial property is due to the cationic domain at N-terminus directly responsible for the bactericidal effect [48].

A

B

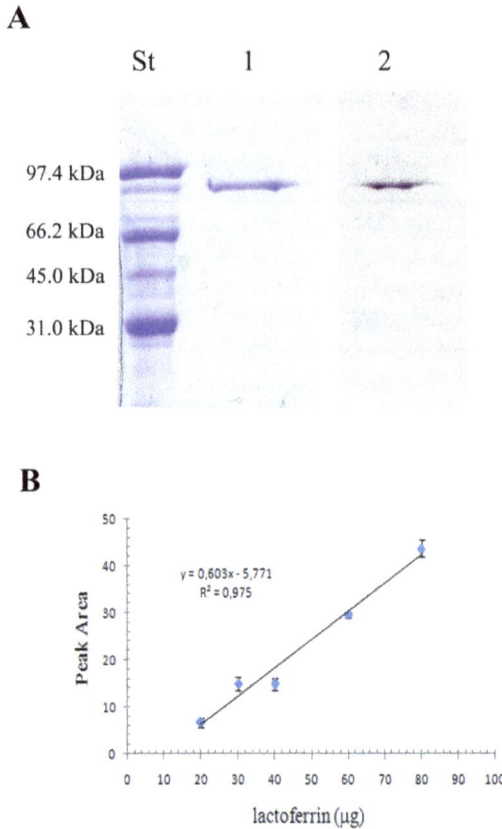

Figure 3. A) 12% SDS-PAGE (lane 1) and immunoblotting using anti-lactoferrin antibodies (lane 2) of lactoferrin purified by Mono S column. St: Bio-Rad low-molecular-weight standard (phosphorylase b, 97.4 kDa; bovine serum albumin, 66.2 kDa; ovalbumin 45.0 kDa; carbonic anhydrase, 31 kDa; soybean trypsin inhibitor, 21.5 kDa; lysozyme, 14.4 kDa.
B) Calibration line of standard solutions of lactoferrin (20-100 final μg).

Lactoperoxidase is a glycoprotein consisting of a single peptide chain with a molecular weight of 78.0 kDa. This enzyme exerts its antimicrobial action through the oxidation of thiocyanate ions (SCN-) by hydrogen peroxide, both present in biological fluids and also in milk. Lactoperoxidase activity in skimmed donkey milk was evaluated by a continuous spectrophotometric rate determination using as substrate 2,2'-Azinobis (3-Ethylbenzthiazoline-6-Sulfonic Acid) [49]. In donkey milk the activity of lactoperoxidase is very low, 4.83±0.35 mU/mL. The enzyme quantification was achieved by a calibration line obtained by plotting the nanograms of peroxidase standard solutions against the enzymatic activity The mean (± SD) concentration of donkey milk lactoperoxidase was calculated to be 0.11±0.027 mg/L, close to the value obtained with human milk (0.77±0.38 mg/L) [50]. In table 5 the concentration of three proteins with antimicrobial effect are compared from donkey, human and cow milk. From these data is evinced that human and donkey milk contain considerable amounts of lysozyme and lactoferrin but lactoperoxidase is present only in small amounts.

Milk	Lactoperoxidase (mg/L)	Lysozyme (g/L)	Lactoferrin (g/L)
Human	0.77	0.12	0.3-4.2
Donkey	0.11	1.0	0.080
Bovine	30-100	trace	0.10

Table 5. Content of lactoperoxidase, lactoferrin and lysozyme from bovine, donkey and human milk [5, 50, 51] (copyright permission obtained).

5.3. Two-dimensional electrophoresis (2-DE) analysis

Whole casein was obtained from skimmed milk by adjusting the pH to 4.6 with 10% (v/v) acetic acid and centrifuging at 3000xg for 10 min in order to obtain a supernatant of whey proteins and the isoelectrically precipitated caseins. In the 2-DE analysis [41], the first dimension was an isoelectric focusing (IEF) performed using a pre-cast immobilized pH gradient gel strip Immobiline DryStrip, (IPG-strip, length 18 cm) in the pH range of 4-7 for the casein fraction and in the pH range of 3-11 for the whey proteins. The second dimension, performed using a Protean II apparatus (Bio-Rad, 180 x 200 x 1.5 mm), consisted of a 13% SDS-PAGE for the casein fraction and a 15% or a 7.5% SDS-PAGE for the whey proteins. After staining, the gel were analyzed using PD-Quest software (Version 7.1.1; Bio-Rad) in order to define spot-intensity calibration, spot detection, background abstraction, calibration, and calculation of molecular mass and pI. The spots considered interesting and quantitatively significant were subjected to N-terminal analysis for their identification. The results of donkey milk 2-DE analysis is shown in figure 4.

The casein fraction (pH range 4-7, figure 4) showed the presence of 13 major protein spots: eight of them were identified as β-caseins (N-terminal sequence: RKEELNVSS) with a pI values ranging from 4.63 to 4.95. Furthermore four β-caseins spots (from spot A to D) showed molecular weights ranging from 33.10 to 33.74 kDa whereas the other four β-caseins spots (from spot E to H) displayed molecular weights ranging from 31.15 to 32.15 kDa, with a difference of about 1000 aminoacids. This results are in good agreement with [52] who demonstrated the presence of a full-length β-casein variant carrying 7, 6, 5 phosphate groups, with a pI of 4.74, 4.82, 4.91 respectively and a spliced variant (-923 aminoacids), carrying 7, 6, 5 phosphate groups with a pI of 4.64, 4.72, 4.80 respectively. On the basis of these observations, and looking at figure 4, it may be evinced that the spots B, C and D may correspond to the full-length forms of β-caseins (pI: 4.72, 4.82, 4.92 respectively) whereas the spots E, F, and G may correspond to the spliced variants of β-caseins (pI: 4.68, 4.80, 4.88). The other remaining five spots (from I to N) were identified as αs1-caseins (N-terminal sequence: RPKLPHRPE) with a pI values ranging from 4.92 to 5.36 (see figure 4). In donkey milk an heterogeneity for the αs1-casein was found [52], assigned to either discrete

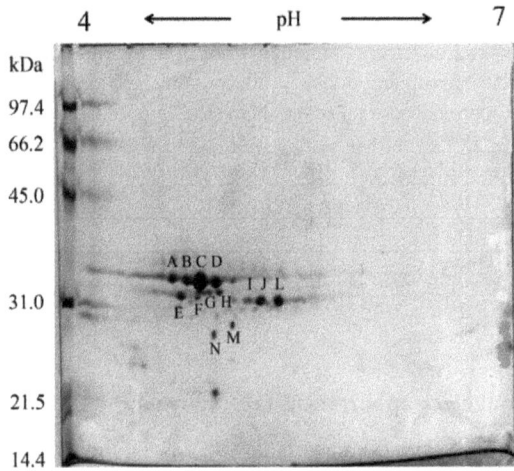

Figure 4. Two-dimensional electrophoresis analysis of donkey milk casein fraction. The first dimension was performed in the pH range of 4-7, the second dimension consisted of a 13%SDS-PAGE. St: Bio-Rad low molecular weight standard (phosphorylase b, 97.4 kDa; bovine serum albumin, 66.2 kDa; ovalbumin 45.0 kDa; carbonic anhydrase, 31 kDa; soybean trypsin inhibitor, 21.5 kDa; lysozyme, 14.4 kDa). [41] (copyright permission obtained).

phosphorylation (5, 6 and 7 phosphate/mole) or non-allelic spliced forms. In our work we found in donkey milk five αs1-caseins: three of them showed a high molecular weight (about 31. 3 kDa) and probably correspond to the full-length phosphorylated forms, whereas two αs1- caseins showed a lowest molecular weight (about 28 kDa) therefore they may correspond to the spliced variants. In our study the presence of αs2-casein and κ-casein were not demonstrated probably because of their low amount in donkey milk. Another group of authors [53] identified in donkey milk the presence of a weak spot identified as αs2-casein and three very weak spots identified as κ-casein. Therefore, the heterogeneity shown in the whole casein analysis by 2-DE may be due to a variable degree of phosphorylation and to spliced forms of αs1- and β-caseins [54-56].

The whey fraction was analyzed by 2-DE in a pH range of 3-11 for the first dimension but with two different polyacrylamide gel percentages in the second dimension in order to have a better differentiation and identification of the low- and high-molecular weight whey proteins as shown in figure 5 A and B.

The separation of low-molecular-weight whey proteins achieved by 2-DE (first dimension: IPG-strip, pH 3-11, second dimension: 15% SDSPAGE) revealed the presence of two isoforms of α-lactalbumin (Figure 5A) corresponding to the spots R and S. This result is in agreement with [57] who observed oxidized methionine forms for α-lactoalbumin (Met 90), due to in vivo oxidative stress that give rise to two α-lactalbumin isoforms. Furthermore from 2-DE, three isoforms of donkey milk β-lactoglobulin (Figure 5A), corresponding to the spots O, P, and Q, were observed. In this case from literature it is known that in donkey

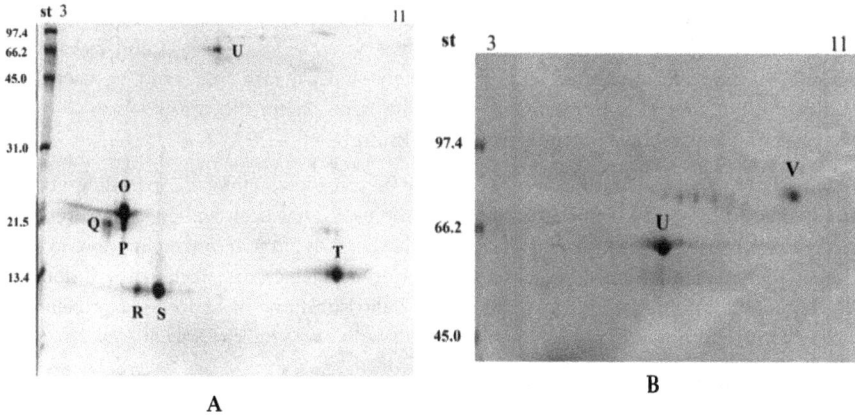

Figure 5. Two-dimensional electrophoresis analysis of donkey milk whey protein fraction. The first dimension was performed in the pH range of 3-11. The second dimension was carried out by: **A)** 15% SDS-PAGE for the identification of the low molecular weight whey proteins, **B)** 7.5 % SDS-Page for the identification of high molecular weight whey proteins. The standard (st) was the same as figure 4. [41] (copyright permission obtained).

milk, this protein exists under two different forms, named β-lactoglobulin I, that constitutes the major form (80%), and β-lactoglobulin II [58, 59] that constitutes the minor form (20%). Successively, a genetic variant for β-lactoglobulin I (named β-lactoglobulin I B) with three amino acid substitutions explained by the degeneracy of the genetic code was found [29], and two genetic variants for β-lactoglobulin II (named β-lactoglobulin II B and C). Successively another β-lactoglobulin II variant (named D) was detected as minor component in the whey fraction of donkey milk [57]. Finally, after 2-DE experiments, only one spot corresponding to donkey milk lysozyme was detected (Figure 5A, spot T) even if from literature the presence of two donkey milk lysozyme genetic variants that arise from an oxidized methionine residue at position 79 [29, 57]. Finally, Figure 5B shows the electropherogram for the donkey milk high molecular-weight whey proteins separated by 2-DE (first dimension: IPG-strip pH 3-11, second dimension: 7.5% SDS-PAGE). By N-terminal sequencing it was possible to assign the spot U to serum albumin (kDa/ pI: 62.7/7.1) and the spot V to lactoferrin (kDa/pI: 77.0/9.8), already discussed in the section 5.2 [41].

6. Conclusion

Recent clinical evidence has renewed the interest in donkey milk because of high tolerability in infants with cows' milk protein allergy. To be successful as a substitute for human milk in infant nutrition, donkey milk must be capable of performing many biological functions associated with human milk. The specific protein fraction in donkey milk can be a good indicators of its potential role. In this study, donkey milk whey proteins were analyzed by 2-DE and were also quantified. From the proteomic map was revealed the presence of two isoforms of α-lactalbumin, three isoforms of β-lactoglobulin, lysozyme, albumin and lactoferrin.

The high lysozyme and α-lactalbumin content found in donkey milk may be responsible for the low bacterial count reported in literature. Lysozyme, lactoperoxidase and lactoferrin have been recognized as antimicrobial and bacteriostatic agents and could be useful to prevent intestine infections in infants. Their action may extend the conservation of fresh donkey milk and the relative potential commercial supply.

On the basis of results obtained donkey milk may be considered suitable for feeding young children affected by severe cow's milk allergy. In the past it has been widely used to replace human milk because its chemical composition and particularly protein content are close to that of human. Great attention must be obviously given to the hygienic characteristics of donkey milk production, in order to consider this milk a valid substitute of hydrolysed proteins or soy-bean derived formulae in the treatment of infants with cow's milk protein allergy.

Food is called functional if it contains one or more components that can provide a benefit to human health, beyond their traditional nutritional role. Donkey milk may be configured as functional food in early childhood and not only.

Author details

Paolo Polidori and Silvia Vincenzetti
Università di Camerino, Italy

7. References

[1] Carroccio A, Cavataio F, Montalto G, D'Amico D, Alabrese L, Iacono G (2000) Intolerance to hydrolysed cow's milk proteins in infants: clinical characteristics and dietary treatment. Clin. Exp. Allergy 30: 1597-1603.

[2] Monti G, Bertino E, Muratore MC, Coscia A, Cresi F, Silvestro L, Fabris C, Fortunato D, Giuffrida MG, Conti A (2007) Efficacy of don key's milk in treating highly problematic cow's milk allergic children: an in vivo and in vitro study. Pediatr. Allergy Immun. 18: 258-264.

[3] Polidori P, Beghelli D, Mariani P, Vincenzetti S (2009) Donkey milk production: state of the art. Ital. J. Anim. Sci. 8(Suppl. 2): 677-683.

[4] Salimei E, Fantuz F, Coppola R, Chiofalo B, Polidori P, Varisco G (2004) Composition and characteristics of ass's milk. Anim. Res. 53: 67-78.

[5] Vincenzetti S, Polidori P, Mariani P, Cammertoni N, Fantuz F, Vita A (2008) Donkey's milk protein fractions characterization. Food Chem. 106: 640-649.

[6] Pacheco MTB, Costa Antunes AE, Sgarbieri VC (2008) New Technological and physiological functional properties of milk proteins. In: Boscoe AB, Listow CR, editors, Protein Research Progress. 117-168, New York: Nova Science Publishers Inc. pp. 117-168.

[7] Ng-Kwai-Hang KF (2003) Milk proteins – Heterogeneity, Fractionation and Isolation. In: Roginski H, Fuquay JW, Fox PF, editors, Encyclopedia of Dairy Sciences. London: Academic Press. pp. 1881-1894.

[8] Schanbacher FL, Talhouk RS, Murray FA, Gherman LI, Willett LB (1998) Milk-borne bioactive peptides. Int. Dairy J. 8: 393-403.

[9] Tanaka T (2007) Antimicrobial activity of lactoferrin and lactoperoxidase in milk. In: Ling JR, editor, Dietary Proteins Research Trends. New York: Nova Science Publishers Inc. pp. 101-115.

[10] Bjorck L (1978) Antibacterial effect of the lactoperoxidase system on psychotrophic bacteria in milk. J. Dairy Res. 45: 109-118.

[11] Vincenzetti S, Polidori P, Vita A (2007) Nutritional characteristics of donkey's milk protein fraction. In: Ling JR, editor, Dietary Proteins Research Trends. New York: Nova Science Publishers Inc. pp. 207-225.

[12] Businco L, Gianpietro PG, Lucenti P, Lucaroni F, Pini C, Di Felice G, Iacovacci P, Curadi C, Orlandi M (2000) Allergenicity of mare's milk in children with cow's milk allergy. J. Allergy Clin. Immunol. 105: 1031-1034.

[13] McClenathan DT, Walker WA (1982) Food allergy. Cow milk and other common culprits. Postgraduate Medicine 72: 233-239.

[14] Park YW, Haenlein GFW (2006) Therapeutic and hypoallergenic values of goat milk and implication of food allergy. In: Park YW, Haenlein GFW, editors. Handbook of milk of non-bovine mammals. Oxford: Blackwell Publishing. pp. 121-135.

[15] Park YW (1994) Hypo-allergenic and therapeutic significance of goat milk. Small Rumin. Res. 14: 151-159.

[16] El-Agamy EI (2007) The challenge of cow milk protein allergy. Small Rumin. Res. 68: 64-72.

[17] Carroccio A, Cavataio F, Iacono G (1999) Cross-reactivity between milk proteins of different animals. Clin. Exp. Allergy 29: 1014-1016.

[18] Docena GH, Ferandez R, Chirdo FG, Fossati CA (1996) Identification of casein as the major allergenic and antigenic protein of cow's milk. Allergy 51: 412-416.

[19] Heyman M, Desjeux JF (1992) Significance of intestinal food protein transport. J. Pediatr. Gastroent. Nutr. 15: 48-57.

[20] Sampson HA (2004) Update on food allergy. J. Allergy Clin. Immunol. 113: 805-819.

[21] Hill DJ, Hosking CS (1996) Cow milk allergy in infancy and early childhood. Clin. Exp. Allergy 26: 254-261.

[22] Stintzing G, Zetterstrom R (1979) Cow's milk allergy, incidence and pathogenetic role of early exposure to cow's milk formula. Acta Paediatr. Scandin. 68: 383-387.

[23] Bock SA (1987) Prospective appraisal of complaints of adverse reactions to foods in children during the first 3 years of life. Paediatrics 79: 683-688.

[24] Host A, Husby S, Osterballe O (1988) A prospective study of cow's milk allergy in exclusively breast-fed infants. Incidence, pathogenic role of early exposure to cow's milk formula, and characterization of bovine milk protein in human milk. Acta Paediatr. Scandin. 77: 663-670.

[25] Host A, Halken S (1990) A prospective study of cow's milk allergy in Danish infants during the first 3 years of life. Clinical course in relation to clinical and immunological type of hypersensitivity reaction. Allergy 45: 587-596.

[26] Restani P, Beretta B, Fiocchi A, Ballabio C, Galli CL (2002) Cross-reactivity between mammalian proteins. Ann. Allergy Asthma Immunol. 89(Suppl.): 11-15.

[27] Guo HY, Pang K, Zhang XY, Zhao L, Chen SW, Dong ML, Ren FZ (2007) Composition, Physiochemical properties, nitrogen fraction distribution, and amino acid profile of donkey milk. J. Dairy Sci. 90: 1635-1643.

[28] Malacarne M, Martuzzi F, Summer A, Mariani P (2002) Protein and fat composition of mare's milk: Some nutritional remarks with reference to human and cow's milk. Intern. Dairy J. 12: 869–877.

[29] Herrouin M, Mollé D, Fauquant J, Ballestra F, Maubois JL, Léonil J (2000) New genetic variants identified in donkey's milk whey proteins. J. Protein Chem. 19: 105–115.

[30] Uniacke-Lowe T, Huppertz T, Fox PF (2010) Equine milk proteins: chemistry, structure and nutritional significance. Intern. Dairy J. 20: 609-629.

[31] Giuffrida MG, Cantisani A, Napoletano L, Conti A, Godovac-Zimmerman J (1992) The amino-acid sequence of two isoforms of α-lactalbumin from donkey (Equus asinus) milk is identical. Biol. Chem. Hoppe-Seyler 373: 931–935.

[32] Zhang M, Yang F Jr, Yang F, Chen J, Zheng CY, Liang Y (2009) Cytotoxic aggregates of alpha-lactalbumin induced by unsaturated fatty acid induce apoptosis in tumor cells. Chem. Biol. Interactions 180: 131-142.

[33] Yamaguchi M, Yoshida K, Uchida M (2009) Novel functions of bovine milk derived alpha-lactalbumin: anti-nociceptive and anti-inflammatory activity caused by inhibiting cyclooxygenase-2 and phospholipase A2. Biol. Pharm. Bull. 32: 366-371.

[34] Miranda G, Mahé MF, Leroux C, Martin P (2004) Proteomic tools to characterize the protein fractions of Equidae milk. Proteomics 4: 2496–2509.

[35] Iacono G, Carroccio A, Cavataio F, Montalto G, Soresi M, Balsamo V (1992) Use of ass's milk in multiple food allergy. J. Pediatric Gastroent. Nutr. 14: 177–181.

[36] Liang L, Tajmir-Riahi HA, Subirade M (2008) Interaction of beta-lactoglobulin with resveratrol and its biological implications. Biomacromolecules 9: 50-56.

[37] Dugo P, Kumm T, Lo Presti M, Chiofalo B, Salimei E, Fazio A, Cotroneo A, Mondello L. (2005) Determination of tryacylglycerols in donkey milk by using high performance liquid chromatography coupled with atmospheric pressure chimica ionization mass spectrometry. J. Sep. Sci. 28: 1023-1030.

[38] Aldamiz-Echevarria L., Bilbao A, Andrade F, Elorz J, Prieto JA, Rodriguez-Soriano J (2008) Fatty acid deficiency profile in children with food allergy managed with elimination diets. Acta Paediatr. 97: 1572-1576.

[39] Wilkins MR, Appel RD, Williams KL, Hochstrasser DF (2007) Proteome research: concepts, technology and application. Berlin: Springer-Verlag, 240 p.

[40] Pennington SR, Wilkins MR, Hochstrasser DF, Dunn MJ (1997) Proteome analysis: from protein characterization to biological function. Trends Cell Biol. 7: 168-173.

[41] Vincenzetti S, , Amici A, Pucciarelli S, Vita A, Micozzi D, Carpi FM, Polzonetti V, Natalini P, Polidori P (2012) A Proteomic Study on Donkey Milk. Biochem. Anal. Biochem. Available: http://dx.doi.org/10.4172/2161- 1009.1000109.

[42] Egito AS, Miclo L, Lopez C, Adam A, Girardet JM, Gaillard JL (2002) Separation and characterization of mare's milk αs_1-, β-, k-caseins, γ-casein-like and proteose peptone component 5-like peptides. J. Dairy Sci. 85: 697-706.

[43] Stelwagen K (2003) Milk protein. In: Roginski H, Fuquay JW, Fox PF, editors. Encyclopedia of Dairy Sciences. London: Academic Press. pp 1835-1842.

[44] Hennart PF, Brasseu DJ, Delogne-Desnoeck JB, Dramaix MM, Robyn CE (1991) Lysozyme, lactoferrin, and secretory immunoglobulin A content in breast milk: Influence of duration of lactation, nutrition status, prolactin status, and parity of mother. Amer. J. Clin. Nutrit. 53: 32-39.

[45] Kanyshkova TG, Buneva VN, Nevinsky GA (2001) Lactoferrin and its biological functions. Biochem. (Moscow) 66: 1-7.

[46] Ward PP, Paz E, Conneely OM (2005) Multifunctional roles of lactoferrin: a critical overview. Cell. Molecular Life Sci. 62: 2540-2548.

[47] Valenti P, Antonini G (2005) Lactoferrin: an important host defense against microbial and viral attack. Cell. Molecular Life Sci. 65: 2576-2587.

[48] Bellamy W, Takase M, Yamauchi K, Wakabayashi H, Kawase K, Tomita M (1992) Identification of the bactericidal domain of lactoferrin. Biochem. Biophys. Acta 1121: 130-136.

[49] Pruitt KM, Kamau DN. (1993) Indigenous Antimicrobial Agents of Milk. In: International Dairy Federation Editions. Bruxelles: IDF. pp. 73-87.

[50] Shin K, Hayasawa H, Lönnerdal B (2001) Purification and quantification of lactoperoxidase in human milk with use of immunoadsorbent with antibodies against recombinant human lactoperoxidase. Amer. J. Clin. Nutrit. 73: 984-989.

[51] Masson PL, Heremans JF (1971) Lactoferrin in milk from different species. Comp. Biochem. Physiol. 39: 119-129.

[52] Chianese L, Calabrese MG, Ferranti P, Mauriello R, Garro G, De Simone C, Quarto M, Addeo F, Cosenza G, Ramunno L (2010) Proteomic characterization of donkey milk "caseome". J. Chrom. A 1217: 4834-4840.

[53] Bertino E, Gastaldi D, Monti G, Baro C, Fortunato D, Perono Garoffo L, Coscia A, Fabris C, Mussap M, Conti A (2010) Detailed proteomic analysis on DM: insight into its hypoallergenicity. Frontieres in Biosciences E2: 526-536.

[54] Visser S, Jenness R, Mullin RJ (1982) Isolation and characterization of β- and γ-caseins from horse milk. Biochem. J. 203: 131-139.

[55] Ochirkhuyag B, Chobert JM, Dalgarrondo M, Haertlè T (2000) Characterization of mare caseins. Identification of $\alpha s1$- and $\alpha s2$-caseins. Lait 80: 223-235.

[56] Criscione A, Cunsolo V, Bordonaro S, Guastella AM, Saletti R, Zuccaro A, D'Urso G, Marletta D (2009) Donkey milk protein fraction investigated by electrophoretic methods and mass spectrometry analysis. Intern. Dairy J. 19: 190-197.

[57] Cunsolo V, Saletti R, Muccilli V, Foti S (2007) Characterization of the protein profile of donkey's milk whey fraction. J. Mass Spectrom. 42: 1162–1174.

[58] Godovac-Zimmermann J, Conti A, James L, Napolitano L (1988) Microanalysis of the amino-acid sequence of monomeric beta-lactoglobulin I from donkey (Equus asinus)

milk. The primary structure and its homology with a superfamily of hydrophobic molecule transporters. Biol. Chem. Hoppe-Seyler 369: 171–179.

[59] Godovac-Zimmermann J, Conti A, Sheil M, Napolitano L (1990) Covalent structure of the minor monomeric beta-lactoglobulin II component from donkey milk. Biol. Chem. Hoppe-Seyler 371: 871–879.

Chapter 8

Allergenicity of Milk Proteins

Simonetta Caira, Rosa Pizzano, Gianluca Picariello, Gabriella Pinto,
Marina Cuollo, Lina Chianese and Francesco Addeo

Additional information is available at the end of the chapter

1. Introduction

Adverse reactions to food are currently classified into toxic and non-toxic reactions. There is a normal range of concentrations of naturally occurring toxic compounds, which can easily increase during food processing. For example, thermal processing can cause the unintended and undesirable formation of toxic compounds, such as acrylamide in fried potato chips and furan in sterilized canned vegetables, together with losses of certain nutrients. The incidence of non-toxic reactions depends on individual susceptibility to a specific food or food ingredient, although these reactions are often dose-dependent. The non-toxic types may be divided further into immune- and non immune-mediated reactions. The term "hypersensitivity" is used for immune-mediated reactions, and the term 'intolerance' is used for non immune-mediated reactions (Figure 1). Immune-mediated reactions may be IgE-mediated (i.e., allergy or type I hypersensitivity) or non-IgE-mediated, whereas food intolerance may be enzymatic, pharmacologic or undefined. The incidence of immune-mediated adverse reactions to foods has increased in recent decades. In healthy subjects, orally ingested dietary proteins induce antigen-specific systemic hyporesponsiveness, termed oral tolerance. This phenomenon is well described in animal models, although the mechanisms remain unknown. Abrogation of oral tolerance or failure to induce oral tolerance may result in the development of food hypersensitivity. Immune reactions that cause tissue damage may be mediated by four reaction types that were defined by Coombs and Gell [1] (Figure 2). Type I, or anaphylactic hypersensitivity, is mediated by the reaction of an antigen with specific IgE antibodies that are strongly bound through their Fc receptor (CD23, IgεR) to the surface of the mast cell. Crosslinking of Igε receptors by divalent hapten leads immediately to the release of mediators, for example, some cytokines [primarily interleukin 4 (IL4)] and histamine, which are both activities of eosinophil chemotactic factor (ECF) and neutrophil chemotactic factor (NCF) (Figure 2A). This type of hypersensitivity occurs within minutes of antigen exposure. Some reaction mediators produce local skin, gastrointestinal and respiratory tract manifestations, and a systemic allergic reaction to an allergen that is associated with a dangerously low blood pressure. Moreover, some of the mediators exhibit chemoattractant activity and induce the infiltration of neutrophils, eosinophils, macrophages, lymphocytes and basophils within 6-12 h after challenge. The localized late-phase inflammatory response may also be mediated partly by cytokines that are released from mast cells. In type II or antibody-dependent cytotoxic hypersensitivity, antibodies recognize antigens on the surface of specific cells or tissues. Once activated, the complement system can initiate a variety of responses that can lyse and destroy cells. Phagocytic and cytotoxic K cells, which have receptors for the Fc-part of IgG or an activation component of complement, i.e., C3b, may also destroy cells (Figure 2B). In type III, or immune complex-mediated hypersensitivity, the soluble antigen can activate the complement and deposited phagocytes. Leucocytes may release tissue-damaging mediators and activate phagocytes, culminating in tissue damage. The complex can also induce thrombin-mediated platelet aggregation and release a vasoactive amine (Figure 2C). Type IV reaction, or delayed-type hypersensitivity, arises more than 24 h after an encounter with the antigen. Type IV reactions are mediated by antigen-sensitized CD4+ T cells (T helper cells) that release inflammatory mediators [e.g., IL2 and interferon-y (lFN-y)], attract phagocytes to site of infection, activate an inflammation response and lyse invading cells (Figure 2D). Because lytic enzymes are secreted from the phagocytic cells into the surrounding tissue, localized tissue destruction can progress.

CLASSIFICATION AND TERMINOLOGY OF ADVERSE REACTION TO FOOD

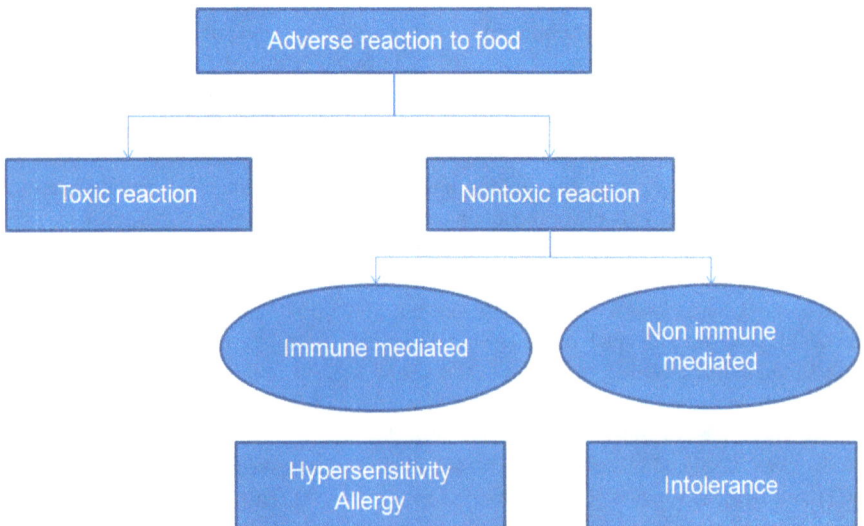

Figure 1.

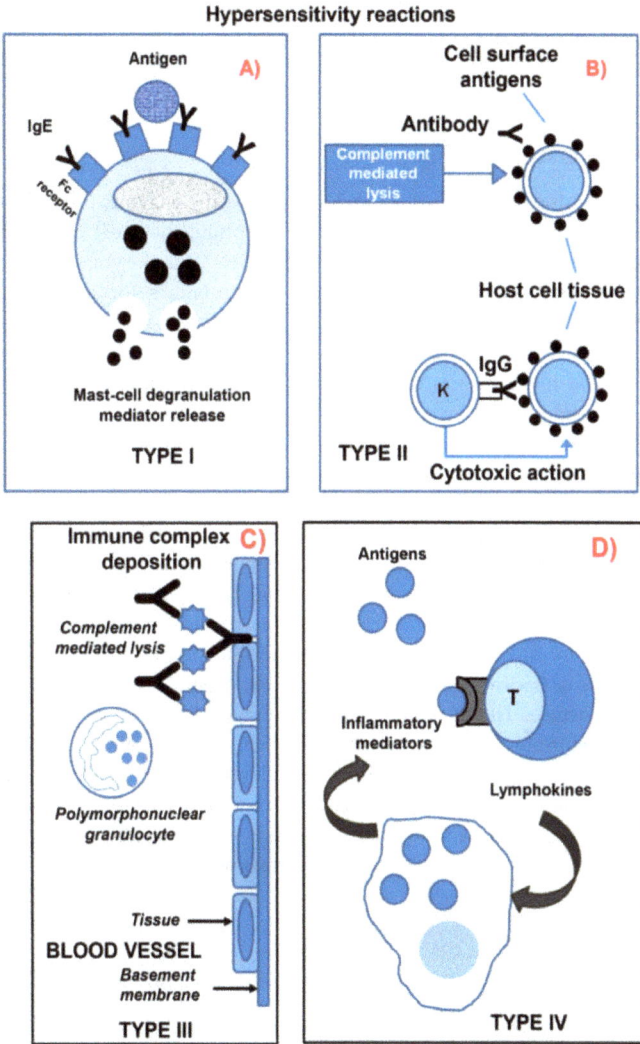

Figure 2. Types of Allergy Mechanisms according to Coombs and Gell [1].

The mechanisms underlying of allergic sensitization to food include genetic susceptibility, aberrant barrier functions of the skin epithelium and gut mucosa and dysregulation of immune functions. Despite a wide range of clinical manifestations, there are at least two common prerequisites for the development of a general food allergy (FA). First, intraluminal antigens must penetrate the mucosal barrier of the intestine. Second, the absorbed antigens must cause harmful immune responses [2].

2. Adverse reaction to milk

Non-toxic adverse reactions to milk are primarily caused by either lactose intolerance or milk allergy.

2.1. Lactose intolerance

Milk intolerance is due to the inherited lack of the specific enzyme, β-galactosidase that is required to hydrolyze lactose. For lactose malabsorption, the most common therapeutic approach excludes lactose-containing milk from the diet. To make yogurt edible, exogenous β-galactosidases that hydrolyze yogurt lactose or probiotics for their bacterial lactase activity are added. However, further studies are required to clarify the role probiotics play in lactose intolerance therapy, which includes considering their well-known beneficial effects on intestinal function, gas metabolism and motility [3]. A prolonged contact time between β-galactosidase and lactose delays the gastrointestinal transit time and chronic lactose ingestion to improve colonic adaptation. It is known that high concentrations of β-galactosidase are physiologically present in neonates, but a genetically programmed and irreversible decline of the activity occurs after weaning [4], which results in primary lactose malabsorption. The secondary hypolactasia because of intestinal mucosa brush border damage that increases the gastrointestinal transit time is a transient and reversible condition [5]. Bloating, flatulence, abdominal pain, the passage of loose and watery stools, excessive flatus and diarrhea are gastrointestinal symptoms of lactose intolerance [6]. However, lactose occurring in the colonic lumen does not necessarily produce gastrointestinal symptoms because of the variable amount of residual intestinal mucosal lactase activity that possibly digests lactose. The availability of recombinant β-galactosidase as an exogenous lactase has resolved problems concerning bacteria that release β-galactosidase during gastric passage. By this means, lactose is partially or fully degraded to glucose and galactose and is therefore easily eliminated by simple dietary adjustments that mediate the effects of lactose intolerance [7]. An accurate diagnosis of lactose intolerance can significantly reduce patient anxiety and avoid unnecessary examinations and treatments [8].

2.2. Milk allergy

Cow's milk allergy (CMA) is a complex disorder that implies an immunologically mediated hypersensitivity reaction with varying mechanisms and clinical presentations. The type I reactions appear to be the most common immune reaction to milk. However, the dominance of IgE reactions (Type I) may be an artifact as the reaction is easy to diagnose because of an immediate IgE measurement, whereas other reaction types are more difficult to diagnose. Non-IgE-mediated hypersensitivity has been increasingly diagnosed, and it is likely that several mechanisms operate in an individual patient. In clinical work, hypersensitivities are classified as IgE- and non-IgE-mediated or as immediate and late reactions based on the appearance of the first types of symptoms [9]. Cow's milk is a member of the "Big-8" food allergens that include egg, soy, wheat, peanuts, tree nuts, fish and shellfish in terms of prevalence [9]. The incidence of CMA varies with age. CMA is prevalent in early childhood with reported incidences between 2 and 6% [10-12] and decreases into adulthood to an incidence of 0.1–0.5% [13-14]. It has been suggested that infants have milk allergies because milk is usually the first source of foreign antigens that they ingest in large quantities, and the infant intestinal system is insufficiently developed to digest and immunologically react to milk proteins. When milk is eliminated, the inflammation response is controlled. After several years, oral tolerance is developed, and milk can again be tolerated [15]. Most milk-allergic children are considered symptom-free by 3 years of age, but several studies have indicated that older children may also have immune reactions to milk. Children whose CMA has been diagnosed at an age older than 3 years do not tend to outgrow the problem. In adults, CMA is less common than lactose intolerance [16-17], even though it has been reported that approximately 1% of the adult population has milk-specific IgE antibodies. However, studies on CMA in adults are scarce. Little is known about the clinical symptoms, eliciting doses, and allergens involved. It has established that CMA in adults is rare but serious [18]. In a study by Stöger et al. [19], the main target organs in adult CMA were the skin and the respiratory tract. Gastrointestinal (mild to moderate) and cardiovascular (severe) symptoms were observed less often in adults compared with children. Milk allergies and hypersensitivity may be more common now than they were several decades ago. Further, the prevalence of atopic diseases has also increased in recent decades. Because genetic diversity has not changed over a short period, environmental factors are believed to have influenced the phenotype. Such factors may include increased air pollution, such as passive smoking, and dietary factors, such as the duration of breastfeeding, amount of antioxidants and the type of dietary fats (favoring saturated fat and n-6 fatty acids). Another approach is the hygiene hypothesis, which states that early exposure to microbial antigens may reduce the risk of having allergies [20]. For as long as milk allergy and hypersensitivity have been experienced, and still may be, these conditions may be misdiagnosed as a disease other than an allergy/hypersensitivity, particularly if symptoms are delayed. Classical IgE-mediated milk allergies with objectively recorded skin reactions may have been diagnosed easily, whereas hypersensitivity with subjective gastrointestinal reactions may have been diagnosed as lactose intolerance, irritable bowel syndrome or some other intestinal disorder. In adults, viral infection, antibiotic treatment or stress may alter intestinal integrity so that its balance is disturbed and the number of protecting agents, such as microflora and mucosal immunity, are altered [21]. This process may result in milk hypersensitivity. Over one-third of women with IgE-mediated reactions to milk proteins have been reported to exhibit their first symptoms of hypersensitivity during or shortly after pregnancy, and one-quarter reported the first symptoms during a period of severe emotional stress [22].

3. Milk allergens

No single major allergen has been identified in cow's milk according to either challenge tests or laboratory procedures; [23]. Indeed, clinical challenge tests demonstrate that most CMA patients react to several protein fractions of cow's milk and each allergenic protein may have several epitopes, which are widely spread along the molecules. The cow milk proteins prevalently implicated in allergic responses in children are the whey proteins α-

Lactalbumin (α-La)(Bos d 4) and β-Lactoglobulin (β-Lg) (Bos d 5), in addition to the casein (CN) fraction (Bos d 8) [24-26]. In adults, the predominant allergen is CN, whereas sensitization to whey proteins is rare. Biochemical characteristic of allergenic cow milk proteins are reported in Table 1. Currently, milk allergen analyses are generally based on immunoassay methods, such as enzyme-linked immunosorbent assay (ELISA) [27]. Commercial test kits are available for the determination of CN, β-Lg, or total milk proteins (CN and β-Lg) [28]. Interlaboratory studies were performed to evaluate the reliability and reproducibility of these kits [29]. Antibody cross-reactivity has been reported for some milk proteins [30]. Confirmatory tests are required to corroborate ELISA detection results and improve the detection specificity of undeclared milk allergens. For the last several decades, mass spectrometry has become the dominant technology for the identification of peptides and proteins. The primary current approaches used for protein identification are top-down [31-33] and bottom-up [34] sequencing. Top-down sequencing involves introducing the intact protein into the gas phase. The protein is identified by measuring either the protein molecular weight or its fragmentation pattern using various techniques [35]. The bottom-up approach is more common. The sample is usually digested with an enzyme, such as trypsin, followed by accurate sequence analysis by tandem mass spectrometry (MS/MS) of the proteolytic fragments. For protein identification, an algorithm is used for database searching based on amino acid sequence assignments.

Cow's Milk Proteins (100%)	Protein	Allergen Name	Allergenicity	Total Protein %	MW (kDa)	pI	Amino Acid Residues	Calcium sensitivity	Phosphate groups
Caseins (80%)	α_{s1}-Casein	Bos d 8	Major	32	26.6	4.9 - 5.0	199	+++	8-9
	α_{s2}-Casein	"	"	10	25.2	5.2 - 5.4	207	++++	10-13
	β-Casein	"	"	28	24,0	5.1 - 5.4		++	4-5
	γ1-Casein	"	"	Traces	20,5	5.5	181	+	1
	γ2-Casein	"	"	Traces	11,9	6.4	104		
	γ3-Casein	"	"	Traces	11.5	5.8	102		
	κ-Casein	"	"	10	19	5.4 - 5.6	169		
Whey proteins (20%)	α-Lactalbumin	Bos d 4	Major	5	14.2	4.8	123		
	β-Lactoglobulin	Bos d 5	Major	10	18.3	5.3	162		
	Immunoglobulins	Bos d 7	–	3	150	–	–	–	1-3
	BSA	Bos d 6	–	1	66.3	4.9 - 5.1	582		
	Lactoferrin	–	–	Traces	80	8.7	703		

Table 1. Chemical characteristics of cow's milk proteins and their inclusion in the official list of allergens

3.1. Casein

The CN fraction is composed of four proteins α_{s1}-, α_{s2}-, β-, and κ-CN, in approximate proportions of 40%, 10%, 40%, and 10%, respectively, and αs1-CN is a major allergen according to IgE and T cell recognition data [36-39]. The specificity of the IgE response to the different purified CNs has been analyzed on 58 sera from patients allergic to whole-CN [40].

Multi-sensitization was observed, which most likely corresponded to a co-sensitization to the different CN components after the disruption of the CN micelles. The CN fractions cross-link to aggregates termed nanoclusters, which combine into micelles. In the new state of aggregation, there is a hydrophobic central part and peripheral hydrophilic one that include phosphorylation sites. Although the primary structure of CNs is known, the micelle structure remains relatively unknown. $\alpha s2$-, $\alpha s1$-, and β-CN bind calcium to form a core that is covered with a κ-CN layer. The κ-CN latter protects CN micelles from precipitation in the presence of milk constitutive calcium ions. During the last five decades, different models for the structure of the bovine CN micelle have been proposed. Walstra has summarized the common structural elements into a "ball-shaped model" [41]. According to Walstra, a well-accepted model considers the CN micelle as follows: (i) it is roughly spherical; although it does not have a smooth surface; (ii) it is built of smaller units, termed sub-micelles, which mainly contain CN and have a mixed composition; (iii) sub-micelles vary in composition and consist of two main types-one primarily consists of αs and β-CNs and the other primarily consists of αs- and κ-CNs; (iv) the sub-micelles can be linked together by small calcium phosphate clusters bridging them; (v) the sub-micelles aggregate until they have formed a micelle in which those with κ-CN are outside; and (vi) consequently, molecular chains of the C-terminal end of κ-CN protrude from the micelle surface to form a "hairy" layer that prevents any further aggregation of sub-micelles by steric and electrostatic repulsion. The hairy layer is also held responsible for the stability of the micelles against flocculation. Destabilization of CN micelles can be made by treating milk with milk clotting enzymes, which limitedly degrade κ-CN affording to CN micelle fusion and formation of a para-κ-CN aggregate. Finally, a variable proportion, up to 5% total, consists of a heterogeneous group of CNs termed γ-CNs (γ-CN) that result from the limited proteolysis of β-CN by plasmin, the native milk protease. Plasmin disperses in low amounts from blood to milk during secretion to generate CN fragments whose structure are shown in Figure 7 with peptides labeled according to the current nomenclature [42]. CNs are highly sensitive to proteolysis and do not maintain a unique folded conformation [43], which has led them to being termed rheomorphic. Because the folded structure is limited, heating does not generally change the structure and hence its IgE binding [44]. $\alpha s1$-CN, $\alpha s2$-CN, and β-CN can chelate Ca^{2+}, Zn^{2+}, and Fe^{3+}, respectively. The four CNs share little sequential homology, but, despite this lack of homology, simultaneous sensitization is often observed.

3.1.1. α_{s1}-CN

$\alpha s1$-CN consists of major and minor polypeptides that include the same amino acid sequence but have different degrees of phosphorylation [45]. Allergenic epitopes were identified by Spuergin et al. [46] in $\alpha s1$-CN regions 19–30, 93–98 and 141–150 as immunodominant epitopes. Some sequential IgE-binding regions were recognized at AA 17–36, 39–48, 69–78, 93–102, 109–120, 123–132, 139–154,159–174, and 173–194 using sera from nine older children (> 9 years old), and the epitopes AA 69–78 and 175–192 were recognized by 60% and 80% of sera, respectively, from older children [47]. A later study by Elsayed et al. [48] has demonstrated that the N- and C-terminal peptides, AA 16–35 and 136–155, respectively, have the highest human IgE-binding affinity, and AA 1–18 and 181–199 exhibited high binding to rabbit IgG. cDNA coding for α_{s1}-CN from a bovine mammary gland cDNA library with allergic patients' IgE Abs has been isolated. IgE epitopes of α_{s1}-CN were determined with recombinant fragments and synthetic peptides that spanned the α_{s1}-CN sequence using microarray components and sera from 66 cow's milk-sensitized patients. The allergenic activity of recombinant α_{s1}-CN and the α_{s1}-CN-derived peptides exhibited IgE reactivity, but mainly the intact recombinant α_{s1}-CN induced strong basophil degranulation. The results indicate that α_{s1}-CN contains several sequential IgE epitopes, but the isolated peptides were less potent than the complete allergen in initiating effector cell degranulation. These results suggest that primarily intact α_{s1}-CN or larger IgE-reactive portions thereof are responsible for IgE-mediated symptoms of FA [49].

3.1.2. α_{s2}-CN

α_{s2}-CN is the most hydrophilic of the four CNs because of an anionic group cluster. α_{s2}-CN consists of two major and several minor components that exhibit varying levels of phosphorylation. α_{s2}-CN contains two cysteines and forms disulfide-linked dimers. Using 99 synthetic decapeptides, 10 regions binding IgE from the sera were identified as allergenic, i.e., 31–44, 43–56, 83–100, 93–108, 105–114, 117–128, 143–158, 157–172, 165–188, and 191–200 [50]. Studies on the presence of α_{s2}-CN epitopes 87–96 and 159–168 with weak binding to 145–154 and 171–180 with originate from individuals with persistent cow's milk allergy. Patients with transient allergies exhibited only weak binding to α_{s2}-CN peptides [51].

3.1.3. β-CN

β-CN is the most hydrophobic component of total CN. Sequence variants are known because of both partial proteolysis and variant genes [42]. β-CN is less phosphorylated than α_{s1}-CN and α_{s2}-CN, with five potential phosphorylated sites located in the N-terminal region. Plasmin cleaves β-CN into γ_1-, γ_2-, and γ_3-CN. Synthetic decapeptides were used to estimate the β-CN region that binds IgE of patients, and peptides 1–16, 45–54, 55–70, 83–92, 107–120, 135–144, 149–164, 167–184, and 185–208 were described as typical for patients with persistent CMA. Sera from eight young patients exhibited a simpler IgE pattern because the sera were bound to peptides 1–16, 45–54, 83–92, 107–120, and 135–144 with weak binding to residues 57-66 and the C-terminal region [52].

3.1.4. κ-CN

κ-CN consists of a major carbonate-free component and a mini-bonded polymer that ranges from dimers to octamers. κ-CN plays an important role in the stability and coagulation properties of milk [89]. κ-CN is most likely more structured than α_s- and β-CN and contains specific disulfide bonds [53-54]. The β-CN can rearrange on heating [55]. κ-CN is sensitive to proteolysis and is hydrolyzed by chymosin to produce para-κ-CN and caseinomacropeptide in the cheese-making process. Allergenic potential sequences remain in cooked cheeses. κ-

CN is essential to the stability of CN micelles [55]. Diagnosing patients with persistent cow's milk allergies with the use of 80 overlapping synthetic decapeptides helped to identify some regions that bind to IgE from sera, specifically sequences 15–24, 37–46, 55–80, 83–92, and 105–116 [37].

3.2. Whey proteins

Whey proteins contain two major allergens, β-Lg and α-La, and minor constituents, e.g., lactoferrin, bovine serum albumin (BSA), and immunoglobulins. In this fraction, proteolytic fragments of CN and fat globule membrane proteins can occur.

3.2.1. β-Lg

β-Lg (Bos d 5) is the major whey protein in ruminant milks, comprising 50% of the total whey protein. β-Lg is found in the milk of other mammals but is missing from the milk from rodents, lagomorphs and humans. Notably, 13-76% of patients are found to react with β-Lg. β-Lg has a molecular weight of approximately 18 kDa [37] and belongs to the lipocalin superfamily. The β-Lg (Bos d 5) allergen is capable of binding lipids, including retinol, β-carotene, saturated and unsaturated fatty acids, and aliphatic hydrocarbons, and transporting hydrophobic molecules, which is an important function [56-57]. Under physiological conditions, β-Lg is an equilibrium mixture of monomeric and dimeric forms. The proportion of monomers increases after heating to 70°C. β-Lg contains five cysteine residues, of which four are engaged in intra-chain disulfide bridges. Because of the single unpaired cysteine, β-Lg predominantly exists as a stable dimer that tends to dissociate into monomers at a pH between 2 and 3 [58]. β-Lg is present in several variants, i.e., A, B, and C, that are found in the Jersey breed. β-Lg is sensitive to thermal processes. Ehn et al. [59] reported that heating β-Lg to 74°C and 90°C reduced IgE binding significantly. Heating to 90 °C reduced IgE binding more extensively. Chen et al. [60] reported that nearly 90% loss and denaturation of β-Lg are observed in processed milk and that high heat is the major cause of protein aggregation. Circular dichroism demonstrated no significant conformational changes at temperatures below 70°C for as long as 480 s. The rapid changes of β-Lg occurred between 80°C and 95°C. Fifty percent of the maximal changes could be reached within 15 s. Guyomarc'h et al. [61] reported that large micellar aggregates, 4×106 Da, are formed upon heating milk that contained 3:1 ratios of β-Lg and α-La together with κ-CN and α_{s2}-CN. Proteolysis and use of monoclonal antibodies proved that β-Lg possesses many allergenic epitopes spread over the β-Lg structure [62]. Major human IgE epitopes for β-Lg amino acid fragments are composed of residues 41–60, 102–124, and 149–162; intermediate 1–8, 25–40, and 92–100; and minor 9–14, 84–91, 125–135, and 78–83 [63]. Similar IgE epitope regions (21–40, 40–60, 107–117, and 148–168) were reported for a rat model of β-Lg allergy [64,65].

3.2.2. α-La

α-La (Bos d 4) is a homologue of C-type lysozymes. It is a member of glycohydrolyase family 22 and Pfam family and weighs 14,186 Da in the mature form and between 15,840 to

16,690 Da for the glycosylated forms. α-La is stabilized by binding to calcium. Polverino de Laureto et al. [66] reported that α-La is cleaved by pepsin at pH 2 in the region of residues 34–57, which produces large fragments. Veprintsev et al. [67] reported that differential scanning calorimetry of α-La at pH 8.1 exhibited transitions at 20°C–30°C with calcium chelator ethylene glycol tetra-acetic acid and near 70°C with the addition of calcium. McGuffey et al. [68] investigated the heating effects of purified α-La and demonstrated that the extent of irreversible aggregation varies at temperatures between 67°C and 95°C. When milk is heated to 95°C, α-La denatures more slowly than β-Lg. The folded α-La structure is destabilized at low pHs with the formation of a molten globule[69]. The stability to denaturation is also strongly lowered by the reduction of disulfides [70]. Disulfide exchange can occur during thermal denaturation, which leads to the formation of aggregates [71]. Although evolved from a lysozyme [72], the function of α-La is to form a complex with galactosyltransferase, which alters the substrate specificity and increase the lactose synthase rate in milk. The galactosyltransferase and α-La is termed lactose synthase [73]. α-La alone does not have any catalytic activity as a lysozyme or a synthase. Several other properties of α-La and possible additional functions have been described [74], which include binding of several ligands and antimicrobial activity, both as a complete molecule [75] or as peptides [76]. The cytotoxic effects against mammalian cells have also been investigated [77]. The major component of α-La is unglycosylated. However, a mass spectrum of α-La contains at least 15 distinct peaks [78], and a minor glycosylated form (approximately 10%) results from asparagine 45 glycosylation. A study concerning the allergenic properties of α-La demonstrated that in 60% of the study patients, allergic sera were specific for intact α-La with only 40% binding to peptides obtained after tryptic hydrolysis. Residue 17–58 was the most frequently recognized in the sequence 59–93, 99–108, and 109–123 [79]. The linear epitopes were identified by using sera of patients suffering from persistent allergies and IgE to cow's milk levels > 100 kU(A)/L. Serum IgE bound most strongly to peptides 1–16, 13–26, 47–58, and 93–102 [80].

3.2.3. Minor allergens

Bovine serum albumin (BSA) (Bos d 6), which is a heat-labile protein, is a major allergen in beef but a minor allergen in milk [81-84]. Accordingly, beef allergic individuals are at risk of being allergic to cow's milk and vice versa. BSA allergies account for 0-88% of sensitization events, whereas clinical symptoms occur in up to 20% of patients. BSA is one of the proteins most frequently involved in binding with circulating IgE [85-86]. Bovine immunoglobulins (Bos d 7) may be also responsible for clinical symptoms in CMA.

4. Post-translational modifications

In evaluating the allergenic potential of a protein, post-translational modifications of amino acid residues should be considered in addition to sequential and conformational IgE binding domains. Notably, such modifications may either generate additional IgE epitopes or induce changes in protein folding that affect IgE-protein interactions. Accordingly, recombinant allergens do not generally have the IgE-binding capacity of their natural counterparts, most likely because of a deficiency in the post-translational events [87]. Regarding milk proteins, selective phosphorylation of serine residues in all of the four CNs, O-glycosylation of threonine residues in κ-CN and N-glycosylation of asparagine residues in α-La have been long described.

4.1. Phosphorylation

The removal of phosphate groups from CNs significantly reduces the CN-binding capacity of IgE from patients who suffer from milk allergies, which indicates that at least part of anti-CN IgE is directed against CN domains that comprise a major phosphorylation site [88, 89]. It has been suggested that currently observed co- and cross-sensitization to the different CNs that are encoded by different genes and display few amino acid sequence homologies can be caused by the occurrence of common highly conserved major sites of phosphorylation, i.e., the Ca^{2+} binding CN sequence SerP-SerP-SerP-Glu-Glu that corresponds to α_{s1}-CN 66-70, β-CN 17-21 and α_{s2}-CN 8-12 and 56-60 [87]. Most likely, sensitization to milk is caused by a large release of phosphopeptides that are resistant to further degradation by digestive enzymes [90] during intestinal proteolysis of milk proteins. However, serine phosphorylation poorly affects the overall antigenic potential of individual CNs. Notably, antisera raised against native β- and α_{s2}-CNs can recognize their targets after dephosphorylation or deletion of a major phosphorylation site [87]. Furthermore, polyclonal antisera that are produced in rabbits using a bovine β-CN 1-28 phosphorylated peptide as an antigen have been utilized to detect all of the tryptic phosphopeptides that originate exclusively from the 1-28 region of β-CN, regardless of the content of the phosphorylated Ser residues, and none of those generated by the other bovine CN fractions [91]. β-CN from human milk contains the phosphopeptide cluster ^{5}Glu-Ser-Leu-SerP-SerP-SerP-Glu-Glu^{12}, also found in the bovine β-CN sequence 14-21; however, a lower level of phosphorylation has been generally observed. For example, according to the phosphopeptide analysis of human milk that is reported in Table 2 and Figure 3, the 2092.8 Da component, which corresponds to β-CN(f2-18)2P, caused the third peak in intensity order in combination with the fully phosphorylated components. This lower phosphorylation level is lacking in its bovine counterpart. The overall higher degree of phosphorylation of bovine CNs can play a role in sensitizing humans to bovine milk.

4.2. Glycosylation

The effect of glycosylation on the allergenic potential of milk proteins has been long disregarded despite efforts to identify the domains responsible for the allergenicity of milk proteins, mostly based on an epitope mapping approach. Notably, the role of carbohydrate epitopes in initiating an allergic reaction is still unclear [92]. Potential glycosylation sites have been identified in major milk proteins, i.e., N^{45} and N^{71} of mature α-La [93] and T^{131}, T^{133}, T^{135}, T^{136}, T^{142}, and S^{141} of mature κ-CN [94]. Approximately 10% of α-La has been found glycosylated at N^{45}, giving rise to at least 14 distinct peaks by electrospray-ionization mass spectrometry analysis [95]. However, these glycosylated forms were not included among the IgE epitopes in a study because they were not detected by matrix-assisted laser desorption/ionization time-of-flight (MALDI-TOF) analysis of α-La; notwithstanding in the same study, IgE reactivity of sera from patients allergic to α-La were proven to be sensitive to periodic acid treatment [93]. As reported in Table 3, the genetic variant A of water buffalo α-La that carries an N^{45}–D^{45} substitution cannot be glycosylated. To assess the effect of glycosylation on the allergenicity of α-La, it might be used as substrate for IgE reactivity testing of sera from patients sensitized to α-La.

Molecular mass (Da)		CPP identification
Theoretical	Measured MH$^+$	
Human β-Casein		
2488.1	2489.1	β-CN (f1–18)5P
2408.1	2409.1	β-CN (f1–18)4P
2328.2	2329.1	β-CN (f1–18)3P
2248.2	2249.3	β-CN (f1–18)2P
2168.2	2169.2	β-CN (f1–18)1P
2252.0	2252.7	β-CN (f2–18)4P
2172.0	2173.0	β-CN (f2–18)3P
2092.0	2093.0	β-CN (f2–18)2P
2012.0	2013.0	β-CN (f2–18)1P
3100.8	3101.0	β-CN (f1–23)5P
3020.8	3021.8	β-CN (f1–23)4P
2940.9	2941.0	β-CN (f1–23)3P
2860.9	2861.9	β-CN (f1–23)2P
2780.9	2781.9	β-CN (1–23)1P
Human α$_{s1}$-Casein		
3077.0	3078.9	α$_{s1}$-CN (f15–38)3P
3068.1	3069.1	α$_{s1}$-CN (f12–36)2P
2488.5	2489.5	α$_{s1}$-CN(f8–27)2P
2408.5	2409.5	α$_{s1}$-CN (f8–27)2P
2118.8	2119.5	α$_{s1}$-CN (f68–83)5P

Table 2. Identification of human milk soluble TCA 12% peptide fractions enriched on hydroxyapatite after MALDI-TOF analysis by FindPept (http://www.expasy.org/tools/findpept.html) software.

Despite some indications that the allergenic character has been identified in the glycosidic moiety of native κ-CN [96], at present this issue remains to be settled. Glycated forms of κ-CN account for approximately 40% of the κ-CN that normally occurs in bovine milk, but glycans are not randomly distributed among potential glycosylation sites. The hierarchy of glycan addition proceeds according to the order T^{131}, T^{142}, T^{133}, whereas the other sites remain latent until these sites are occupied [97]. κ-CN is cleaved by chymosin during the primary stage of cheese making at the peptide bond F^{105}-M^{106}. The C-terminal 106-169 fragment, known as glycopeptide because all of the glucides originally present in κ-CN are retained, is released and lost in the whey. Therefore, cheese is devoid of any glycosylated major component. Potential allergenicity of κ-CN glycoforms has been suggested by analyzing the IgE binding capacity of an individual human serum from an adult atopic patient who had outgrown a cow milk allergy in early childhood. Bovine κ-CN has been selectively recognized by IgE immunostaining of an electrophoretic profile of milk proteins. No additional IgE-reactive proteins other than bovine κ-CN have been found in either bovine cheese, regardless of the cheese making technology and time ripening, or in ewe, goat and water buffalo milk. Moreover, chemical removal of glucide chains from bovine κ-CN has not impaired IgE binding, thus proving a primary involvement of the glycoside moiety of the protein in IgE recognition. According to the specificity displayed by IgE, N-acetylneuraminic acid as a terminal unit of a tetrasaccharide chain has been argued to be an IgE epitope [98].

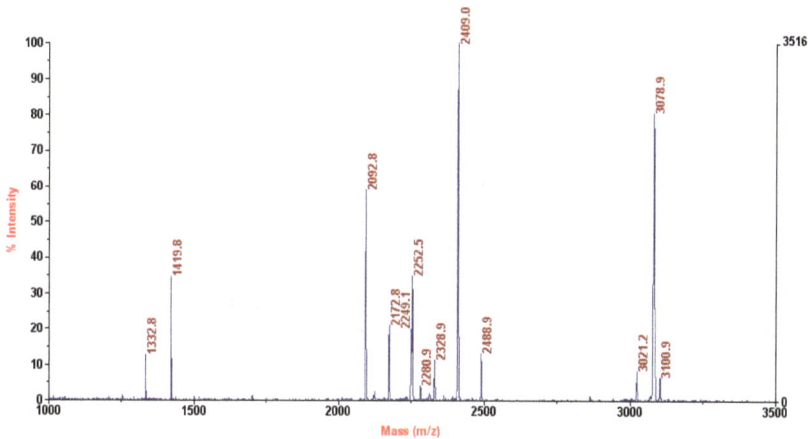

Figure 3. Mass spectrum of human milk soluble TCA 12% fraction enriched on hydroxyapatite by MALDI-TOF.

Site	10	17	45
Species			
Bovine A	Gln	Gly	Asn
Bovine B	Arg	Gly	Asn
Water buffalo A	Arg	Asp	Asp
Water buffalo B	Arg	Asp	Asn
Caprine	Gln	Asp	Asn
Ovine	Gln	Asp	Asn

Table 3. Position and amino acid differences among the genetic variant of α-La from four animal species.

5. "Allergenomics"

The application of proteomic methodologies for the analysis of food allergens has been termed "allergenomics" [99]. For "type I" FA, IgE-binding indicates that the target carries the risk to be an allergen. MS-based proteomic methods have identified many proteins and allergens. MALDI [100] and electrospray ionization (ESI) [101] and MS/MS sequencing are the techniques most widely used to produce high-quality spectra of post-translationally modified peptides [102-104] or intact proteins (see [105]). The characterization of glycosylated allergens has been partly overcome by specific enrichment using lectin or hydrophilic resins (HILIC) prior to MS analysis [106,107]. Native and de-glycosylated peptides are analyzed by MALDI or ESI-MS. Because of the difficulty of profiling oligosaccharides released by glycoprotein [108-110], glycan profiles are obtained after permethylation of the oligosaccharide chains according to the procedure of Das et al. [111]. By this means, glycosylated site(s) are identified together with the peptide backbone. Although widespread, several studies are dedicated to milk protein analysis for the detection of allergenic proteins or peptides in dairy products. Thus is determined by the concentration of allergenic compounds in food products that are often secondarily masked by dominant non-allergenic proteins. Among the various methods currently used to detect allergens in food products, immunochemical techniques that rely on antibody-binding properties have been developed. Commercially based kits are used for rapid screening, and enzyme-linked immunosorbent assays (ELISA) provide evaluations. Limits of detection (LOD) attained by ELISA tests are in the range of 1–5 ppm. Because the epitopes to be detected and their possible cross-reactivity with matrix components are unknown, detection reliability strongly depends on various factors that include the thermal changes of whey proteins, which are of primary importance. Furthermore, in several foods, linear epitopes can be released by parent protein hydrolysis, whereas retain their allergenic potential. As cited above, MS measurements can be finalized to evaluate the molecular mass of proteins and derived-peptides (MS1), determine the amino acid sequence and identify post-translational modifications (MS/MS or MS^n). Two-dimensional electrophoresis (2DE) separates proteins according to the pI (i.e., first dimension, isoelectric focusing, IEF) and subsequently molecular weight (SDS-PAGE) in an orthogonal dimension. By this means, the separation of thousands of proteins has been achieved using highly specific stains that visualize specific protein classes, i.e., phosphoproteins, or nonspecific stains that simultaneously target total proteins without particular functional groups. Protein spots are localized, excised from the gel and subjected to an in-gel tryptic digestion. Mass spectrometry analysis either by MALDI reflectron TOF or microcapillary liquid chromatography MS-MS detects the proteins based on the expected masses of peptides available in databases and other plant proteins in pollen diffusates. More directly, tandem mass spectrometry is used to identify the peptide sequence and search for allergens in databases. More elegantly, allergens are localized on the gels after a one- or two-dimensional electrophoretic separation followed by a nitrocellulose transfer of the proteins (i.e., western blotting), which is stained with sera from allergic patients as a source of specific IgE. Combined with the analysis by mass spectrometry of electrophoretically separated allergens, immunoblotting is useful for the rapid determination of allergen identities. Allergen–IgE complexes are also detected using conjugated anti-human IgE as a secondary antibody. Once localized in a 2DE map, the allergen can be monitored using allergen specific antibodies [112,113]. Immune-reactive allergenic protein(s) are identified along the immunoblots by comparison with a reference electrophoretic map. All of the major milk proteins are allergen candidates because sera of allergic patients contain various percentages of immune-reactive proteins that are recognized by IgE [114]. The order of milk protein allergenicity is as follows: α_{s2}-CN>α_{s1}-CN>β-CN>κ-CN>β-Lg> serum albumin > IgG-heavy chain>Lf. This list contains allergenic proteins that have been identified by experiments on several MS platforms [115]. Thus far, it is quite difficult to find a separation method that can accommodate the diversity of proteins equally. Therefore, modern separation techniques have been performed off-line or by online ion-exchange/reversed phase liquid chromatography prior to MS analysis. ESI-MS is currently the interface most frequently used to perform an LC separation of intact proteins. The protein identification is most commonly achieved after a proteolytic digestion and molecular weight determination of the LC-separated peptides. "Shotgun" proteomics is the most effective LC/MS-based strategy because a trypsin-digested protein sample generates thousands of peptides that are subsequently separated by LC prior to MS/MS sequencing [103,116]. Proteins with at least one matching peptide are candidates to occur in the sample. However, with CNs, it is difficult to determine which fractions are present in the sample if they share the same set of phosphopeptides or have only one constituent peptide detected. LC-Q/TOF MS/MS has been used to detect wine CN as a fining agent. Two peptides were identified from α_{s1}-CN and four peptides from the tryptic digestion of α- and β-CNs [117]. A similar strategy could be applied to monitoring allergens in processed milk products. Signature peptides could be identified as CN or whey protein allergens by submitting protein concentrates to trypsinolysis. In this manner, information on the molecular weight of the intact allergen is lost, but cross-reactive immunogenic peptides can be discriminated. Because of the higher sensitivity of MS in the detection of peptides, MS expands the dynamic range of the protein species detected. MS is a method for discovering "hidden" or traces of allergens. Proteomics has become pivotal to the development of modern structural immunology and to the understanding of interacting systems that are involved in immune responses, regardless of FA status.

5.1. Allergen quantification methods

Difference gel electrophoresis (DIGE) is utilized to compare multiple proteins in samples migrating in parallel in the same chamber. The proteins are labeled with three distinct fluorescent dyes on the same 2D gel and differentially visualized via fluorescence at different wavelengths. This methodology enables the detection of a differential presence of proteins and small differences in protein abundance. Allergens can be quantified by LC–MS. An accurate evaluation of the protein/peptide requires a suitable standard. In the direct quantification of intact proteins, the intensity of multi-charged analyte ions is compared with that of an internal or external standard. For example, quantification of cow's milk allergens in fruit juice samples [118] and whey drink [119] was performed by simultaneously monitoring several multiple-charged ions of whey protein components.

With a similar approach, internal standard β-Lg was used to quantify non-bovine β-Lg in different milk-derived products [120]. The use of "bottom-up" methods, such as SILAC, ICAT, and iTRAQ, for quantitative analysis in proteomics has progressively increased [121,122]. Quantification of allergens in complex samples requires simple and precise methods of analysis, such as selected reaction monitoring (SRM) [123]. SRM is presently considered the gold standard for absolute quantification, whereas multiple reaction monitoring (MRM) can monitor the masses of selected signature peptides. For an allergen evaluation, internal reference peptides are required for food product monitoring by LC–SRM MS [124]. Hydrolyzed protein samples are spiked with known amounts of synthetic peptides and monitored by LC–MS in the SRM mode. Absolute amounts of peptide(s) are determined by the ratio of the ion intensities of natural and synthetic peptides (Figure 4).

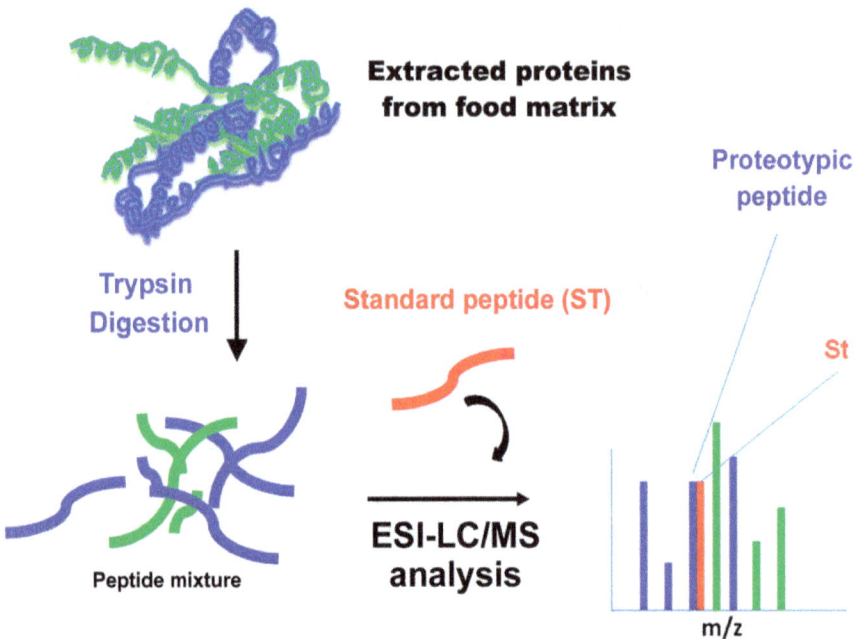

Figure 4. Mass Spectrometry procedure analysis for protein absolute quantification. Proteotypic peptides are selected with a preliminary fullscan. For quantitative analysis whole protein extracts are trypsinized and the peptide mixture is spiked with external standard peptide. Proteotypic peptides (blue colored) selected as analytical probes of the target protein(s), are quantified by comparing the ionic intensities.

This strategy has demonstrated its validity for using signature peptides as analytical surrogates to measure allergens in crude protein extracts. One advantage of the SRM procedure is the possibility of one-step monitoring of a variety of allergens. Recently LC–triple quadrupole MS operating in an MRM mode has been effectively demonstrated to simultaneously detect allergens from seven different potentially allergenic matrices, such as milk, eggs, soy, hazelnuts, peanuts, walnuts and almonds. The detection limits were in the 10-1000 µg/g range. However, prior knowledge of allergens was required to monitor the most suitable allergenic peptides [125]. Based on the above-specified considerations, allergen evaluation requires the following: (i) allergen extraction from the food; (ii) enzyme proteolysis, usually trypsinolysis; and (iii) identification of signature peptides that are characteristic of food proteins or food ingredients. The signature peptides should be determined experimentally by prior LC/MS analysis of food-derived digested protein extracts.

5.2. Standardization of allergen preparations

Although the search for clinically relevant allergens has progressed, the characterization of allergens still requires studies on milk proteins as starting material. Pure native and recombinant allergens are needed as reference materials to calibrate methods among different laboratories. Recently, a panel of 46 food plant and animal allergens [126,127,128] has been made available within an EU-funded research project. In a recent 2DE application, calibration has been utilized for microbial complex protein systems using data obtained by MS [129]. Developing more allergen standards could be realized in the near future. Moreover, the search for allergenic sequence stretches would comprise only those immuno-dominant produced during digestion that can to translocate the gut barrier and reach the mucosal immune system. Among the digestion/adsorption models of food protein stability, pepsin digestion has been included in the Food and Agriculture Organization/World Health Organization to assess food safety [130]. A model study has established milk-derived peptide candidate-mediated resistance to proteases to display allergic effects. The survival of milk protein epitopes [131,114] requires structure determination. To this end, a simulated digestion of bovine milk proteins in vitro that includes the sequential use of pepsin, pancreatic proteases, and extracts of human intestinal brush border membranes, has allowed the identification of produced peptides by MS. The presence of characteristic β-Lg resistant peptides could implicate β-Lg in the case of a cow's milk allergy [132]. The identification by MS of peptides arising from simulated digestion is complicated by a lack of enzyme specificity. Currently, there is no treatment to fully resolve or provide long-term remission from FA allergies. The research for therapy is mainly focused on the introduction of anti-IgE antibodies and specific oral tolerance induction. Immunotherapy appears to be an attractive approach; however, the risk of anaphylaxis should be considered. To this purpose, engineered proteins have been designed, i.e., anaphylaxis-initiating epitopes have been removed within these proteins, while preserving the tolerance-inducing epitopes [133-135]. It appears clear that to successfully pursue similar strategies, the precise identification of epitopes is necessary. It is expected that such approaches will be extended to an increasing number of food sources, whereby MS will play a key role for characterizing novel protein entities. The accurate characterization of the offending sequences could also be the starting point for developing less allergenic food products through the use of enzymatic, microbiological and technological processes to effectively remove allergens [136–139].

6. Dairy research versus CMA

6.1. Milk and dairy products from mammals different from cow

According to the current clinical approach to FA and intolerance based on an elimination diet, the treatment of choice is complete avoidance of cow milk. Although of moderate importance in an adult diet, cow milk elimination has a significant nutritional significance in the infant diet, especially during early childhood. Milk from other mammals has been suggested as a possible alternative to cow milk. At first, goat milk had been proposed as a hypoallergenic infant food or cow milk substitute in human diet, but much of this thesi has no credible scientific evidence. Despite the immunological cross-reactivity between cow and goat milk proteins, due to the close biochemical similarity associated with the same phylogenetic origin [140], it has been estimated that from 40 to 100% of patients allergic to cow milk proteins can tolerate goat milk intake [141]. However, clinical and immunochemical studies aimed at evaluating goat milk safety for cow milk allergic subjects have demonstrated that goat milk cannot be a substitute for cow milk without risk of anaphylactic reactions [142,143]. It has been suggested that evidence for goat milk tolerance in clinical trials can be due, at least in part, to a higher number of genetic polymorphisms in goat CNs, especially for α_{s1}-CN [144]. Null or reduced expression of α_{s1}-CN in individual goats; consequently, the overall α_{s1}-CN content in goat bulk milk is lower than that found in cow bulk milk. According to this general finding, and taking into account that little β-Lg persists in cheese, fresh cheese produced from raw milk has been suggested to be a promising hypoallergenic protein source [145]. Unexpectedly, water buffalo milk yogurt has successfully been employed as an alternative food for children with cow milk allergies [146] despite homologous proteins from cow and water buffalo milk [147]. In contrast, several studies have reported the existence of allergies to goat and sheep milk [148-151] and cheese [152,153] in patients with tolerance to cow milk proteins. Overall, this type of allergy is less common and occurs later than that initiated by cow milk proteins, which is likely because goat and sheep dairy products are not usually included in an infant diet. Moreover, IgE epitopes have been widely recognized in the CN components of goat and sheep milk. Differences in the degree of CN phosphorylation, on average lower in goat and sheep milk than in cow milk, rather than differences in IgE epitope sequences, may be involved in initiating selective allergies to goat and sheep milk, as observed in recent cases. In addition to the four ruminant species of dairy interest (i.e., cow, water buffalo, sheep and goat), other monogastric mammals produce milk for human consumption, such as mares and donkeys. Mare milk, which is more similar to human than cow milk, has been proven to be an acceptable substitute of cow milk for children with severe IgE-mediated cow milk allergy; although the evidence of its tolerability by a supervised oral challenge test is recommended [154]. However, mare milk availability is limited, and its collection is difficult. Donkey milk provides nutritional adequacy and excellent palatability similar to that of mare milk but is more readily available. The composition of donkey milk is more similar to human milk than cow milk because of the higher lactose content (6.5 vs. 5 g/100 mL), lower protein content (1.2 vs. 3.2 g/100 mL), lower CN/whey protein ratio (approximately 1 vs. 4) and a higher non-protein nitrogen fraction level (0.29 vs. 0.18%) [155]. These features have prompted clinicians to propose donkey milk as a valuable breast milk substitute. Additionally, donkey milk intake has demonstrated positive effects in the diet therapy of patients allergic to cow milk proteins [156]. Although the mechanism of this tolerance is unclear, the reduced allergenic properties of donkey milk can be related to the structural differences compared with bovine milk. Because of scientific and clinical interest in donkey milk, characterization of the whey protein fraction [157], caseome [158], and the minor protein components [159] of donkey milk have been recently provided. Presently, milk from mammals with a geographically restricted distribution area, such as reindeer living in Northern Europe, has been utilized to overcome immunological cross-reactivity among proteins from mammals other than cows. In particular, β-Lg from reindeer milk, although belonging to the lipocalin family and similar to its homologous bovine protein, lacks several IgE epitopes of bovine β-Lg that are involved in CMA [160]. Recently, camel milk, mainly available in the Gulf area and Mauritania, is of growing interest to both nutritionists and pediatricians because of its high nutritive value and unique electrophoretic protein patterns, which strongly suggest a different immunological reactivity of camel milk proteins with respect to the bovine counterparts [161].

6.2. Gut microflora

It has been suggested that gut flora may be involved in the etiology of atopic diseases. It has been demonstrated that the gut microflora differs in children with high or low rates of allergy. Commensal gut flora play a role in inducing an oral tolerance, and the importance of the intestinal microbiota in developing food allergies is essential at early ages when the mucosal barrier and immune system are still immature. Probiotics interact with the mucosal immune system by the same pathways as commensal bacteria. A recent study has demonstrated that probiotic bacteria induced in vivo increased plasma levels of IL-10 and total IgA in children with allergic predisposition. Many clinical studies have reported significant benefits by probiotics supplementation in FA prevention and management. However, not everyone agrees on the effectiveness of probiotics supplementation. The differences are most likely related to the selected populations and probiotic strains used. The hygiene hypothesis proposes that disturbances in the gastrointestinal microbiota are associated with increased prevalence of allergic and autoimmune diseases [162]. Changes in the establishment of gut microbiota have been observed in Western infants [163,164]. This is most likely because of improved hygiene and cleanliness in Western countries and excessive use of antibiotics, which causes a reduced bacterial stimulus. Several clinical studies have reported differences in the composition of bacterial communities in the feces of children with and without allergic diseases. Many of those studies have highlighted the involvement of *Bifidobacterium* and *Bacteroides* in the protection against the development of atopy [165-168], but this observation remains a matter of debate [162]. Moreover, the mechanisms underlying such protective effects remain elusive. There is increasing evidence that T-regulatory cells derived from the thymus or induced in the periphery including the gut mucosa [169,170] are key players of immune regulation [171-173]. Using a single strain mouse model and defined bacterial communities and conventional mice, it has been recently demonstrated that the gut microbiota plays a protective role against allergen sensitization and allergic response in a mouse model of FA [174]. The difference between healthy and allergic children may be in their microflora. At 3 weeks of age, infants in whom atopy developed then had more Clostridia and fewer *bifidobacteria* in their feces compared with infants who remained healthy. Moreover, fecal *bifidobacteria* microflorae were different between healthy and allergic children; the healthy infants' microflora was mainly *Bifidobacterium bifidum*, whereas the microflora was mainly *Bifidobacteria adolescentis* in the allergic infants. It can be hypothesized that individual species, rather than an entire genus, can affect the manifestation of allergy. In a recent study, the microflorae of milk-hypersensitive and control adults before and after a 4-week supplementation with probiotic bacteria (*Lactobacillus rhamnosus* GG, ATCC 53103) have been studied. The anaerobic microflora before supplementation was comparable between the healthy and hypersensitive subjects, whereas the response after supplementation was different. The number of *bifidobacteria* in the healthy subjects increased significantly after supplementation. However, this did not occur with the supplementation in milk-hypersensitive subjects; this may be because of altered intestinal integrity. However, other studies have suggested a beneficial effect of probiotic bacteria in milk-hypersensitive subjects. In one study, symptoms of hypersensitivity abated along with an elimination diet in 28% of the patients in 4 years. It can be hypothesized that with milk elimination and long-lasting probiotic treatment, the intestinal severity of IgE-mediated hypersensitivity reactions may increase the intestinal microflora or even eliminate them.

6.3. Reduction of allergenicity of milk proteins by hydrolysis

The main objective of the milk industry is to supply products while preserving both the nutritive value and safety against developing allergies. Nutritional value is preserved by exposing liquid or powdered milk to low heat treatments to reduce heat susceptible amino acid side chain modifications and preserve the integrity of triacylglycerols, native vitamins and other milk components. As noted above, infants can develop milk allergies because of increased gut permeability to large molecules, in addition to other causes [175]. This result is supported by measurements of unmodified proteins or partially modified proteins in the sera of infants and adults [176]. Milk proteins have a molecular mass between 14 and 80 kDa. To reduce allergenicity, milk proteins can be submitted to different hydrolysis procedures. Attempts to classify products by protein hydrolysis include "extensive" or "high degree" hydrolysis and "partial" or "low degree" hydrolysis. The rationale of such a classification is the spectrum of peptide molecular weights or the ratio of α amino acids to total nitrogen. For quality assurance, *in vitro* product characterization requires size measurements of the peptides that are generated by protein hydrolysis and then an in vivo allergenicity determination. The in vivo step would include evaluating immunogenic or allergenic effects in a recipient infant. The European Union regulates that infant formulas contain immunoreactive proteins in quantities lower than 1% of nitrogen compounds [177] to reduce allergenicity knowing that only pure amino acids are strictly non-allergenic. This criterion could be encountered by milk proteins that have undergone extensive hydrolysis partially to cleave amino acids [178-180]. In contrast, formulas with moderately reduced allergenicity (partially hydrolyzed) are not recommended for the treatment of allergies because of the high amounts of residual allergens [181]. The low quantity of native proteins or residual high molecular mass peptides may produce adverse effects in highly sensitive patients. Therefore, a milk hydrolysate can be considered safe and non-allergenic if the nitrogen fraction does not contain unmodified milk proteins or high molecular mass peptides [178-180]. In the latter case, the product could be classified in the "low degree" protein hydrolysate category. However, the antigenic properties of protein hydrolysates may not be dependent on the molecular size of the peptide components alone [182]. By comparing protein structures with known allergens and allergen epitopes, protein allergenicity has been predicted [183]. Although this is true for crystallized proteins, such as α-La, β-Lg and Lf, this procedure cannot be applied to uncrystallized CNs. Because infants who are diagnosed for milk protein allergies must ingest foods that exclude the causal protein, including those ingested by the mother and filtered in breast milk, extensively hydrolyzed milk formulas are used for the development of appropriate dietary and management strategies. Preclinical testing of infant formulas is necessary to characterize the molecular properties and residual antigenicity of proteins [184-186]. Stringent criteria specify that extensively hydrolyzed CN with a molecular weight below 5000 Da should be reduced by at least 99.99%. There is a need for accurate diagnostic methods to confirm the amount of extensively hydrolyzed CN. Milk for allergic infants would consist of extensively hydrolyzed CN and whey proteins of which at least 99.99% of the hydrolysis products have molecular weights below 5000 Da. The crucial criterion is for the level of allergens to be sufficiently low as to cause no significant reaction, even in infants who are highly allergic to cow's milk. There have been no reports of adverse reactions because of whey [187, 188] and CN hydrolysates [189-191]. Therefore, caution must be maintained that milk formulas destined to infants with milk allergies contain correctly hydrolyzed proteins. This generic indication requires that molecular properties and residual antigenicity of proteins would be characterized [184-186]. In vitro incubation of milk proteins with pepsin, trypsin, and chymotrypsin causes the cleavage of numerous peptides of various sizes. Bacterial, fungal and plant proteases may also act as hydrolyzing agents. Various enzyme combinations, such as alcalase, pancreatin and enzymes from fungal sources, have been utilized to produce protein hydrolysates. Commercial hypoallergenic products are currently characterized by an average degree of hydrolysis (DH) of the protein components. The DH 19 milk protein value is calculated from the increase of the number of primary amino groups compared to that of native proteins. In practice, the DH value could vary from 1 to 100% in the case of total hydrolysis of the proteins. In the case of partly hydrolyzed proteins, intact and partly hydrolyzed proteins are visible bands along an electrophoretic pattern of the products. Two commercial formula preparations with DH values of 6.3 and 1.3% contained some intact β-Lg and peptides with an Mr between 6000 and 8000 Da [192]. Using gel permeation chromatography, quantitative results on peptides with an Mr larger than 10 kDa were obtained [192]. Regardless of the technique used, descriptive information was obtained on either the molecular mass or the origin of the peptides. As a result, consumption of infant formulas by allergic patients cannot be attributed to one specific protein or high molecular mass peptide. To suppress or reduce the antigenicity of peptides, natural enzyme cleaving of many or most of the peptide bonds is required. In this manner, epitopes that determine the antigenicity of the protein molecules are destroyed. This result proves that evaluation of the adequacy of infant formula composition in preventing or delaying antigenicity is not based solely on DH or Mr determination. In highly sensitized infants with IgE-mediated cow's milk allergies, life-threatening anaphylactic shock usually develops shortly after the consumption of claimed hypoallergenic milk products in which a number of epitopes would have survived in the highly proteolytic environment. The possibility of using well-characterized monoclonal antibodies in ELISA tests can be used for assessing the origin of immunoreactive bovine milk proteins. Because the clinical significance of residual antigenicity requires prior molecular approaches, hypoallergenic products may first be screened for peptide identification in hydrolyzed milk products. Although protein hydrolysates can provide a positive effect, they can contain undefined peptide components, which are undesirable for pharmaceutical production purposes. In many cases, hydrolysates are produced by methods that are not well-controlled. Other complications arise from the raw starting material and differences in processing that lead to lot-to-lot hydrolysate composition variability. For these reasons, constant chemically defined products are needed. The data presented here represent the initial steps that have been taken to identify peptides treated with pepsin (P) and trypsin (T) and were used in succession to hydrolyze commercial milk protein powder (PT hydrolysate). To mimic commercial milk hydrolysates, the protein powder was treated with enzymes after a thermal shock deactivation treatment. Subsequently, the peptides are identified. RP-HPLC fractionation was used to aid with the peptide separation. In Figure 5, the hydrolysate was analyzed to demonstrate a correlation between proteolytic enzymes and the presence of peptides. Commercial milk protein hydrolysates may contain trace amounts of allergenic proteins whose molecular weights were determined by MALDI-TOF analysis. Among the number of peptides present in the hydrolyzed sample, the proteins/peptides exhibited molecular masses less than or equal to 2431 Da (Figure 6). This means that the CNs and whey proteins were digested by pepsin and trypsin into peptides with masses less than 3000 Da. This type of hydrolysate is not expected to elicit allergic reactions in already sensitized allergic patients (neither anaphylactic shock nor positive passive cutaneous anaphylaxis), as verified in experimental animals [193]. LC-ESI-MS/MS analysis was performed on the hydrolysate to identify peptides occurring herein. No sequence peptides with 3 or 4 residues were detected because the MALDI signals were acquired at a mass gate of m/z 400. Because some short peptides were in the hydrolysate, milk proteins were hydrolyzed by P and T into oligopeptides with different biological activities. In Figure 7, the amino acid sequence of the four bovine CN fractions and β-Lg are reported with a subscript that indicates the number of amino acid residues in MS/MS-identified fragment.

The proteins in the milk powder sample, which contained modified amino acid residues that may indicate the quality of the protein in milk powder, were not examined within the present work.

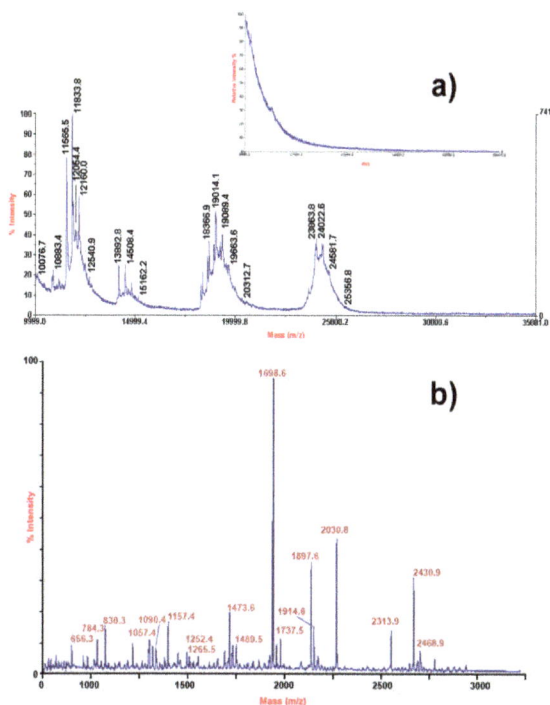

Figure 5. MALDI spectra of a sample of milk powder before and after sequential hydrolysis with pepsin and trypsin. A search for residual intact proteins and high molecular derived peptides (a) and measurements of molecular mass value of oligopeptides in the mixture (b). No peptide at a molecular mass higher than 3430.97 Da was observed in the MALDI spectrum.

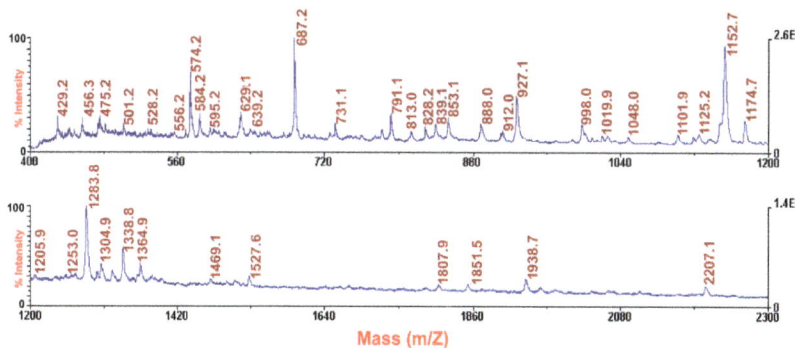

Figure 6. MALDI analysis of a milk sample after sequential hydrolysis with pepsin and trypsin. A molecular mass value corresponds to that of oligopeptides in mixture. No peptide at a molecular mass higher than 2207.1 Da was observed in the MALDI spectrum.

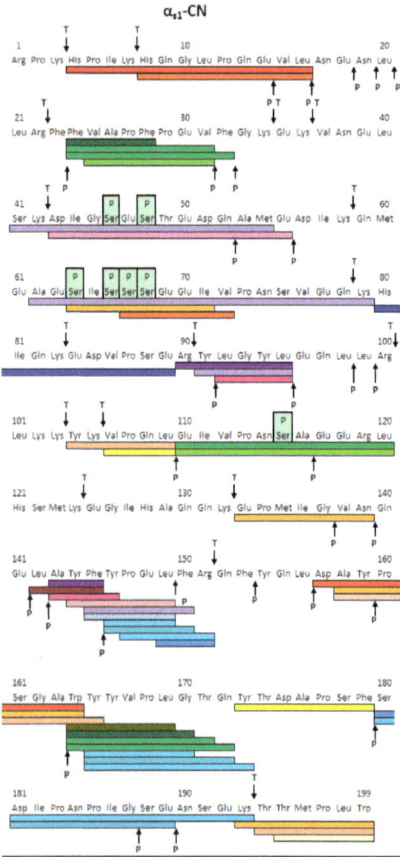

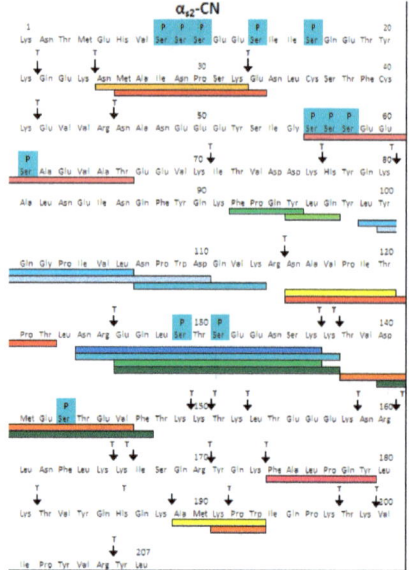

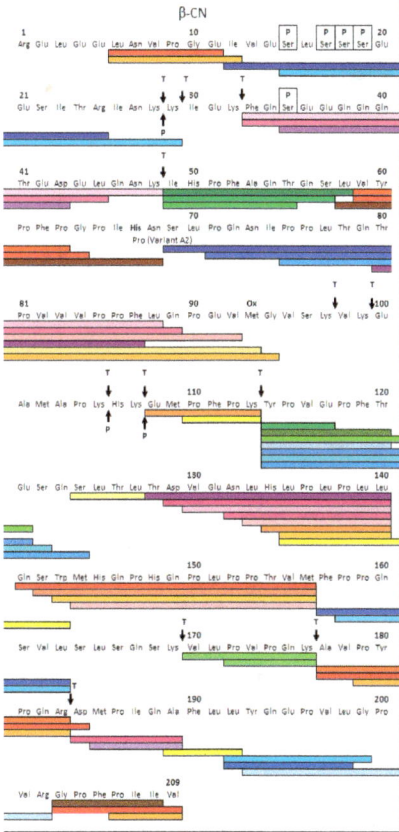

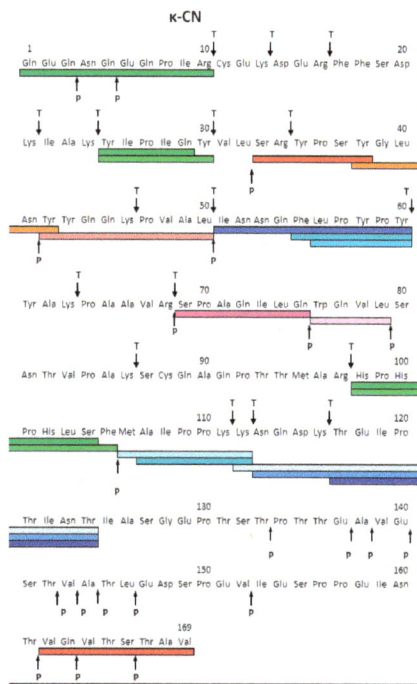

β-Lg

```
                          T              T
  1                       ↓        10    ↓                        20
 Leu Ile Val Thr Gln Thr Met Lys Gly Leu Asp Ile Gln Lys Val Ala Gly Thr Trp Tyr
                                    30                            40 T
 Ser Leu Ala Met Ala Ala Ser Asp Ile Ser Leu Leu Asp Ala Gln Ser Ala Pro Leu Arg ↓

                     T
                     ↓              50                            60 T
 Val Tyr Val Glu Glu Leu Lys Pro Thr Pro Glu Gly Asp Leu Glu Ile Leu Leu Gln Lys ↓

                                  T   T               T       T
                                  ↓70 ↓    70         ↓       ↓        80
 Trp Glu Asn Asp Glu Cys Ala Gln Lys Lys Ile Ile Ala Glu Lys Thr Lys Ile Pro Ala
         T                              T
         ↓                              ↓  90                      100 T
 Val Phe Lys Ile Asp Ala Leu Asn Glu Asn Lys Val Leu Val Leu Asp Thr Asp Tyr Lys ↓

     T
     ↓                    110                         120
 Lys Tyr Leu Leu Phe Cys Met Glu Asn Ser Ala Glu Pro Glu Gln Ser Leu Val Cys Gln

             T                            T           T
             ↓            130             ↓           ↓   140
 Cys Leu Val Arg Thr Pro Glu Val Asp Asp Glu Ala Leu Glu Lys Phe Asp Lys Ala Leu

     T                        T
     ↓                        ↓   150                     160
 Lys Ala Leu Pro Met His Ile Arg Leu Ser Phe Asn Pro Thr Gln Leu Glu Glu Gln Cys

     162
 His Ile
```

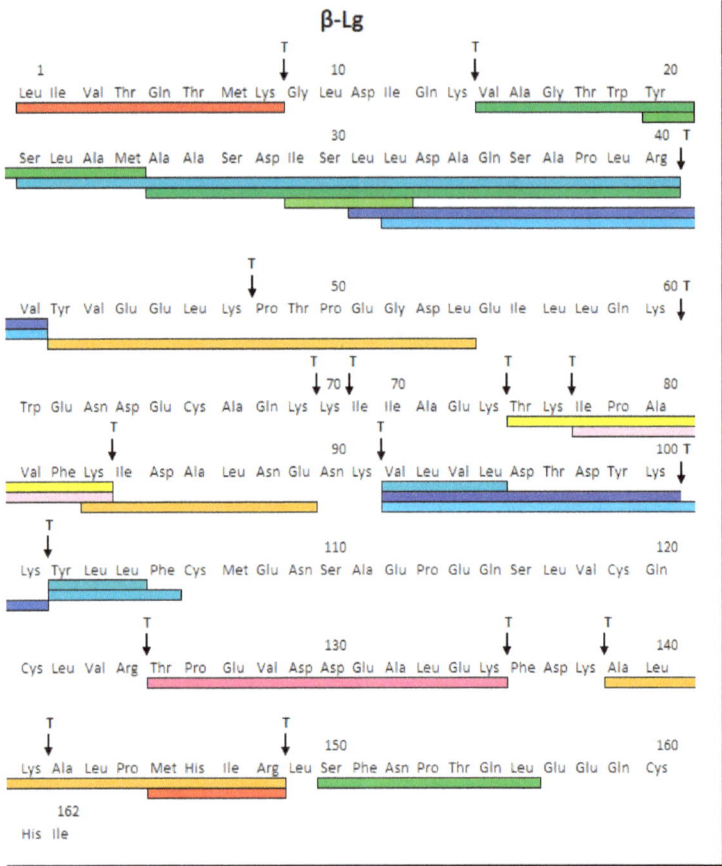

Figure 7. Primary structure of bovine β-, α_{s1}-, α_{s2}-, κ-CN and β-Lg together with identified peptides in commercial milk protein powder sample hydrolysed by sequential action of pepsin and trypsin. Protein/enzyme ratio 100/1 (w/w). Incubation with each enzyme was carried out for 16h at appropriate pH. Ends of horizontal bars indicate first and last amino acid residue of peptides isolated within this work. Arrows indicate casein peptide bonds broken by pepsin (P) and trypsin (T) as reported by Pelissier [194].

7. Oral immunotherapy as future perspective in CMA management

The primary treatment for managing food allergies is eliminating the offending food from the diet. In the case of milk, it is extremely difficult to achieve complete elimination because milk can be masked in any number of foods, which may lead to unwanted severe reactions. The natural course of a cow's milk protein allergy is the acquisition of tolerance spontaneously through an elimination diet, and 85% of patients overcome CMA by the time they are 4-5 years old [195-198]. In recent years, a number of studies have been published regarding desensitization or oral tolerance to food antigens, particularly to cow's milk. Major advances in understanding the immunological processes involved in the development of CMA have revealed a considerable number of allergenic epitopes and the heterogeneity of allergic responses. Importantly, an elimination diet of dairy foods has negative consequences in terms of inadequate calcium and vitamin intake. In the literature, several conflicting studies have reported on possible desensitizing therapies in the treatment of FA allergies. The possibility to obtain an oral desensitization is now gaining acceptance widely, even if the mechanism is still unclear. Oral tolerance or desensitization is the active non-response of the immune system to an antigen through sublingual, oral administration. Tolerance or the long-term loss of allergic reactivity follows a desensitization treatment. In the literature, several studies have reported on possible physiopathogenetic mechanisms of oral desensitization, but the exact mechanism is still unknown. That tolerance may be involved in the mechanism of desensitization is still uncertain [199-201]. A growing understanding of the molecular and cellular mechanisms of oral tolerance is reinforcing advances in potential therapies for food allergies and is pivotal to eventually curing allergies in sensitized individuals. Oral desensitization should be taken into consideration in the management of food-allergic patients even if the physiopathogenetic mechanisms are still unexplained. Moreover, this treatment should be considered, particularly for children, because elimination from the diet of some foods (e.g., milk and eggs) for these patients could cause psychological and/or nutritional problems. Oral immunotherapy (oral desensitization) may be a promising treatment strategy for cow's milk allergy in children and valuable for other foods, such as eggs or peanuts. Although the mechanisms of IgE-mediated allergies are fairly well understood, the immunology and variety of non-IgE-mediated reactions remains largely unknown. A better understanding of these allergy mechanisms is a prerequisite to the development of improved diagnostics, which in turn will facilitate an improved understanding of the epidemiology of CMA, particularly for non-IgE-mediated reactions.

8. Conclusions

A better understanding will also aid the development of hypoallergenic dairy products, especially for adults with CMA for whom there is currently a dearth of suitable low-allergenic dairy products. Some of the risk factors for developing CMA have been identified; a familiar history of atopy is one of the main determinants. However, the mechanisms of allergic sensitization and the precise interactions between genetics and various environmental factors that lead to CMA remain unclear. The first few months of life, during which the immune system is still maturing, appear to be a critical risk period for allergic sensitization. For at-risk infants with at least one atopic parent, breastfeeding during this period is currently the best identified preventative strategy; the use of hydrolyzed formulas is recommended for babies who cannot be breastfed. The use of immunomodulatory dietary adjuvants, such as probiotics, is an emerging approach with considerable promise for primary prevention. For CMA sufferers, the avoidance of dietary milk proteins remains the only effective management strategy but carries with it nutritional implications, particularly for adequate vitamin and calcium intake as well as protein and energy where unorthodox alternative diets are implemented. Increasing knowledge of the molecular and cellular mechanisms of oral tolerance reinforces the advances in potential FA therapies and is pivotal to eventually curing allergies in sensitized individuals. Unraveling the links between innate and adaptive immunity and characterizing the roles of dendritic cells and T cells in directing immune responses and homeostasis to environmental antigens are likely to remain a focus of fundamental FA research in the coming years.

Author details

Simonetta Caira, Rosa Pizzano, Gianluca Picariello
Food Science Institute of the National Research Council (C.N.R.),Avellino, Italy

Gabriella Pinto, Marina Cuollo, Lina Chianese and Francesco Addeo
Department of Food Science, University of Naples "Federico II", Portici (Naples), Italy

Acknowledgement

The Authors gratefully acknowledge the American Journal Experts Association for the text revision (http://www.journalexperts.com/). This work was partly supported by the financial aid to C.L. from MIUR, Program PRIN-2008 HNHAT7-004.

9. References

[1] Gell, PGH, RRA Coombs. The classification of allergic reactions underlying disease. In: Coombs, R.R.A., Gells, P.G.H. (Eds.) Clinical Aspects of Immunology. Blackwell, Oxford 1963.

[2] Johansson SGO, Bieber T, Dahl R, Friedmann PS, Lanier BQ, Lockey RF, Motala C, Ortega Martell JA, TAE Platts-Mills, Ring J, Thien F, Van Cauwenberge P, Williams HC. Revised nomenclature for allergy for global use: Report of the Nomenclature Review Committee of the World Allergy Organization, October 2003. The Journal of Allergy and Clinical Immunology 2004;113(5) 832-836.

[3] Montalto M, Curigliano V, Santoro L, Vastola M, Cammarota G, Manna R, Gasbarrini A, Gasbarrini G. Management and treatment of lactose malabsorption. World Journal of Gastroenterology 2006;12(2) 187-191.

[4] Wang Y, Harvey CB, Hollox EJ, Phillips AD, Poulter M, Clay P, Walker-Smith JA, Swallow DM. The genetically programmed down-regulation of lactase in children. Gastroenterology 1998;114(6) 1230-1236.

[5] Labayen I, Forga L, González A, Lenoir-Wijnkoop I, Nutr R, Martínez JA. Relationship between lactose digestion, gastrointestinal transit time and symptoms in lactose malabsorbers after dairy consumption. Alimentary Pharmacology & Therapeutics 2001;15(4) 543-549.

[6] Shaw AD, Davies GJ. Lactose intolerance: problems in diagnosis and treatment. Journal of Clinical Gastroenterology 1999;28(3) 208-216.

[7]   Rosado JL, Solomons NW, Lisker R, Bourges H. Enzyme replacement therapy for primary adult lactase deficiency. Effective reduction of lactose malabsorption and milk intolerance by direct addition of beta-galactosidase to milk at mealtime. Gastroenterology 1984;87(5) 1072-1082.

[8]   Swagerty DL Jr, Walling AD, Klein RM. Lactose intolerance. American Family Physician 2002;65(9) 1845-1850.

[9]   Crittenden RG, Bennett LE. Cow's milk allergy: a complex disorder. Journal of the American College of Nutrition 2005;24(6 Suppl) 582S-91S.

[10] Hill DJ, Hosking CS, Zhie CY, Leung R, Baratwidjaja K, Iikura Y, Iyngkaran N, Gonzalez-Andaya A, Wah LB, Hsieh KH: The frequency of food allergy in Australia and Asia. Environmental Toxicology and Pharmacology 1997;4(1-2) 101-110.

[11] Exl BM, Fritsché R. Cow's milk protein allergy and possible means for its prevention. Nutrition 2001;17(7-8) 642-51.

[12] Garcia-Ara MC, Boyano-Martinez MT, Diaz-Pena JM, Martin-Munoz MF, Martin-Esteban M. Cow's milk-specific immuno-globulin E levels as predictors of clinical reactivity in the follow-up of the cow's milk allergy infants. Clinical & Experimental Allergy 2004;34(6) 866-870.

[13] Woods RK, Thien F, Raven J, Walters EH, Abramson MA. Prevalence of food allergies in young adults and their relationship to asthma, nasal allergies, and eczema. Annals of Allergy, Asthma & Immunology 2002;88(2) 183-189.

[14] Bindels JG, Hoijer M. Allergens: latest developments, newest techniques. Bulletin of the International Dairy Federation 2000; 351 31-32.

[15] Høst A. Dietary products used in infants for treatment and prevention of food allergy. Joint Statement of the European Society for Paediatric Allergology and Clinical Immunology (ESPACI) Committee on HypoallergenicFormulas and the European Society for Paediatric Gastroenterology, Hepatology and Nutrition (ESPGHAN) Committee on Nutrition. Archives of Disease in Childhood 1999;81(1)80-84.

[16] Distler JW. Food allergy Children. An update on diagnosis and treatment. ADVANCE for NPs & Pas 2010;1(2) 26-30.

[17] Jansen JJ, Kardinaal AF, Huijbers G, Vlieg-Boerstra BJ, Martens BP, Ockhuizen T. Prevalence of food allergy and intolerance in the adult Dutch population. Journal of Allergy and Clinical Immunology 1994;93(2) 446-456.

[18] Lam H-Y, Van Hoffen E, Michelsen A, Guikers K, Van Der Tas CHW, Bruijnzeel-Koomen CAFM, Knulst AC. Cow's milk allergy in adults is rare but severe: both casein and whey proteins are involved. Clinical & Experimental Allergy 2008;38(6) 995-1002.

[19] Stöger P, Wuthrich B. Type I allergy to cow milk proteins in adults. A retrospective study of 34 adult milk and cheese allergic patients. International Archives of Allergy and Immunology 1993;102(4) 399-407.

[20] Rautava S, Ruuskanen O, Ouwehand A, Salminen S, Isolauri E. The hygiene hypothesis of atopic disease-an extended version. Journal of Pediatric Gastroenterology and Nutrition 2004;38(4) 378-88.

[21] Strannegård O, Strannegård IL. The causes of the increasing prevalence of allergy: is atopy a microbial deprivation disorder? Allergy 2001;56(2) 91-102.

[22] Bito T, Kanda E, Tanaka M, Fukunaga A, Horikawa T, Nishigori C. Cows milk-dependent exercise-induced anaphylaxis under the condition of a premenstrual or ovulatory phase following skin sensitization. Allergology International 2008;57(4) 437-439.

[23] Savilahti E, Kuitunen M. Allergenicity of cow milk proteins. Journal of Pediatrics 1992;121(5 Pt 2) S12-20.

[24] International Union of Immunological Societies Allergen Nomenclature Sub-Committee. Allergen Nomenclature. Retrieved from
http://www.allergen.org/Allergen.aspx. Accessed 2009.

[25] Wal J-M. Cow's milk proteins/allergens. Annals of Allergy, Asthma & Immunology 2002;89(Suppl 9) 3-10.

[26] Restani P, Ballabio C, Di Lorenzo C, Tripodi S, Fiocchi A. Molecular aspects of milk allergens and their role in clinical events. Analytical and Bioanalytical Chemistry 2009;395(1) 47-56.

[27] Plebani A, Restani P, Naselli A, Galli CL, Meini A, Cavagni G, Ugazio AG, Poiesi C. Monoclonal and polyclonal antibodies against casein components of cow milk for evaluation of residual antigenic activity in 'hypoallergenic' infant formulas. Clinical & Experimental Allergy 1997;27(8) 949-956.

[28] Poms RE, Klein CL, Anklam E. Methods for allergen analysis in food: a review. Food Additives and Contaminants 2004;21(1) 1-31.

[29] Rozenfeld P, Docena GH, Anon MC,Fossati CA. Detection and identification of a soy protein component that cross-reacts with caseins from cow's milk. Clinical & Experimental Immunology 2002;130(1) 49-58.

[30] Loo JA, Edmonds CG, Smith RD. Primary sequence information from intact proteins by electrospray ionization tandem mass spectrometry. Science 1990; 248(4952) 201-204.

[31] Senko MW, Speir JP, McLafferty FW. Collisional activation of large multiply charged ions using Fourier transform mass spectrometry. Analytical Chemistry 1994;66(18) 2801-2808.

[32] Little DP, Speir JP, Senko MW, O'Connor PB, McLafferty FW. Infrared multiphoton dissociation of large multiply charged ions for biomolecule sequencing. Analytical Chemistry 1994;66(18) 2809-2815.

[33] Oh H, Breuker K, Sze SK, Ge Y, Carpenter BK, McLafferty FW. Secondary and tertiary structures of gaseous protein ions characterized by electron capture dissociation mass spectrometry and photofragment spectroscopy. Proceedings of the National Academy of Sciences 2002;99(25) 15863-15868.

[34] Yates, JR, III. Database searching using mass spectrometry data. Electrophoresis 1998; 19(6) 893-900.

[35] Wysocki, VH, Resing KA, Zhang Q, Cheng, G. Mass spectrometry of peptides and proteins. Methods 2005;35(3) 211-222.

[36] Spuergin P, Mueller H, Walter M, Schiltz E, and Forster J. Allergenic epitopes of bovine αs1-casein recognized by human IgE ad IgG. Allergy 1996;51(5) 306-312.

[37] Chatchatee P, Jarvinen KM, Bardina L, Beyer K, and Sampson HA. Identification of IgE- and IgG-binding epitopes on αs1-casein: differences in patients with persistent and

transient cow's milk allergy. Journal of Allergy and Clinical Immunology 2001;107(2) 379-383.

[38] Ruiter B, Tregoat V, M'Rabet L, Garssen J, Bruijnzeel-Koomen CA, Knol EF, and Hoffen E. Characterization of T cell epitopes in αs1-casein in cow's milk allergic, atopic and non-atopic children. Clinical & Experimental Allergy 2006;36(3) 303-310.

[39] Ruiter B, Rozemuller EH, van Dijk AJ, Garssen J, Bruijnzeel-Koomen CA, Tilanus MG, Knol EF, and van Hoffen E.Role of human leucocyte antigen DQ in the presentation of T cell epitopes in the major cow's milk allergen αs1-casein. International Archives of Allergy and Immunology 2007;143(2) 119-126.

[40] Bernard H, Creminon C, Yvon M, Wal JM. Specificity of the human IgE response to the different purified caseins in allergy to cow's milk proteins. International Archives of Allergy and Immunology 1998;115(1) 235-244.

[41] Walstra P. Casein sub-micelles: do they exist? International Dairy Journal 1999; 9(3-6) 189-192.

[42] Farrell HM Jr, Jimenez-Flores R, Bleck GT, Brown EM, Butler JE, Creamer LK, Hicks CL, Hollar CM, Ng-Kwai-Hang KF, Swaisgood HE. Nomenclature of the proteins of cows' milk-sixth revision. Journal of Dairy Science 2004;87(6) 1641-1674.

[43] Horne, D.S.Casein structure, self-assembly and gelation. Current Opinion in Colloid & Interface Science 2002; 7(5-6) 456-461.

[44] Kohno Y, Honma K, Saito K, Shimojo N, Tsunoo H, Kaminogawa S, and Niimi H. Preferential recognition of primary protein structures of alpha-casein by IgG and IgE antibodies of patients with milk allergy. Annales of Allergy 1994; 73(5) 419-422.

[45] Monaci L, Tregoat V,van Hengel AJ, Anklam E. Milk allergens, their characteristics and their detection in food: A review. European Food Research And Technology 2006; 223(2) 149-179.

[46] Spuergin P, Walter M, Schiltz E, Deichmann K, Forster J, Mueller H. Allergenicity of alpha-caseins from cow, sheep, and goat. Allergy 1997;52(3) 293-298.

[47] Vila L, Beyer K, Järvinen KM, Chatchatee P, Bardina L, Sampson HA. Role of conformational and linear epitopes in the achievement of tolerance in cow's milk allergy. Clinical & Experimental Allergy 2001;31(10) 1599-1606.

[48] Elsayed S, Hill DJ, Do TV. Evaluation of the allergenicity and antigenicity of bovine-milk alphas1-casein using extensively purified synthetic peptides. Scandinavian Journal of Immunology 2004;60(5) 486-93.

[49] Schulmeister U, Hochwallner H, Swoboda I, Focke-Tejkl M, Geller B, Nystrand M, Härlin A, Thalhamer J, Scheiblhofer S, Keller W, Niggemann B, Quirce S, Ebner C, Mari A, Pauli G, Herz U, Valenta R, Spitzauer S. Cloning, expression, and mapping of allergenic determinants of alphaS1-casein, a major cow's milk allergen. The Journal of Immunology 2009;182(11) 7019-7029.

[50] Busse PJ, Järvinen KM, Vila L, Beyer K, Sampson HA. Identification of sequential IgE-binding epitopes on bovine alpha(s2)-casein in cow's milk allergic patients. International Archives of Allergy and Immunology 2002;129(1) 93-96.

[51] Järvinen KM, Beyer K, Vila L, Chatchatee P, Busse PJ, Sampson HA. B-cell epitopes as a screening instrument for persistent cow's milk allergy. Journal of Allergy and Clinical Immunology. 2002;110(2) 293-297.

[52] Chatchatee P, Järvinen KM, Bardina L, Vila L, Beyer K, Sampson HA. Identification of IgE and IgG binding epitopes on beta- and kappa-casein in cow's milk allergic patients. Clinical & Experimental Allergy 2001;31(8) 1256-1262.

[53] Rasmussen LK, Højrup P, Petersen TE.The multimeric structure and disulfide-bonding pattern of bovine kappa-casein. European Journal of Biochemistry 1992;207(1) 215-222.

[54] Rasmussen LK, Højrup P, Petersen TE. Disulphide arrangement in bovine caseins: localization of intrachain disulphide bridges in monomers of kappa- and alpha s2-casein from bovine milk. Journal of Dairy Research 1994;61(4) 485-493.

[55] Creamer LK, Plowman JE, Liddell MJ, Smith MH, Hill JP. Micelle stability: kappa-casein structure and function. Journal of Dairy Science 1998;81(11) 3004-3012.

[56] Godovac-Zimmermann J, Conti A, Liberatori J, Braunitzer G The amino-acid sequence of beta-lactoglobulin II from horse colostrum (Equus caballus, Perissodactyla): beta-lactoglobulins are retinol-binding proteins. Biological chemistry Hoppe-Seyler 1985;366(6) 601-608.

[57] Kontopidis G, Holt C, Sawyer L. Invited review: beta-lactoglobulin: binding properties, structure, and function. Journal of Dairy Science 2004;87(4) 785-796.

[58] McKenzie HA, Sawyer WH, Smith MB. Optical rotatory dispersion and sedimentation in the study of association-dissociation: bovine beta-lactoglobulins near pH 5. Biochimica and Biophysica Acta 1967;147(1) 73-92.

[59] Ehn BM, Ekstrand B, Bengtsson U, Ahlstedt S. Modification of IgE binding during heat processing of the cow's milk allergen beta-lactoglobulin. Journal of Agricultural and Food Chemistry 2004;52(5) 1398-1403.

[60] Chen, WL, Hwang MT, Liau CY, Ho JC, Hong KC, and Mao SJ. Beta-lactoglobulin is a thermal marker in processed milk as studied by electrophoresis and circular dichroic spectra. Journal of Dairy Science 2005;88(5) 1618-1630.

[61] Guyomarc'h F, Law AJ, and Dalgleish DG. Formation of soluble and micelle-bound protein aggregates in heated milk. Journal of Agricultural Food Chemistry 2003;51(16) 4652-4660.

[62] Clement G, Boquet D, Frobert Y, Bernard H, Negroni L, Chatel J-M, Adel-Patient K, Creminon C, Wal J-M, and Grassi J. Epitopic characterization of native bovine b-lactoglobulin. Journal of Immunological Methods 2002;266(1-2) 67-78.

[63] Sélo I, Clément G, Bernard H, Chatel J, Créminon C, Peltre G, Wal J. Allergy to bovine beta-lactoglobulin: specificity of human IgE to tryptic peptides. Clinical & Experimental Allergy 1999;29(8) 1055-1063.

[64] Fritsché R, Adel-Patient K, Bernard H, Martin-Paschoud C, Schwarz C, Ah-Leung S, Wal JM. IgE-mediated rat mast cell triggering with tryptic and synthetic peptides of bovine beta-lactoglobulin. International Archives of Allergy and Immunology 2005;138(4) 291-297.

[65] Miller K, Meredith C, Selo I, Wal JM. Allergy to bovine beta-lactoglobulin: specificity of immunoglobulin E generated in the Brown Norway rat to tryptic and synthetic peptides. Clinical & Experimental Allergy 1999;29(12) 1696-1704.

[66] Polverino de Laureto P, Frare E, Gottardo R, Van Dael H, Fontana A. Partly folded states of members of the lysozyme/lactalbumin superfamily: a comparative study by circular dichroism spectroscopy and limited proteolysis. Protein Science 2002;11(12) 2932-2946.

[67] Veprintsev DB, Permyakov SE, Permyakov EA, Rogov VV, Cawthern KM, Berliner LJ. Cooperative thermal transitions of bovine and human apo-alpha-lactalbumins: evidence for a new intermediate state. FEBS Letters 1997;412(3) 625-628.

[68] McGuffey MK, Epting KL, Kelly RM, Foegeding EA. Denaturation and aggregation of three alpha-lactalbumin preparations at neutral pH. Journal of Agricultural and Food Chemistry 2005;53(8) 3182-3190.

[69] Redfield C.Using nuclear magnetic resonance spectroscopy to study molten globule states of proteins. Methods 2004;34(1) 121-32.

[70] Chang JY. Evidence for the underlying cause of diversity of the disulfide folding pathway. Biochemistry 2004;43(15) 4522-4529.

[71] Livney YD, Verespej E, Dalgleish DG. Steric effects governing disulfide bond interchange during thermal aggregation in solutions of beta-lactoglobulin B and alpha-lactalbumin. Journal of Agricultural and Food Chemistry 2003;51(27) 8098-8106.

[72] McKenzie HA. alpha-Lactalbumins and lysozymes. EXS 1996;75 365-409.

[73] Ramakrishnan B, Qasba PK. Crystal structure of lactose synthase reveals a large conformational change in its catalytic component, the beta1,4-galactosyltransferase-I. Journal of Molecular Biology 2001;310(1) 205-218.

[74] Permyakov EA and Berliner LJ. alpha-Lactalbumin: Structure and function. FEBS Letters 2000;473(3) 269-274.

[75] Hakansson A, Svensson M, Mossberg AK, Sabharwal H, Linse S, Lazou I, Lonnerdal B, and Svanborg C. A folding variant of alpha-lactalbumin with bactericidal activity against Streptococcus pneumoniae. Molecular Microbiololgy 2000;35(3) 589-600.

[76] Pellegrini A, Thomas U, Bramaz N, Hunziker P, von Fellenberg R. Isolation and identification of three bactericidal domains in the bovine alpha-lactalbumin molecule. Biochimica and Biophysica Acta 1999;1426(3) 439-448.

[77] Permyakov SE, Pershikova IV, Khokhlova TI, Uversky VN, Permyakov EA. No need to be HAMLET or BAMLET to interact with histones: binding of monomeric alpha-lactalbumin to histones and basic poly-amino acids. Biochemistry 2004;43(19) 5575-5582.

[78] Slangen CJ, Visser S. Use of mass spectrometry To rapidly characterize the heterogeneity of bovine alpha-lactalbumin. Journal of Agricultural and Food Chemistry 1999;47(11) 4549-4556.

[79] Maynard F , Jost R, Wal JM. Human IgE binding capacity of tryptic peptides from bovine alpha-Lactalbumin. International Archives of Allergy and Immunology 1997;113(4) 478-488.

[80] Wróblewska B, and Jędrychowski L. Milk allergens. In Lucjan Jędrychowski edtr Chemical and Biological Properties of Food Allergens Lucjan Jędrychowski, Harry Wichers Editors CRC Press, 2010 ISBN 142005855X, 9781420058550

[81] Szepfalusi Z, Ebner C, Urbanek R, Ebner H, Scheiner O, Boltz-Nitulescu G, Kraft D. Detection of IgE antibodies specific for allergens in cow milk and cow dander. International Archives of Allergy and Immunology 1993;102(3) 288-294.

[82] Han GD, Matsuno M, Ito G, Ikeucht Y, Suzuki A. Meat allergy: investigation of potential allergenic proteins in beef. Bioscience, Biotechnology, and Biochemistry 2000;64(9) 1887-1895.

[83] Fiocchi A, Restani P, Riva E. Beef allergy in children. Nutrition 2000;16(6) 454-457.

[84] Fiocchi A, Restani P, Riva E, Mirri GP, Santini I, Bernardo L, Galli CL. Heat treatment modifies the allergenicity of beef and bovine serum albumin. Allergy 1998;53(8) 798-802.

[85] Restani P, Fiocchi A, Beretta B, Velona T, Giovannini M, Galli CL. Meat allergy: III--Proteins involved and cross-reactivity between different animal species. Journal of the American College of Nutrition 1997;16(4) 383-389.

[86] Werfel S, Cooke SK, Sampson HA. Clinical reactivity to beef in children allergic to cow's milk. Journal of Allergy and Clinical Immunology 1997;99(3) 293-300.

[87] van Ree R, van Leeuwen WA, Aalberse RC. How far can we simplify in vitro diagnostics for grass pollen allergy?: A study with 17 whole pollen extracts and purified natural and recombinant major allergens. Journal of Allergy and Clinical Immunology 1998;102(2) 184-190.

[88] Bernard H, Meisel H, Creminon C, Wal JM. Post-translational phosphorylation affects the IgE binding capacity of caseins. FEBS Lett. 2000;467(2-3) 239-44.

[89] Molina ACT., Alli I, Boye JI. Effects of dephosphorylation and proteolysis on immunogenicity of caseins. Milchwissenschaft-Milk Science International 2007;62(4) 371-375.

[90] Schlimme E and Meisel H. Bioactive peptides derived from milk proteins. Structural, physiological and analytical aspects. Nahrung/Food 1995;39 (1) 1-20.

[91] Pizzano R, Nicolai MA, Padovano P, Ferranti P, Barone F, Addeo F. Immunochemical evaluation of bovine beta-casein and its 1-28 phosphopeptide in cheese during ripening. Journal of Agricultural and Food Chemistry 2000;48(10) 4555-4560.

[92] Mari A. IgE to Cross-Reactive Carbohydrate Determinants: analysis of the distribution and appraisal of the in vivo and in vitro reactivity. International Archives of Allergy and Immunology 2002;129(4) 286-295.

[93] Hochwallner H, Schulmeister U, Swoboda I, Focke-Tejkl M, Civaj V, Balic N, Nystrand M, Härlin A, Thalhamer J, Scheiblhofer S, Keller W, Pavkov T, Zafred D, Niggemann B, Quirce S, Mari A, Pauli G, Ebner C, Papadopoulos NG, Herz U, van Tol EA, Valenta R, Spitzauer S. Visualization of clustered IgE epitopes on alpha-lactalbumin Journal of Allergy and Clinical Immunology 2010;125(6) 1279-1285.e9.

[94] Tran NT, Daali Y, Cherkaoui S, Taverna M, Neeser JR, Veuthey JL. Routine o-glycan characterization in nutritional supplements--a comparison of analytical methods for the

monitoring of the bovine kappa-casein macropeptide glycosylation. Journal of Chromatography A 2001; 929(1-2) 151-63.

[95] Slangen CJ, Visser S. Use of mass spectrometry To rapidly characterize the heterogeneity of bovine alpha-lactalbumin. Journal of Agricultural and Food Chemistry 1999; 47(11) 4549-4556.

[96] Heine W, Wutzke KD, Radke M. Decreasing the immunogenicity of milk proteins by desialinization. Infusionstherapie 1989;16(6) 264-266.

[97] Holland JW, Deeth HC, Alewood PF. Proteomic analysis of kappa-casein micro-heterogeneity. Proteomics 2004;4(3) 743-752.

[98] Pizzano R, Nicolai MA, Manzo C, Giannattasio M, Addeo F. Human IgE binding to the glycosidic moiety of bovine kappa-casein. Journal of Agricultural and Food Chemistry 2005;53(20) 7971-7975.

[99] Akagawa M, Handoyo T, Ishii T, Kumazawa S, Morita N, Suyama K. Proteomic analysis of wheat flour allergens. Journal of Agricultural and Food Chemistry 2007;55(17) 6863-6870.

[100] Karas M, Hillenkamp F. Laser desorption ionization of proteins with molecular masses exceeding 10,000 daltons. Chemistry 1988;60(20) 2299-2301.

[101] Fenn JB, Mann M, Meng CK, Wong SF, Whitehouse CM. Electrospray ionization for mass spectrometry of large biomolecules. Science 1989;246(4926) 64-71.

[102] Aebersold R, Mann M. Mass spectrometry-based proteomics. Nature 2003;422(6928) 198-207.

[103] Yates JR 3rd. Mass spectral analysis in proteomics. Annual Review of Biophysics and Biomolecular Structure 2004;33 297-316.

[104] Domon B, Aebersold R. Mass spectrometry and protein analysis. Science 2006;312(5771) 212-217.

[105] Han X, Aslanian A, Yates JR 3rd. Mass spectrometry for proteomics. Current Opinion in Chemical Biology 2008;12(5) 483-490.

[106] Hirabayashi J. Lectin-based structural glycomics: glycoproteomics and glycan profiling. Glycoconjugate Journal 2004;21(1-2) 35-40.

[107] Wuhrer M, de Boer AR, Deelder AM. Structural glycomics using hydrophilic interaction chromatography (HILIC) with mass spectrometry. Mass Spectrometry Review 2009;28(2) 192-206.

[108] Harvey DJ. Analysis of carbohydrates and glycoconjugates by matrix-assisted laser desorption/ionization mass spectrometry: an update for the period 2005-2006. Mass Spectrometry Review 2011;30(1) 1-100.

[109] An HJ, Peavy TR, Hedrick JL, Lebrilla CB. Determination of N-glycosylation sites and site heterogeneity in glycoproteins. Analytical Chemistry 2003;75(20) 5628-5637.

[110] Zaia J. Mass spectrometry of oligosaccharides. Mass Spectrometry Review 2004;23(3) 161-227.

[111] Das BC, Gero SD, Lederer E. N-methylation of N-acyl oligopeptides. Biochemical and Biophysical Research Communications 1967;29(2) 211-215.

[112] De Angelis M, Di Cagno R, Minervini F, Rizzello CG, Gobbetti M. Two-dimensional electrophoresis and IgE-mediated food allergy. Electrophoresis 2010;31(13) 2126-2136.

[113] Petersen A. Two-dimensional electrophoresis replica blotting: a valuable technique for the immunological and biochemical characterization of single components of complex extracts. Proteomics 2003;3(7) 1206-1214.

[114] Natale M, Bisson C, Monti G, Peltran A, Garoffo LP, Valentini S, Fabris C, Bertino E, Coscia A, Conti A. Cow's milk allergens identification by two-dimensional immunoblotting and mass spectrometry. Molecular Nutrition & Food Research 2004;48(5) 363-369.

[115] Monaci L, Losito I, Palmisano F, Visconti A. Reliable detection of milk allergens in food using a high-resolution, stand-alone mass spectrometer. Journal of AOAC International 2011;94(4) 1034-1042.

[116] Lu B, Xu T, Park SK, McClatchy DB, Liao L, Yates JR 3rd. Shotgun protein identification and quantification by mass spectrometry in neuroproteomics. Methods in Molecular Biology 2009;566 229-259.

[117] Weber D, Raymond P, Ben-Rejeb S, Lau B. Development of a liquid chromatography-tandem mass spectrometry method using capillary liquid chromatography and nanoelectrospray ionization-quadrupole time-of-flight hybrid mass spectrometer for the detection of milk allergens. Journal of Agricultural and Food Chemistry 2006;54(5) 1604-1610.

[118] Kuppannan K, Albers DR, Schafer BW, Dielman D, Young SA. Quantification and characterization of maize lipid transfer protein, a food allergen, by liquid chromatography with ultraviolet and mass spectrometric detection. Analytical Chemistry 2011;83(2) 516-524.

[119] Huber CG, Premstaller A. Evaluation of volatile eluents and electrolytes for high-performance liquid chromatography-electrospray ionization mass spectrometry and capillary electrophoresis-electrospray ionization mass spectrometry of proteins. I. Liquid chromatography. Journal of Chromatography A 1999;849(1) 161-73.

[120] Czerwenka C, Maier I, Potocnik N, Pittner F, Lindner W. Absolute quantitation of beta-lactoglobulin by protein liquid chromatography-mass spectrometry and its application to different milk products. Analytical Chemistry 2007;79(14) 5165-5172.

[121] Wu WW, Wang G, Baek SJ, Shen RF.Comparative study of three proteomic quantitative methods, DIGE, cICAT, and iTRAQ, using 2D gel- or LC-MALDI TOF/TOF. Journal of Proteome Research 2006; 5(3) 651-658.

[122] Gingras AC, Gstaiger M, Raught B, Aebersold R. Analysis of protein complexes using mass spectrometry. Nature Reviews Molecular Cell Biology 2007;8(8) 645-654.

[123] Picotti P, Rinner O, Stallmach R, Dautel F, Farrah T, Domon B, Wenschuh H, Aebersold R. High-throughput generation of selected reaction-monitoring assays for proteins and proteomes. Nature Methods. 2010;7(1) 43-46.

[124] Careri M, Costa A, Elviri L, Lagos JB, Mangia A, Terenghi M, Cereti A, Garoffo LP. Use of specific peptide biomarkers for quantitative confirmation of hidden allergenic peanut proteins Ara h 2 and Ara h 3/4 for food control by liquid chromatography-tandem mass spectrometry. Analytical and Bioanalytical Chemistry 2007;389(6) 1901-1907.

[125] Heick J, Fischer M, Kerbach S, Tamm U, Popping B. Application of a liquid chromatography tandem mass spectrometry method for the simultaneous detection of seven allergenic foods in flour and bread and comparison of the method with commercially available ELISA test kits. Journal of AOAC International 2011;94(4) 1060-1068.

[126] Mann M, Hendrickson RC, Pandey A. Analysis of proteins and proteomes by mass spectrometry. Annual Review of Biochemistry 2001;70 437-473.

[127] Vieths S, Hoffmann-Sommergruber K.EuroPrevall food allergen library. Molecular Nutrition & Food Research 2008;52 Suppl 2:S157-8.

[128] Sancho AI, Hoffmann-Sommergruber K, Alessandri S, Conti A, Giuffrida MG, Shewry P, Jensen BM, Skov P, Vieths S. Authentication of food allergen quality by physicochemical and immunological methods. Clinical & Experimental Allergy 2010;40(7 973-986.

[129] Maass S, Sievers S, Zühlke D, Kuzinski J, Sappa PK, Muntel J, Hessling B, Bernhardt J, Sietmann R, Völker U, Hecker M, Becher D. Efficient, global-scale quantification of absolute protein amounts by integration of targeted mass spectrometry and two-dimensional gel-based proteomics. Analytical Chemistry 2011;83(7).

[130] FAO/WHO, Evaluation of Allergenicity of Genetically Modified Foods, FAO, Rome, 2001, p. 1.

[131] Williams SC, Badley RA, Davis PJ, Puijk WC, Meloen RH. Identification of epitopes within beta lactoglobulin recognised by polyclonal antibodies using phage display and PEPSCAN. Journal of Immunological Methods 1998;213(1) 1-17.

[132] Picariello G, Ferranti P, Fierro O, Mamone G, Caira S, Di Luccia A, Monica S, Addeo F. Peptides surviving the simulatedgastrointestinal digestion of milk proteins: biological and toxicological implications. Journal of Chromatography B: AnalyticalTechnologies in the Biomedical and Life Sciences 2010;878(3-4) 295-308.

[133] Bannon GA, Cockrell G, Connaughton C, West CM, Helm R, Stanley JS, King N, Rabjohn P, Sampson HA, Burks AW. Engineering, characterization and in vitro efficacy of the major peanut allergens for use in immunotherapy. International Archives of Allergy and Immunology 2001;124(1-3) 70-72.

[134] Burks AW, King N, Bannon GA. Modification of a major peanut allergen leads to loss of IgE binding.International Archives of Allergy and Immunology 1999;118(2-4) 313-314.

[135] Swoboda I, Bugajska-Schretter A, Linhart B, Verdino P, Keller W, Schulmeister U, Sperr WR, Valent P, Peltre G, Quirce S, Douladiris N, Papadopoulos NG, Valenta R, Spitzauer S. A recombinant hypoallergenic parvalbumin mutant for immunotherapy of IgE-mediated fish allergy. The Journal of Immunology 2007;178(10) 6290-6296.

[136] Prioult G, Pecquet S, Fliss I. Stimulation of interleukin-10 production by acidic beta-lactoglobulin-derived peptides hydrolyzed with Lactobacillus paracasei NCC2461 peptidases. Clinical and Diagnostic Laboratory Immunology 2004;11(2) 266-271.

[137] Prioult G, Pecquet S, Fliss I. Allergenicity of acidic peptides from bovine beta-lactoglobulin is reduced by hydrolysis with Bifidobacterium lactis NCC362 enzymes International Dairy Journal 2005; 15(5) 439-448

[138] Di Cagno R, Rizzello CG, De Angelis M, Cassone A, Giuliani G, Benedusi A, Limitone A, Surico RF, Gobbetti M. Use of selected sourdough strains of Lactobacillus for removing gluten and enhancing the nutritional properties of gluten-free bread. Journal Of Food Protection 2008;71(7) 1491-1495.

[139] Gianfrani C, Siciliano RA, Facchiano AM, Camarca A, Mazzeo MF, Costantini S, Salvati VM, Maurano F, Mazzarella G, Iaquinto G, Bergamo P, Rossi M. Transamidation of wheat flour inhibits the response to gliadin of intestinal T cells in celiac disease. Gastroenterology 2007;133(3) 780-789.

[140] Restani P, Beretta B, Fiocchi A, Ballabio C, Galli CL. Cross-reactivity between mammalian proteins. Annals of Allergy, Asthma & Immunology 2002;89(6) 11-15.

[141] Park YW. Hypo-allergenic and therapeutic significance of goat milk. Small Ruminant Research 1994;14(2) 151-159.

[142] Bellioni-Businco B., Paganelli R., Lucenti P., Giampietro P. G., Perborn H., Businco L. Allergenicity of goat's milk in children with cow's milk allergy. Journal of Allergy and Clinical Immunology 1999;103(6) 1191-1194.

[143] Pessler F, Nejat M. Anaphylactic reaction to goat's milk in a cow's milk-allergic infant. Pediatric Allergy and Immunology 2004;15(2) 183-185.

[144] Ballabio C, Chessa S, Rignanese D, Gigliotti C, Pagnacco G, Terracciano L, Fiocchi A, Restani P, Caroli AM. Goat milk allergenicity as a function of αs_1-casein genetic polymorphism. Journal of Dairy Science 2011;94(2) 998-1004.

[145] Tomotake H, Katagiri M, Fujita M, Yamato M. Preparation of fresh cheese from caprine milk as a model for the reduction of allergenicity. Journal of Nutritional Science and Vitaminology (Tokyo) 2009;55(3) 296-300.

[146] Sheehan WJ, Phipatanakul W. Tolerance to water buffalo milk in a child with cow milk allergy. Annals of Allergy, Asthma & Immunology 2009;102(4) 349.

[147] D'Auria E, Agostoni C, Giovannini M, Riva E, Zetterström R, Fortin R, Greppi GF, Bonizzi L, Roncada P. Proteomic evaluation of milk from different mammalian species as a substitute for breast milk. Acta Paediatrica 2005;94(12) 1708-1713.

[148] Lamblin C, Bourrier T, Orlando JP, Sauvage C, Wallaert B. Allergie aux laits de chèvre et de brebis sans allergie associée au lait de vache. Revue Française d'Allergologie et d'Immunologie Clinique 2001; 41(2) 165-168.

[149] Bidat E, Rancé F, Baranès T, Goulamhoussen S. Goat's milk and sheep's milk allergies in children in the absence of cow's milk allergy. Revue Française d'Allergologie et d'Immunologie Clinique 2003;43(4) 273–277.

[150] Muñoz Martín T, de la Hoz Caballer B, Marañón Lizana F, González Mendiola R, Prieto Montaño P, Sánchez Cano M. Selective allergy to sheep's and goat's milk proteins. Allergologia et Immunopathologia (Madr) 2004;32(1) 39-42.

[151] Ah-Leung S, Bernard H, Bidat E, Paty E, Rancé F, Scheinmann P, Wal JM. Allergy to goat and sheep milk without allergy to cow's milk. Allergy. 2006;61(11) 1358-1365.

[152] Wüthrich B, Johansson SG. Allergy to cheese produced from sheep's and goat's milk but not to cheese produced from cow's milk. Journal of Allergy and Clinical Immunology 1995;96(2) 270-273.

[153] Umpiérrez A., Quirce S, Marañón F, Cuesta J, García-Villamuza Y, Lahoz C, Sastre J. Allergy to goat and sheep cheese with good tolerance to cow cheese. Clinical & Experimental Allergy 1999;29(8) 1064-1068.
[154] Businco L, Giampietro PG, Lucenti P, Lucaroni F, Pini C, Di Felice G, Iacovacci P, Curadi C, Orlandi M. Allergenicity of mare's milk in children with cow's milk allergy. Journal of Allergy and Clinical Immunology. 2000;105(5) 1031-1034.
[155] Guo HY, Pang K, Zhang XY, Zhao L, Chen SW, Dong ML, Ren FZ. Composition, physiochemical properties, nitrogen fraction distribution, and amino acid profile of donkey milk. Journal of Dairy Science. 2007;90(4) 1635-1643.
[156] Monti G, Bertino E, Muratore MC, Coscia A, Cresi F, Silvestro L, Fabris C, Fortunato D, Giuffrida MG, Conti A. Efficacy of donkey's milk in treating highly problematic cow's milk allergic children: an in vivo and in vitro study. Pediatric Allergy and Immunology 2007;18(3) 258-264.
[157] Cunsolo V, Saletti R, Muccilli V, Foti S. Characterization of the protein profile of donkey's milk whey fraction. Journal of Mass Spectrometry 2007;42(9) 1162-1174.
[158] Chianese L, Calabrese MG, Ferranti P, Mauriello R, Garro G, De Simone C, Quarto M, Addeo F, Cosenza G, Ramunno L.Proteomic characterization of donkey milk "caseome". Journal of Chromatography A 2010;1217(29) 4834-4840.
[159] Cunsolo V, Muccilli V, Fasoli E, Saletti R, Righetti PG, Foti S. Poppea's bath liquor: the secret proteome of the-donkey's milk. Journal of Proteomics 2011;74(10) 2083-99.
[160] Suutari TJ, Valkonen KH, Karttunen TJ, Ehn BM, Ekstrand B, Bengtsson U, Virtanen V, Nieminen M, Kokkonen J. IgE cross reactivity between reindeer and bovine milk beta-lactoglobulins in cow's milk allergic patients. Journal of Investigational Allergology and Clinical Immunology 2006;16(5) 296-302.
[161] El-Agamy EI, Nawar M, Shamsia SM, Awad S, Haenlein GFW. Are camel milk proteins convenient to the nutrition of cow milk allergic children? Small Ruminant Research 2009;82(1) 1-6.
[162] Okada H, Kuhn C, Feillet H, and Bach J-F. The 'hygiene hypothesis' for autoimmune and allergic diseases: an updatClinical and Experimental Immunology 2010; 160(1) 1-9.
[163] Campeotto F, Waligora-Dupriet AJ, Doucet-Populaire F, Kalach N, Dupont C, Butel MJ. Establishment of the intestinal microflora in neonates. Gastroenterologie Clinique et Biologique 2007;31(5) 533-542.
[164] Adlerberth I, Wold AE. Establishment of the gut microbiota in Western infants. Acta Paediatrica 2009;98(2) 229-238.
[165] Stsepetova J, Sepp E, Julge K, Vaughan E, Mikelsaar M, de Vos WM. Molecularly assessed shifts of Bifidobacterium ssp. and less diverse microbial communities are characteristic of 5-year-old allergic children. FEMS Immunology & Medical Microbiology. 2007;51(2) 260-269.
[166] Vael C, Nelen V, Verhulst SL, Goossens H, Desager KN. Early intestinal Bacteroides fragilis colonisation and development of asthma. BMC Pulmonary Medicine. 2008; 26 8-19.
[167] Sjögren YM, Tomicic S, Lundberg A, Böttcher MF, Björkstén B, Sverremark-Ekström E, Jenmalm MC. Influence of early gut microbiota on the maturation of childhood mucosal

and systemic immune responses. Clinical & Experimental Allergy. 2009;39(12) 1842-1851.

[168] Sjögren YM, Jenmalm MC, Böttcher MF, Björkstén B, Sverremark-Ekström E. Altered early infant gut microbiota in children developing allergy up to 5 years of age. Clinical & Experimental Allergy 2009;39(4) 518-26.

[169] Chen W, Jin W, Hardegen N, Lei KJ, Li L, Marinos N, McGrady G, and Wahl SM. Conversion of peripheral CD4+CD25-naive T cells to CD4+CD25+regulatory T cells by TGF-beta induction of transcription factor Foxp3. The Journal of Experimental Medicine. 2003; 198, 1875–1886.

[170] Karim M, Kingsley CI, Bushell AR, Sawitzki BS, Wood KJ. Alloantigen-induced CD25+CD4+ regulatory T cells can develop in vivo from CD25-CD4+ precursors in a thymus-independent process. The Journal of Immunology 2004;172(2) 923-928.

[171] O'Mahony C, Scully P, O'Mahony D, Murphy S, O'Brien F, Lyons A, Sherlock G, MacSharry J, Kiely B, Shanahan F, O'Mahony L. Commensal-induced regulatory T cells mediate protection against pathogen-stimulated NF-kappa B activation. PLoS Pathogens 2008;4(8) e1000112.

[172] Lyons A, O'Mahony D, O'Brien F, MacSharry J, Sheil B, Ceddia M, Russell WM, Forsythe P, Bienenstock J, Kiely B, Shanahan F, O'Mahony L. Bacterial strain-specific induction of Foxp3+ T regulatory cells is protective in murine allergy models. Clinical & Experimental Allergy 2010;40(5) 811-819.

[173] Round JL, Mazmanian SK. Inducible Foxp3+ regulatory T-cell development by a commensal bacterium of the intestinal microbiota. Proceedings of the National Academy of Sciences 2010;107(27) 12204-12209.

[174] Rodriguez B, Prioult G, Bibiloni R, Nicolis I, Mercenier A, Butel MJ, Waligora-Dupriet AJ. Germ-free status and altered caecal subdominant microbiota are associated with a high susceptibility to cow's milk allergy in mice. FEMS Microbiology Ecology 2011;76(1) 133-144.

[175] Høst A, Koletzko B, Dreborg S, Muraro A, Wahn U, Aggett P, Bresson JL, Hernell O, Lafeber H, Michaelsen KF, Micheli JL, Rigo J, Weaver L, Heymans H, Strobel S, Vandenplas Y. Dietary products used in infants for treatment and prevention of food allergy. Joint Statement of the European Society for Paediatric Allergology and Clinical Immunology (ESPACI) Committee on Hypoallergenic Formulas and the European Society for Paediatric Gastroenterology, Hepatology and Nutrition (ESPGHAN) Committee on Nutrition. Archives of Disease in Childhood 1999;81(1) 80-84.

[176] Høst A. Cow's milk protein allergy and intolerance in infancy. Some clinical, epidemiological and immunological aspects. Pediatric Allergy and Immunology 1994;5(5 Suppl) 1-36.

[177] Commission of the European Communities. 1996, Official Journal of the European Commission, 39, 12.

[178] de Boissieu D, Matarazzo P, Dupont C. Allergy to extensively hydrolyzed cow milk proteins in infants: identification and treatment with an amino acid-based formula. The Journal of Pediatrics 1997;131(5) 744-747.

[179] Halken S, Høst A. How hypoallergenic are hypoallergenic cow's milk-based formulas? Allergy. 1997;52(12) 1175-1183.

[180] Vanderhoof JA, Murray ND, Kaufman SS, Mack DR, Antonson DL, Corkins MR, Perry D, Kruger R. Intolerance to protein hydrolysate infant formulas: an underrecognized cause of gastrointestinal symptoms in infants. The Journal of Pediatrics 1997;131(5) 741-744.

[181] Ragno V, Giampietro PG, Bruno G, Businco L. Allergenicity of milk protein hydrolysate formulae in children with cow's milk allergy. European Journal of Pediatrics 1993;152(9) 760-762.

[182] Takase M, Fukuwatari Y, Kawase K, Kiyosawa I, Ogasa K, Suzuki S, Kuroume T. Antigenicity of casein enzymatichydrolysate. Journal of Dairy Science 1979;62(10) 1570-1576.

[183] Sharma S, Kumar P, Betzel C, Singh TP. Structure and function of proteins involved in milk allergies. Journal of Chromatography B: Biomedical Sciences and Applications 2001;756(1-2) 183-187.

[184] Richter WO, Jacob B, Schwandt P. Molecular weight determination of peptides by high-performance gel permeation chromatography. Analytical Biochemistry 1983;133(2) 288-291.

[185] Hames BD (ed): Get Electrophoresis of Proteins: a Practical Approach, IRL Pr. Mclean, Va. 1981

[186] Coombs RRA, Devey ME, Anderson KJ. Refractoriness to anaphylactic shock after continuous feeding of cow's milk to guinea pigs. Clinical and Experimental Immunology 1978;32(2) 263-271.

[187] Ellis MH, Short JA, Heiner DC. Anaphylaxis after ingestion of a recently hydrolyzed whey protein formula. The Journal of Pediatrics 1991;118 74-77.

[188] Businco L, Cantani A, Longhi Ma, Giampietro Pg. Anaphylactic reactions to a cow's milk whey protein hydrolyzate (Alfa-Ré, Nestlé) in infants with cow's milk allergy. Annals of Allergy 1989;62(4) 333-335.

[189] Saylor Jd, Bahna Sl. Anaphylaxis to casein hydrolysate formula. Journal Of Pediatrics 1991; 118(1) 71-74.

[190] Bock SA. Probable allergic reaction to casein hydrolysate formula. Journal Of Allergy And Clinical Immunology 1989; 84(2) 272-272.

[191] Sampson HA. Immunological mechanisms in adverse reactions to foods. Immunology and Allergy Clinics of North America 1991;11(4) 701-716.

[192] Van Beresteijn ECH, Meijer RJ, Schmidt DG. Residual antigenicity of hypoallergenic infant formulas and the occurrence of milk-specific IgE antibodies in patients with clinical allergy. Journal of Allergy and Clinical Immunology 1995;96(3) 365-374.

[193] Van Beresteijn, ECH, Peeters RA, Kaper JM, Ron JGM, Robben AJPM, Schmidt DG. Molecular Mass Distribution, Immunological Properties and Nutritive Value of Whey Protein Hydrolysates. Journal of Food Protection 1994;57(7) 619-625.

[194] Pelissier J-P, Mercier J-C and Ribadeau Dumas B. Etude de la protéolyse des caséines αs1 et β bovines par la présure spécificité d'action. peptides amers libérés. Annales des Biologie Animales Biochimie Biophysic 1974;14(2) 343-362.

[195] Høst A. Clinical course of cow's milk protein allergy and intolerance. Pediatric Allergy and Immunology 1998;9(11) 48-52.

[196] Host A, Halken S, Jacobsen HP, Estmann A, Mortensen S, Mygil S. The natural course of cow's milk protein allergy/intolerance. Journal of Allergy and Clinical Immunology 1997;99(1) S490

[197] Bishop JM, Hill DJ, Hosking CS Natural history of cow milk allergy: Clinical outcome. The Journal of Pediatrics. 1990;116(6) 862-867.

[198] Alvarado MI, Alonso E, GÁlvarez M, Ibáñez MD Laso MT. Persistencia de sensibilización a proteínas de leche de vaca: estudio clínico. Allergologia et Immunopathologia 2000;28(3) 189.

[199] Lowney ED. Immunological unresponsiveness to a contact sensitizer in man. Journal of Investigative Dermatology 1968.51(6) 411-417.

[200] Bellanti JA. Prevention of food allergies. Annals of Allergy 1984;53(6 Pt 2) 683-688.

[201] Tomasi TB, Barr WG, Challacombe SJ, Curran G. Oral tolerance and accessory-cell function of Peyer's patches. Annals of the New York Academy of Sciences 1983;409 145-63.

Genetic Diversity

Milk Proteins' Polymorphism in Various Species of Animals Associated with Milk Production Utility

Joanna Barłowska, Anna Wolanciuk, Zygmunt Litwińczuk and Jolanta Król

Additional information is available at the end of the chapter

1. Introduction

In the recent years a significant progress has occurred in the understanding of many of the complex processes in cells of the body at a molecular level. To clarify the basis of this phenomenon the development in the field of molecular biology was a great contribution. This area has provided many hitherto unknown tools that enabled the understanding of fundamental life processes at a basic level. The utilization of the achievements of biological and zootechnical sciences led to a significant increase in animal productivity.

The first research studies concerning the detection of genetic diversity among farm animals were based on morphological, chromosomal and biochemical markers and therefore, they are not free of defects. Most of the morphological markers are limited by gender or age and are influenced by the environment. The biochemical markers, however, contribute to low levels of polymorphism (Boichard et al., 2003; Naqvi, 2007). Currently a key role in animal genetics is connected with molecular markers. Molecular markers are specific pieces of DNA that can be identified within the genome and are inherited according to Mendel's laws (Teneva & Petrović, 2010). They can be successfully utilized to detect and track the inheritance mechanisms of polymorphic traits that contribute to genetic diversity (Khatkar et al., 2004). Molecular markers which enable the detection of genetic variants at DNA sequence level are devoid of these limitations typical for morphological, chromosomal and protein markers. They also have unique properties that make them more useful than other markers. They are ubiquitous throughout the genome, often multi-allelic, giving an average heterozygosity of more than 70%. What is more, molecular markers are not influenced by environment and usually do not have pleiotropic effect on quantitative traits loci (QTL) (Teneva, 2009).

Up to date, numerous techniques of studying DNA variations at the molecular level are known. The most famous include RFLP (restriction fragment length polymorphism), SNP (single nucleotide polymorphism) and STR (microsatellite DNA polymorphism) (Beuzen et al., 2000; Teneva & Petrović, 2010).

RFLP (restriction fragment length polymorphism)

Restriction fragment length polymorphism (RFLP) is associated with the occurrence of differences in nucleotide sequences in the gene. It is a result of point mutations, which may manifest themselves phenotypically. They appear through the use of restriction enzymes that recognize sequences specific to their nucleotide. The mutations result in the creation of new places that are identified by restriction enzymes that cut DNA into fragments of various lengths. This technique consists of amplification of specific parts of the genome and the amplicon is digested by one or more restriction enzymes. The obtained DNA fragments are distributed on an agarose gel and, depending on their size, migrate at different speed rates. Smaller fragments tend to move faster in the comparison to larger ones (Beuzen et al., 2000; Bogdzińska, 2011; Marle-Köster & Nel, 2003).

SNP (single nucleotide polymorphism)

The system enables detection of single nucleotide polymorphism within the examined DNA sequences. It is based on the amplification of specific parts of the genome in the PCR reaction and sequencing of the product obtained. A comparison between electrophoresis images of amplification products is conducted, which allows determining whether a mutation in a given region had occurred. Point mutation leads to the formation of a polymorphic SNP if each of the alleles appear in the population with the frequency not lesser than 1 %. What is more, these markers are present in both coding and non-coding parts of the genome (Stoneking, 2001). SNP polymorphism is usually associated with the presence of only two alleles in the gene pool of the population (Beuzen et al., 2000). On the one hand, a great advantage of this polymorphism is its university in the genome of different species and highly efficient identification of polymorphism within the tested sequence while on the other hand, the high cost of the analysis makes it a disadvantage. The high density of SNP markers in the genome leads to their extensive utilization in the genetic analysis (Charon & Świtoński, 2000).

Microsatellite DNA

Genetic diversity may be also determined by utilization of microsatellite sequences. This group of markers is polymorphic in more than 90 % of the cases. Microsatellites are also known as simple sequence repeats (SSR) or short tandem repeats (STR). These repeats consist of several nucleotides (1-6) sequences also referred to as motifs. They are repeated from 20 to 50 times and their lengths range from 60 to 300 bp (Ellegren, 2004; Liu & Cordes, 2004). They occur mainly in non-coding regions of the genes, thus they can be also identified in flanking sequences or more rarely in coding sequences. What is more, they are characterized by uniform dispersion at 6 000 to 10 000 bp (Li et al., 2004; Liu et al., 2001). The function of microsatellites is not yet fully understood (Li et al., 2002). Probably through the dispersion across the genome they have an impact on increasing or decreasing the expression of genes (Pisarchik & Kartel, 2000). Features of microsatellites like high level of polymorphism, high frequency of occurrence, ease of identification and uniform distribution across the genome contributed to their common usage. They are used in the estimation of the genetic variability of animals, in the research on the control of origin, to characterize the structure and degree of inbreeding of the population and also to identify the genes of quantitative traits (QTL). What is more, they are used to conduct the selection based on genetic markers (MAS- *Marker Assisted Selection*) (Citek et al., 2006).

In the evaluation procedure of animals breeding value the knowledge of genome organization and polymorphism is increasingly utilized due to the fact of vast and easy access to many molecular technics. What is more, many mutations directly affecting the phenotype were recognized. On the other hand, thousands of anonymous genetic markers, because of their potential linkage with novel mutations of large scale of activity, may be utilized for estimation of the breeding values and selection based on genetic markers (MAS).

2. Location of genes and the frequency of alleles conditioning the synthesis of selected milk proteins in most species of animals used for milk productivity

The major milk proteins include casein: α_{s1}-, α_{s2}-, β-, κ- and two whey proteins: α-lactalbumin and β-lactoglobulin. These fractions, in most species, are polymorphic.

2.1. Cattle

Polymorphism of milk proteins has been widely explored in the case of cattle. Cattle casein *loci* is located on chromosome 6 (6/BTA 6q31-33) and occupies a total DNA fragment of 200 kb. Genes are arranged in order: CSN1S1, CSN2, CSN1S2, CSN3 and encode, respectively: α_{s1}-casein, α_{s2}-casein, β-casein and κ-casein. These genes are closely linked and form a cluster (Bai et al., 2008; Caroli et al., 2009).

α_{s1}-casein is a fraction which forms up to 40 % of bovine caseins in milk. It consists of one major and one minor component. Both of these proteins are composed of a single polypeptide chain of the same amino acid sequence. The reference protein for this family is α_{s1}-CN B-8P, a single chain protein with no cysteine residues. It consists of 199 amino acids residues: Asp_7, Asn_8, Thr_5, Ser_8, Ser P_8, Glu_{25}, Gln_{14}, Pro_{17}, Gly_9, Ala_9, Val_{11}, Met_5, Ile_{11}, Leu_{17}, Tyr_{10}, Phe_8, Lys_{14}, His_5, Trp_2, and Arg_6 with a molecular mass of 23.615 (Mercier et al., 1971). So far 8 alleles were identified within the α_{s1}-casein: A, B, C, D, E, F, G and H (Farrell et al., 2004). The most frequent, however, are alleles B and C found both in dairy and beef cattle (Litwińczuk et al., 2006). For European cattle breeds allele B is the most common – it exceeds the frequency of 0.9 (Kučerova et al., 2006; Luhken et al., 2009). Allele B at the 192 position of the polypeptide chain encodes glutathione, whereas the allele C encodes glycine. Variant A occurs sporadically. Alleles C, D and E were created due to the mutation of allele B. Table 1 shows the frequencies of as1-casein alleles in various breeds of cattle.

Cattle breed	Allelic frequencies of α_{s1}-casein	References
Holstein-Friesian	B = 0.981 C = 0.019	Çardak, 2005
Simmentaler	B = 0.932 C = 0.068	
Reggiana	B = 0.750 C = 0.250	Caroli et al., 2004
Uruguayan Creole	B = 0.8654 C = 0.1346	Rincón et al., 2006
Polish Red	B = 0.928 C = 0.042 D = 0.030	Erhardt et al., 1998
Red Danish Dairy Cattle	A = 0.003 B = 0.994 C = 0.003	
German Red	B = 0.975 C = 0.025	
Czech Fleckvieh	B = 0.893 C = 0.107	Kučerova et al., 2006
Hereford and crossbreeding F₁ and R₁ (Black and White x Hereford)	A = 0.025 B = 0.710 C = 0.136 D= 0.130	Litwińczuk & Król, 2002
Limousine and crossbreeding F₁ and R₁ (Black and White x Limousine)	A = 0.020 B = 0.617 C = 0.214 D = 0.148	
Simmentaler	B = 0.750 C = 0.250	
Estonian Native	B = 0.915 C = 0.085	Jõuru et al., 2007
Western Finncattle	B = 0.939 C = 0.061	
Danish Jersey	B = 0.781 C = 0.219	
Brazilian Zebu cattle (Gyr, Guzerat, Sindi, Nelore)	B = 0.000 – 0.136 C = 0.864 – 1.000	Da Salva & Del Lama, 1997

Table 1. The frequencies of α_{s1}-casein alleles in various breeds of cattle

The gene CSN1S2 encoding α_{s2}-casein has the length of 18 438 nucleotides and is divided into 18 exons ranging from 21 to 266 nucleotides (Ramunno et al., 2001). So far only 4 genetic variants of bovine α_{s2}-casein were described: A, B, C and D. Allele A which was created by mutation of allele D exists in most European breeds up to date. Alleles B and C are specific, respectively, for zebu and yaks (Ibeagha-Awemu et al., 2007).

β-casein is a fraction which forms up to 45 % of bovine caseins in milk. The native milk protease plasmin leads to the creation of γ_1-, γ_2- and γ_3-CN, which are actually fragments of β-casein consisting of the following sections of the chain 29-209, 106-209 and 108-209 (Farrell et al., 2004). 12 genetic variants that determine the synthesis of this protein were identified so far, i.e. A^1, A^2, A^3, B, C, D, E, F, G, H^1, H^2 and I (Farrell et al., 2004). The reference protein for this family is β-casein A2-5P consisting of a single chain protein, assembled from 209 amino acids residues without cysteine residues. The most common forms of β-casein in dairy cattle are A^1 and A^2, which differ by only one amino acid (Farrell et al., 2004). At the 67 position of amino acid chain, respectively, variant A^1 contains histidine and variant A^2 proline. Variant A^2, however, is the original form and is identified in old breeds of cattle (Zebu, Guernsey), whereas variant A^1 evolved much later and is characteristic to contemporary breeds (Hanusova et al., 2010). Variant B is less common, and A^3 and C exist rarely (Farrell et al., 2004). These alleles are most common for European breeds of cattle.

Allele E was only identified for the Italian Piemontese breed. The occurrence of genetic variants is based on nucleotide substitutions within exon VII (A1, A2, A3 and B) and exon VI (C and E) (Jann et al., 2002). Table 2 shows the frequencies of β-casein alleles in various breeds of cattle.

The κ-CN family consists of a major carbohydrate-free component and a minimum of 6 minor components. The 6 minor components are detected by PAGE in urea with 2-mercaptoethanol. In the basic structure of the reference protein of κ-CN family is a major carbohydrate-free component of κ-CN-1P. It consists of 169 amino acid residues arranged in the following order: Asp_4, Asn_8, Thr_{15}, Ser_{12}, Ser P_1, $Pyroglu_1$, Glu_{12}, Gln_{14}, Pro_{20}, Gly_2, Ala_{14}, Cys_2, Val_{11}, Met_2, Ile_{12}, Leu_8, Tyr_9, Phe_4, Lys_9, His_3, Trp_1 and Arg_5, with a molecular weight of 19.307 (Farrell et al., 2004). What is more, the length of κ-casein is less than 13 kb, though most of the coding sequences are comprised in the exon IV. κ-casein differs significantly from other caseins, both in terms of properties and structure. It is highly homologous to the fibrinogen gamma chain. What is more it serves a similar function, while being a stabilizing factor during the formation of the clot (Azevedo et al., 2008). So far 12 genetic variants of κ-casein were identified: A, A^1, B, C, E, F^1, F^2, G^1, G^2, H, I and J (Chen et al., 2008; Farrell et al., 2004). The differences between them are caused by two point mutations involving a substitution of threonine with isoleucine at the 136 position of polypeptide chain and aspartic acid with alanine at the 148 position (Azevedo et al., 2008). What is more, at the C-terminal part of κ-casein molecule a macropeptide residue can be found (106-169 chain fragment) (Farrell et al., 2004). Numerous studies focused on the analysis of genes that control the frequency of κ-CN polymorphism exhibit a superior frequency of allele A above allele B - for dairy, meat and also high productivity and local breeds (Król, 2003; Tsiaras et al., 2005). Azewedo et al. (2008) in their research on Brazilian cattle stated that the frequency of allele B of κ-casein ranges from 0.01 to 0.30. Kučerova et al. (2006) have also demonstrated a high frequency of allele B (0.38) and genotype BB (13 %) in the population of Czech Fleckvieh cattle. Table 3 shows the frequencies of κ -casein alleles in various breeds of cattle.

Cattle breed	Allelic frequencies of β-casein	References
Holstein-Friesian	$A^1 = 0.472$ $A^2 = 0.496$ $B = 0.026$	Ehrmann et al., 1997
Jersey	$A^1 = 0.093$ $A^2 = 0.721$ $B = 0.186$	
Jersey	$A^1 = 0.123$ $A^2 = 0.591$	Winkelman & Wickham, 1997
Polish Red	$A^1 = 0.617$ $A^2 = 0.321$ $B = 0.062$	Erhardt et al., 1998
Red Danish Dairy Cattle	$A^1 = 0.710$ $A^2 = 0.230$ $B = 0.060$	
German Red	$A^1 = 0.628$ $A^2 = 0.302$ $A^3 = 0.013$ $B = 0.057$	
Estonian Native	$A^1 = 0.318$ $A^2 = 0.644$ $B = 0.038$	Jõuru et al., 2007
Western Finncattle	$A^1 = 0.292$ $A^2 = 0.671$ $B = 0.037$	
Danish Jersey	$A^1 = 0.094$ $A^2 = 0.688$ $B = 0.219$	

Table 2. The frequencies of β-casein alleles in various breeds of cattle

Cattle breed	Allelic frequencies of κ-casein	References
Jersey	$A = 0.110$ $B = 0.880$	Ren et al., 2011
Holstein	$A = 0.690$ $B = 0.310$	
Romanian Spotted	$A = 0.650$ $B = 0.350$	Ilie et al., 2007
Brown of Maramures	$A = 0.375$ $B = 0.625$	

Cattle breed	Allelic frequencies of κ-casein	References
Holstein-Friesian	A = 0.940 B = 0.060	Tsiaras et al., 2005
Holstein	A = 0.760 B = 0.240	Bonvillani et al., 2010
Simmentaler	A = 0.531 B = 0.469	Feleńczak et al., 2008
Polish Red	A = 0.690 B = 0.310	Erhardt et al., 1998
Red Danish Dairy Cattle	A = 0.810 B = 0.190	
German Red	A = 0.642 B = 0.288 E = 0.070	
Hereford and crossbreeding F₁ and R₁ (Black and White x Hereford)	A = 0.722 B = 0.277	Litwińczuk & Król, 2002
Limousine and crossbreeding F₁ and R₁ (Black and White x Limousine)	A = 0.607 B = 0.393	
Simmentaler	A = 0.681 B = 0.302 E = 0.017	
Estonian Native	A = 0.695 B = 0.305	Jõuru et al., 2007
Western Finncattle	A = 0.671 B = 0.305 E = 0.024	
Danish Jersey	A = 0.512 B = 0.488	
Estonian Holstein	A = 0.790 B = 0.138 E = 0.072	
Estonian Red	A = 0.642 B = 0.324 E = 0.034	

Table 3. The frequencies of κ-casein alleles in various breeds of cattle

β-lactoglobulin is a major whey protein. Bovine BLG gene is located on chromosome 11 (11q28) and is responsible for coding the main whey protein which is β-lactoglobulin. It consists of 7 exons and its length is approximately 6 700 bp.

Cattle breed	Allelic frequencies of β-lactoglobulin	References
Jersey	A = 0.320 B = 0.680	Ren et al., 2011
Holstein	A = 0.320 B = 0.680	
Czech Fleckvieh	A = 0.511 B = 0.489	Kučerova et al., 2006
Holstein-Friesian	A = 0.480 B = 0.520	Tsiaras et al., 2005
Uruguayan Creole	A = 0.4938 B = 0.5062	Rincón et al., 2006
Polish Red	A = 0.188 B = 0.740 C = 0,058 I = 0,014	Erhardt et al., 1998
Red Danish Dairy Cattle	A = 0.110 B = 0.890	
German Red	A = 0.157 B = 0.775 C = 0.068	
Hereford and crossbreeding F$_1$ and R$_1$ (Black and White x Hereford)	A = 0.278 B = 0.722	Litwińczuk & Król, 2002
Limousine and crossbreeding F$_1$ and R$_1$ (Black and White x Limousine)	A = 0.520 B = 0.480	
Simmentaler	A = 0.379 B = 0.621	
Estonian Native	A = 0.314 B = 0.686	Jõuru et al., 2007
Western Finncattle	A = 0.098 B = 0.902	
Danish Jersey	A = 0.463 B = 0.537	
Estonian Holstein	A = 0.421 B = 0.579	
Estonian Red	A = 0.254 B = 0.746	

Table 4. The frequencies of β-lactoglobulin alleles in various breeds of cattle

The first who discovered its polymorphism were Aschaffenburg and Drewry in 1955 (as cited in El-Hanafy et al., 2010). So far 12 genetic variants of β-lactoglobulin were identified: A, B, C, D, E, F, G, W, H, I, J and X; they encode different forms of proteins (Farrell et al., 2004). For most cow breeds, both variant A and B are most common and occur with high frequency (Heidari et al., 2009). Mutations in the nucleotide sequence resulting in substituting of amino acids are distributed on 3 exons: exon 2 (allele D), exon 3 (alleles A and B) and exon 4 (alleles A, B and I) (Kamiński, 2001). The differences between variant A and B occur because of the existence of different amino acids at position 64: aspartic acid for variant A and glycine for variant B. What is more, at position 118 variant A has valine and variant B alanine (Kučerova et al., 2006). The presence of one of the variants (A or B) significantly influences the physicochemical properties of milk and also raises the actual contents of β-lactoglobulin (Farrell et al., 2004). Tsiaras et al. (2005) noted the presence of 3 genotypes with the frequencies ranging from 0.25 to 0.47. What is more, Heidari et al. (2009) calculated the frequencies of β-lactoglobulin genotypes at the levels of: 0.257 for AA, 0.544 for AB and 0.198 for BB. Kučerova et al. (2006) observed that the proportion of AB β-lactoglobulin genotype ranged from 49.5 to 66.6 % corresponding to other genotypes: for AA from 18.2 to 26.4 % and for BB from 15.2 to 24.1 %. Table 4 shows the frequencies of β-lactoglobulin alleles in various breeds of cattle.

α-LA gene encoding α-lactalbumin is localized on chromosome 5 (BTA5q12-13). α-lactalbumin has the length of 3 061 bp and it consists of 4 exons. The polymorphism of this gene is revealed in cattle breeds deriving directly from *Bos indicus*. What is more α-lactalbumin appear most commonly in 2 forms: A and B (Kamiński, 2001). In bovine milk the concentration of α-lactalbumin varies between 1.2 and 1.5 g/l (Farrell et al., 2004). This whey protein has a precise role in the mammary gland. It reacts with the enzyme β-1,4-galactosyltransferase to form the lactose synthase complex within the Golgia apparatus. It allows the formation of lactose from glucose and UDP-galactose thank to the modifications in the substrate specificity of β-1,4-galactosyltransferase. The function of lactose as a main osmolyte of milk and its production demonstrate the importance of this fraction (Farrell et al., 2004).

2.2. Goats

Goat milk contains 4 caseins (αs1-, αs2-, β- and κ-) linked with each other encoded respectively by autosomal genes: CSN1S1, CSN1S2, CSN2 and CSN3. They are located on chromosome 4 in the following order: αs1-, β-, αs2- and κ-. The CSN1S1 gene encoding αs1-casein has the most complex construction and has a number of polymorphic sites. It has the size of 16.7 kb and it consists of 19 exons with a length of 24 to 385 bp and 18 introns with a length of 90 to 1 685 bp (Supakorn, 2009).

The gene CSN1S1 in goats presents the highest level of variability of all the casein genes among all species of ruminants that have been analysed. So far 16 genetic variants of αs1-casein were identified: A[1], B[1], B[2], B[3], B[4], C, E, F, G, H, I, L, M, N, 0[1] and 0[2]. They probably evolved from 4 original alleles: A, B[1], B[2] and W. What is more, these different alleles are associated with 4 levels of protein synthesis in milk. A high level of αs1-casein (3.5-3.6 g/l) synthesis is connected with "strong" alleles: A, B[1], B[2], B[3], B[4], C, H and L. "Medium" alleles determine the protein synthesis at the levels of 1.1-1.6 g/l, while "weak" alleles are associated with the synthesis only at the amounts of 0.45-0.6 g/l. "Null" alleles (0[1] and 0[2]) account for trace amounts or complete absence of this casein fraction in milk (Caravaca et al., 2008; Ibeagha-Awemu et al., 2005; Moatsu et al., 2006; Veress et al., 2004). Table 5 shows the frequencies of αs1-casein alleles in various breeds of goats.

In the local breeds of goats the frequency of A, B and C alleles (so called "strong" alleles) is generally higher, whereas in typical dairy breeds the frequency of so called "medium" and "weak" alleles exceeds (Barłowska et al., 2007a; Jordana et al., 1996; Torres-Vázquez et al., 2008). Moatsou et al. (2008) analysed the contribution of individual genetic variants of αs1-casein in goats of local Greek and international breeds (Alpine and Saanen). In their research they presented that in local breeds of goats "strong" alleles exceeded and there were no "weak" and "null" alleles. In the international breeds, however, "medium" variants were in majority, and nearly ⅓ had "weak" and "null" alleles. Table 5 shows the frequencies of αs1-casein alleles in various goat breeds.

Goat's CSN1S2 locus is characterized by a much higher genetic diversity compared to cattle or sheep. So far 8 alleles associated with different level of synthesis of αs2-casein have been identified: alleles A, B, C, E and F are the so called "normal" alleles connected with the synthesis of αs2-casein at the level of 2.5 g/l; allele D is an intermediate allele associated with the synthesis of αs2-casein at the level of 1.25 g/l; "null" alleles linked to the absence or synthesis of αs2-casein at trace levels. Chessa et al. (2003) presented in their research that the population of Italian dairy goats (Maltese breed) are characterized by the presence of 5 alleles of CSN1S2: A, B, C, G and 0 with the frequencies: 0.548, 0.062, 0.319, 0.067 and 0.005 respectively.

So far 5 alleles of goat's CSN2 gene were identified that are associated with different levels of β-casein in milk. Alleles A, B and C are connected with normal contents of this protein, while alleles 0 and 0' are linked with undetectable or trace amounts (Ibeagha-Awemu et al., 2005). Sztankóová et al. (2005) in goats from two local breeds White Short- Haired (WSH) and Brown Short-Haired (BSH) maintained in Czech Republic identified allele A associated with "normal" synthesis of β-casein and allele 0 connected with a lack of synthesis of β-casein. Goats in the studied population, however, were not associated with a variant 0 of CSN2.

Goat κ-casein gene is composed of 5 exons, out of which the portion encoding the mature protein is located in the exon 3 (9 amino acids) and exon 4 (162 amino acids). It is assumed that the κ-casein gene is not evolutionary correlated to milk proteins sensitive to calcium, although it is linked with them. There is a hypothesis that it is related to the fibrinogen (Strzelec & Niżnikowski, 2009). CSN3 was not considered a multiallelic gene until 1990, when Di Luccia et al. (1990) reported variability (alleles A and B) in an unspecified Italian breed. Since then, 11 following variants of this protein were identified: C, D, E, F, G, H, I, J, K, L and M (Ibeagha-Awemu et al., 2005). Chessa et al. (2003) in the Maltese goat breed

Breed	Country	Allel								References
		A	B	C	D	N	E	F	D+O	
Toggenburg	Mexico	0.141	0.212			0.179	0.147	0.321		Torres-Vázquez et al., 2008
Appearance of Murciana-Granadina	Mexico	0.250	0.135			0.058	0.442	0.115		
Mosaico Lagunero	Mexico	0.183	0.350			0.183	0.050	0.233		
Saanen	Mexico	0.031	0.108			0.072	0.418	0.371		
Alpine	Mexico	0.185	0.142			0.154	0.241	0.278		
Alpine	Spanish	0.14	0.05	0.01	0.34			0.41	0.05	Jordana et al., 1996
Murciano-Granadina	Spanish	0.08	0.23		0.59			0.08	0.02	
Malagueña	Spanish	0.09	0.09		0.65			0.04	0.13	
Payoya	Spanish	0.05	0.19		0.76					
Canaria	Spanish	0.28	0.32		0.20				0.20	
Palmera	Spanish	0.68	0.23		0.09					
Majorera	Spanish	0.07	0.38		0.24				0.31	
Tinerfeña	Spanish	0.15	0.35		0.32				0.18	
Boer	Malaysia	0.15	0.10	0.71				0.04		Marini et al., 2011
Boer-Feral	Malaysia	0.22		0.68				0.03		
Katjang	Malaysia	0.61	0.11	0.17						
Jamnapari	Malaysia	0.51	0.10	0.28				0.10		
White improved	Poland	0.252	0.285				0.463			Barłowska et al., 2007a
Colored improved	Poland	0.194	0.319				0.487			
White non-improved	Poland	0.185	0.233				0.582			
Colored non-improved	Poland	0.182	0.196				0.622			

Table 5. The frequencies of αs1-casein alleles in various breeds of goats

characterized 3 alleles of CSN3, i.e. A, B and D with the frequency: 0.089, 0.230 and 0.708 respectively. Sztankóová et al. (2005) in White Short-Haired (WSH) and Brown Short-Haired (BSH) breeds of goats also identified 3 alleles of CSN3, although these were alleles A, B and C. The frequencies of A, B and C alleles were, respectively: 0.15, 0.80 and 0.05 for the WSH breed and 0.52, 0.40 and 0.03 for BSH breed. Bemji et al. (2006) in Nigerian Red Sokoto breed characterized the presence of 3 alleles of CSN3, i.e. A, B and M with the following frequencies: 0.453, 0.523 and 0.023.

Breed	-60C allele	-60T allele	References
Muriciano Grandana	0.86	0.14	Yahyaoui et al., 2000
Canaria	1	-	
Payoya	0.73	0.27	
Malguena	0.75	0.25	
Saanen	0.73	0.27	
Hungarian Milk	0.88	0.12	Veress et al., 2004

Table 6. The frequencies of β-lactoglobulin in various breeds of goats

Goat's LAA and LBG genes encoding 2 major whey proteins, respectively, α-lactalbumin and β-lactoglobulin, are characterized with much lower genetic variation in comparison to the genes observed in cattle. Goat's LGB gene is located on chromosome 11q28. What is more there is a high homogeneity of β-lactoglobulin in bovine, ovine and goat (at the level of 95 %). Sheep and goat β-lactoglobulin differs itself from bovine β-lactoglobulin only in 6 positions (Strzelec & Niżnikowski, 2009; Yahyaoui et al., 2000). Table 6 shows the frequencies of β-lactoglobulin in various breeds of goats.

2.3. Sheep

The research on the analysis of milk protein polymorphism in sheep is limited to the study of the gene polymorphism of αs1-casein and β-lactoglobulin.

BLG gene in sheep is located on chromosome 3 and three alleles can be found within its area: A, B and C. Alleles A and B (present in all breeds) differ in the substitution of 1 amino acid at position 20 of the polypeptide chain, i.e. for variant A it is tyrosine, while for variant B it is histidine. Variant C of β-lactoglobulin, which is rare, is a subtype of allele A and it was recognized within the German and Spanish Merino breed. It differs from variant A with a substitution of arginine with glutathione at position 148 of the amino acid chain (Arora et al., 2010; El-Shazly et al., 2012; Mohammadi et al., 2006).

Sheep milk contains 4 casein fractions: αs1-, αs2-, β- and κ- encoded by the genes: CSN1S1, CSN1S2, CSN2 and CSN3. These genes are localized on chromosome 4. So far 8 alleles of CSN1S1 were identified: A, B, C, D, E, F, G and H (Giambra et al., 2010). Vlaic et al. (2011) evaluating 282 sheep from 5 breeds maintained in Romania: Turcana, Carabasa, Tigaie (white and rusty varieties), Cluj Merinos, Botosani Karakul (black, dark grey, brown, light grey, pink and white varieties) have reported the presence of 2 alleles of BLG, i.e. A and B. The frequencies of alleles ranged: for allele A from 0.422 to 0.800 and for allele B from 0.133 to 0.578. Mroczkowski et al. (2004) in Polish Merino identified: 4 alleles of CSN1S1 (A, B, C and D), 3 alleles for CSN2 (A, B and C) and 2 alleles for β-lactoglobulin (A and B). They indicated the existence of alleles, respectively: for α-lactalbumin only 1 allele A with a frequency of 1.00; for αs1-casein 4 alleles with frequencies of A=0.078, B=0.007, C=0.905 and D=0.010; for β-casein 3 alleles with frequencies of A=0.944, B=0.051 and C=0.005; for β-lactoglobulin 2 alleles with frequencies of A=0.498 and B=0.502.

2.4. Buffalo

Buffalo milk is characterized by the presence of all 4 casein fractions (αs_1-, β-, αs_2- and κ-) encoded by 4 closely linked autosomal genes (CSN1S1, CSN2, CSN1S2 and CSN3) that are mapped on chromosome 7 (Iannuzzi et al., 2003). These casein fractions are distributed in buffalo milk respectively: β- (53.45 %), αs_1- (20.61 %), αs_2- (14.28 %) and κ- (11.66 %) (Cosenza et al., 2011). What is more, BLG gene is mapped on chromosome 12 (as cited in El Nahas et al., 2001). Ren et al. (2011) assessing the population of 48 water buffaloes identified only the existence of allele B of CSN3 and LBG. Allele A was not present at all for these fractions of protein. Similar results were established by Shende at al. (2009) who were studying the polymorphism of κ-casein in 20 buffaloes – only allele B of CSN3 was identified.

3. Milk protein genes as markers of production traits

The research on determining the relationship between the presence of different genetic markers and production traits of animals is being conducted for many years now. In the livestock farming the emphasis was put on milk protein genes.

3.1. The association between selected genetic variants with the chemical composition of milk

3.1.1. Cattle

A major milk protein which is considered to be an important genetic marker of quantitative traits is β-lactoglobulin. Analysing the association of genetic variants of is β-lactoglobulin with the chemical composition of milk, most authors link the variant B of β-lactoglobulin with the higher contents of total protein, casein, fat and dry matter in milk (Barłowska, 2007; Barłowska et al., 2007b; Ng-Kwai-Hang, 2002). Lodes et al. (1997) presented in their research that the gene B of β-lactoglobulin led to the increase of total protein in milk (including casein) with the simultaneous decrease of contents of whey proteins. The protein contents, depending on the genotype of β-lactoglobulin was reduced, respectively: for total protein – BC>AD/BB>AB>AA/BD; for casein – BC>BB>AD/AB/BD>AA; for whey protein – AA>AD>AB>BC>BD>BB. A similar relationship was ascertained by Bonfatti et al. (2010) who conducted research on 2 167 cows of the Simmental breed maintained in 47 herds in Northern Italy. They have also reported the positive influence of allele B of β-lactoglobulin on the percentage contribution of all casein fractions and α-lactalbumin and negative effect on the contents of β-lactoglobulin. Contrary to that research, Hallen et al. (2008) presented a negative association of variant B of β-lactoglobulin with the concentration of total protein.

Ryniewicz et al. (1998) in their research showed that the physical structure, i.e. the compactness of the casein curd connected with the ease and speed of its digestion, is associated with the polymorphic forms of milk proteins. What is more, milk with AB heterozygotes of β-lactoglobulin has the most digestible protein which may be associated with better results of rearing of offspring from cows with this genotype. This fact is confirmed by the studies of Litwińczuk & Król (2002) and Król (2003) who evaluated the cows of the Limousine and Hereford breeds; the research focused on calves to the age of approximately 8 months. The average daily gains of calves reared by cows of the Hereford breed with AB genotype of β-lactoglobulin (861.3 g) were 115.7 g higher in comparison to the increments of calves reared by cows with AA genotype and 33 g higher compared to BB homozygotes of β-lactoglobulin. Higher daily gain of calves with AB genotype of β-lactoglobulin led to higher body weight at the age of 210 days (207.4 kg) in comparison to BB (192.7 kg) and AA (184.1 kg) homozygotes. Similar results were presented in Henderson & Marshall (1996) studies on the multi-breed population of beef cattle. They stated that the best results of rearing calves were achieved by cows with the AB and BB genotypes of β-lactoglobulin.

A milk protein which arouses the greatest interest as a genetic marker is κ-casein. On the one hand (Barłowska et al., 2007b; Litwińczuk et al., 2006; Winkelman & Wickham, 1997) in their studies link AA genotype of κ-casein with a higher yield of milk. On the other hand Creamer & Harris (1997) present that the highest daily yield of milk is obtained from cows with AB genotype of κ-casein. Król (2003) who conducted research on beef cattle of the Hereford breed has also concluded that the highest milk yield was associated with the AB genotype of κ-casein. Henderson & Marshall (1996), who analysed the characteristics of milk yield in the multi-breed population of beef cattle in South Dakota, stated that the cows with AB and AA genotypes of κ-casein produced the most milk and BB homozygotes of κ-casein the lowest quantity. Lodes et al. (1997) within 7 genotypes of κ-casein, i.e. AA, AB, AC, AE, BB, BC and BE did not describe significant differences in the productivity of dairy cows. Some authors point to a higher concentration of total protein (including casein) and fat in milk from cows of BB homozygotes of κ-casein (Barłowska, 2007; Karima et al., 2010). Henderson & Marshall (1996) also indicate the relationship between genetic variants of κ-casein with the results of rearing calves.

The progeny of cows with AB genotype of κ-casein achieved the highest daily gain (932 g) and the highest body weight at the peak (242 kg). Similar results were presented in the research of Król (2003) who carried it out on cows of the Hereford and Limousine breed.

In most studies focused on the analysis of genotype of α_{s1}-casein and β-casein no clear significant relation of these genetic variants on cow productivity (and contents of protein and casein in milk) was presented (Huang et al., 2012; Ikonen et al., 2001; Litwińczuk & Król, 2002). Only the variant C of α_{s1}-casein is associated with a slightly higher contents of protein and casein in milk in comparison to allele B of α_{s1}-casein (Winkelman & Wickham, 1997). McLean (1984) indicates the connection of variant C of α_{s1}-casein with a higher concentration of α_{s1}-casein and also with a simultaneous decrease of κ-casein in milk.

In the case of β-casein the research is mainly focused on the subfraction A of β-casein, i.e. A^1, A^2, A^3. Most studies (Nilsen et al., 2009; Olenski et al., 2010; Winkelman & Wickham, 1997) indicate the positive association of variant A^2 of β-casein on milk yield and the contents of protein. According to Nilsen et al. (2009) and Olenski et al. (2010) A^2 allele of β-casein is an effective genetic marker for the productivity of proteins. They suggest conducting a selection of cattle to raise its frequency. Table 7 shows the association of genetic variants of cattle milk proteins with milk yield and its composition.

Item	Protein fraction		
	β-LG	κ-CN	αs1-CN
Milk yield	AA>BB>AB (Creamer & Harris, 1997; Walawski et al., 1994) AB>AA>BB (Henderson & Marshall, 1996; Król, 2003)	AB>AA>BB (Creamer & Harris, 1997; Henderson & Marshall, 1996; Król, 2003) AA>AB>BB (Litwińczuk et al., 2006)	no distinct tendency between genotypes
Total protein	BB>AB>AA (Lunden et al., 1997; Ng-Kwai-Hang, 2002) AA/AB>BB (Creamer & Harris, 1997; Hallen et al., 2008)	BB>AB>AA (Barłowska et al., 2007b; Creamer & Harris, 1997; Feleńczak et al., 2004, Karima et al., 2010)	CC/BC>BB (Winkelman & Wickham, 1997)
Casein	BB>AB>AA (Bonfatti et al., 2010, Creamer & Harris, 1997; Heck et al., 2009)	BB>AB>AA (Creamer & Harris, 1997; Feleńczak et al., 2004, Karima et al., 2010)	CC/BC>BB (Winkelman & Wickham, 1997)
Whey protein	AA>AB>BB (Creamer & Harris, 1997; Lunden et al., 1997)	AA>AB>BB (Creamer & Harris, 1997)	AB>BB>BC (Creamer & Harris, 1997)
Fat	BB>AB>AA (Ng-Kwai-Hang, 2002)	BB>AB>AA (Barłowska et al., 2007b; Karima et al., 2010)	no distinct tendency between genotypes

Table 7. The association of genetic variants of cattle milk proteins with milk yield and its composition

3.1.2. Goats

From all polymorphic proteins in goats the most known is definitely αs1-casein. In most studies, no significant correlation between the variants of αs1-casein and milk productivity was ascertained linking this with a strong influence of various factors such as age, stage of lactation etc. (Barłowska et al., 2007a; Litwińczuk et al., 2007; Remeuf, 1993). Some authors, however (Ryniewicz et al., 2000; Vassal & Manfredi, 1994), indicate a lower daily milk yield from goats with a variant E of αs1-casein. Most of the studies on goat milk were focused on the analysis of the association of polymorphic variants of αs1-casein with the chemical composition of milk and its technological properties (Barłowska et al., 2007a; Clark & Sherbon, 2000; Remeuf, 1993; Schmidely et al., 2002). A positive influence of "strong" variants (A, B and C) of α_{S1}-casein on the concentration of particular milk components (especially proteins) was demonstrated (Barłowska et al., 2007a; Krzyżewski et al., 2000; Litwińczuk et al., 2004; 2007b; Remeuf, 1993; Schmidely et al., 2002; Vassal & Manfredi, 1994). Krzyżewski et al. (2000) analysing the population of White improved breed of goats indicated a higher contents of protein (about 0.15 % more) in the milk of goats characterized by "strong" polymorphic variants of α_{S1}-casein in comparison to animals with "medium" variants. In the studies conducted by Litwińczuk et al. (2004) on goats originating from the region of Wielkopolska and Podkarpacie differences in total protein contents between goats with "strong" and "medium" variants ranged from 0.29 to 0.39 %. Schmidely et al. (2002) compared the chemical composition of homozygotes AA and FF of α_{S1}-casein of the Alpine and Sannen breeds and showed that the animals with AA genotype produced milk with a higher contents of fat (about 1.2 g/kg more) and protein (about 6.3 g/kg more) in comparison to goats with FF homozygotes.

A propitious relationship between "strong" α_{S1}-casein genotypes and contents of fat and protein in milk was also described in the research of Barłowska et al. (2007a and 2007c) conducted on goats of 4 breeds. Moreover, a clear association of α_{S1}-casein genotypes with the concentration of casein was indicated. In a group of goats with the highest contents of casein in milk (≥2.4 %) in each breed goats with "strong" genotypes (AA, AB, BB, AE and BE) were predominant – in comparison to animals with "medium" genotypes the prevalence ranged from 7.2 % (for coloured non-improved breed) to 29.1 % (for white improved breed of goats). The opposite situation was observed for samples containing the lowest amounts of casein (≤1.99 %). In every breed goats with "medium" genotypes of α_{S1}-casein dominated. Their prevalence ranged from 7.9 % for white non-improved breed to 19.0 % for coloured improved breed. According to Vassal & Manfredi (1994) the contents of casein in milk depending on the different alleles of α_{S1}-casein varies, respectively: for "strong" alleles, i.e. A, B and C- 3.6 g/l; for "medium" alleles, i.e. E and G – 1.6 g/l; for "weak" alleles, i.e. D and F – approximately 0.6 g/l. The presence of "null" variant is related to the synthesis of trace amounts or the complete absence of this fraction.

Genetic variants of α_{S1}-casein are not only associated with the protein contents in milk but also with their quantitative proportions both within α_{S1}-casein and the rest of proteins from casein group. What is more, milk from goats with "strong" alleles of α_{S1}-casein contains more Ca and Zn (Krzyżewski et al., 2002).

Ryniewicz et al. (1998) also presented a relationship between polymorphic variants of α_{S1}-casein and susceptibility of total protein to hydrolysis. Significantly higher degree of hydrolysis of protein was determined in the milk of goats characterized by "weak" variants of α_{S1}-casein.

3.1.3. Sheep

Most studies conducted on ovine milk concern the association of polymorphic variants of α_{S1}-casein, β-casein and β-lactoglobulin on the chemical composition of milk. It was indicated that sheep with CC genotype of α_{s1}-casein are characterized by a higher contents of base composition. In the studies conducted on the population of the Polish Merino and Polish Merino x Prolific sheep (Mroczkowski et al., 2002) it was determined that milk obtained from CC homozygotes of α_{s1}-casein contained significantly more protein and dry matter in comparison to AC and BC heterozygotes. What is more, BC genotype of α_{s1}-casein was associated with a higher milk production.

There are not many studies concerning the relationship between polymorphic variants of β-casein and the composition of milk. In the research of Mroczkowski et al. (2004) sheep with AA genotype of β-casein were characterized by a higher contents of protein and dry matter in comparison to individuals with AB genotype of β-casein. However, AB heterozygotes of β-casein corresponded to a higher milk production.

The results of Mroczkowski et al. (2004) indicate that the AA and BB homozygotes of β-lactoglobulin were characterized by higher yields of milk and contents of protein and casein than AB heterozygotes. In the research of Nudda et al. (2000) who analysed the milk of Sarda sheep, it was stated that the genetic variants of β-lactoglobulin have an association with only the milk yield, i.e. animals with AB genotype of β-lactoglobulin were considered to have the highest daily production. The results of other authors are presented in the Table 8.

Item	Protein fraction	
	β-LG	α_{s1}-CN
Milk yield	AA/BB>AB (Mroczkowski et al., 2004) AB>AA>BB (Nudda et al., 2000)	BC>CC>CD (Chianesse et al., 1996) BC>AC>CC (Mroczkowski et al., 2002)
Total protein	BB>AB >AA (Krukovics et al., 1998; Mroczkowski et al., 2002)	CC>CD>BC (Chianesse et al., 1996) CC>AC>BC (Mroczkowski et al., 2002)
Casein	BB>AB>AA (Krukovics et al., 1998)	CC>CD>BC (Chianesse et al., 1996)
Fat	no distinct tendency between genotypes	CC>AC>BC (Mroczkowski et al., 2002; Mroczkowski et al., 2004)

Table 8. The association of genetic variants of ovine milk proteins with the milk yield and chemical composition

3.2. The association between selected genetic variants with the parameters of technological suitability of milk

3.2.1. Cattle

For cattle, polymorphic forms of κ-casein and β-lactoglobulin are strongly associated with the parameters of technological suitability of milk.

Many studies emphasize the distinct association of genetic variants of β-lactoglobulin with the chemical composition of milk and cheese yield. Most of the authors combine BB variant of β-lactoglobulin with the higher contents of fat in milk (Barłowska, 2007; Barłowska et al.,

2007b; Ng-Kwai-Hang, 1997), protein (Ng-Kwai-Hang, 2002), casein (Lunden et al., 1997), dry matter and also with the higher cheese yield as well as better thermal stability of milk (Imafidon & Ng-Kwai-Hang, 1991). Lodes et al. (1997) analysing the chemical composition of milk obtained from 801 cows demonstrated that among 7 genotypes of β-lactoglobulin the highest protein contents was associated with animals with BC (3.76 %) genotype of β-lactoglobulin, and the highest casein contents with BC (2.97 %) and BB (2.85 %) genotypes of β-lactoglobulin. What is more, BW genotype of β-lactoglobulin was characterized with the lowest amount of protein (3.44 %) and casein (2.68 %). In their research Vătăşescu-Balcan et al. (2007) presented that the BB homozygotes of β-lactoglobulin are associated with the production of milk rich in fat and protein and therefore very valuable in the manufacture of cheese. According to Creamer & Harris (1997) variant A of β-lactoglobulin contributes (in contrary to variant B) to the formation of more concise clot at approximate pH of 7. Variant B, in turn, is associated with faster thermal coagulation of milk. Imfidon & Ng-Kwai-Hang (1991) also claim that the milk obtained from cows with AA genotype of β-lactoglobulin is more resistant to high temperatures. In studies of Barłowska (2007) it was presented that the best properties in terms of contents of protein (including casein), fat and dry matter was found in milk obtained from cows with genotypes AA or AB of β-lactoglobulin (which means that the variant A was present). Milk from cows with BB genotype of β-lactoglobulin indicated the shortest (p≤0.05) time of rennet coagulation. However, milk from animals with heterozygotes (β-LG AB) was characterized by the longest heat treatment stability, i.e. about 29 s longer related to AA homozygotes of β-lactoglobulin (p≤0.05).A more explicit studies are formulated by authors presenting an association of genetic variants of κ-casein with milk composition and its suitability for cheese production. Most of them (Azevedo et al., 2008; Feleńczak et al., 2004; FitzGerald, 1996; Imafidon & Ng-Kwai-Hang, 1991; Lunden et al., 1997; Tsiaras et al., 2005) indicate the fact that cows with BB genotype of κ-casein produce milk with a higher contents of total protein and casein. In addition, this genotype increases the fat contents in milk, provides better stability of casein micelles, shorter time of flocculation, firmer clot formation and higher cheese yield. So that, it can be concluded that genes coding the synthesis of κ-casein have an influence on technological processes in cheese-making. Barłowska et al. (2007b) stated in their research that the presence of B variant of κ-casein in cows (of the Polish Red and White-back breed) favourably affects the contents of dry matter, total protein, casein and also it reduces the time of enzymatic coagulation and prolongs the colloidal stability of milk. According to Robitaille et al. (2001) the genetic polymorphism in the gene expression of κ-casein may influence the physicochemical properties of casein micelles (it is a component stabilizing micelles) and thus the technological properties of milk. What is more, BB genotype not only affects the increase of stability of casein micelles but also quantities of casein (in which it increases the contents of κ-casein fraction within all caseins). FitzGerald (1996) in his research presented that from milk with BB genotype of κ-casein it is possible to achieve higher yields of cheese (Edam, Gouda, Cheddar, Mozzarella). This may be associated with a higher fat recovery from cheeses manufactured from that milk. In the research of Imafidon & Ng-Kwai-Hang (1991) it was presented that milk with BB genotype of κ-casein is more resistant to heat treatment.

Four casein *loci* (CSN1S1, CSN2, CSN1S2 and CSN3) are closely linked and conduct as one genetic unit, thus form different combinations of alleles (haplotypes). According to Matajicek et al. (2007) 15 combinations of κ-casein and β-lactoglobulin were identified and the most frequent were AB/AB (21.0 %) and AA/AB (18.3 %). What is more, BB/AA genotype was determined to have the highest positive association with the evaluated properties of milk. AB/BB, BB/BB, BB/AB and AB/AB genotypes also had a positive correlation with the quality of milk and its clotting properties, whereas genotypes with allele E of κ-casein negatively affected these parameters. It was presented that the distribution of allele A in the combination of κ-casein and β-lactoglobulin genotypes resulted in the increase of milk yield, while the presence of allele B was associated with increased contents of protein and fat in milk. Comin et al. (2008) indicated that κ-casein and β-lactoglobulin genotypes had a strong relationship with parameters of milk coagulation but not with fat and protein contents and other parameters of milk quality. The best results affecting the clotting of milk were these combinations of κ-casein and β-lactoglobulin that contained at least 1 allele B in both *loci*. κ-casein *locus* was more strongly associated with milk coagulation parameters, whereas β-lactoglobulin was more connected with milk yield and proteins.

3.2.2. Goats

Many studies (Barłowska et al., 2007a,c; Devold et al., 2010; Remeuf, 1993; Mahé et al., 1994; Strzałkowska et al., 2004) indicate a strong relationship between genetic variants of CSN1S1 with the chemical composition of milk, mainly the contents of casein, but also with its technological parameters important for cheese production. Homozygous goats with "strong" alleles of CSN1S1 produce milk with significantly higher percentage of protein, fat, calcium and a small diameter of casein micelles. It is possible to acquire more cheese from this type of milk and the curd is more concise in comparison to milk obtained from homozygous goats with "medium" and "weak" alleles. Sacchi et al. (2005) reported that goats with "strong" alleles of α_{s1}-casein (A, B^1, B^2, B^3, B^4, C, H, L, M) synthetize this protein fraction on the level of 3.5 g/l of milk. For "medium" (E, J) alleles this value is lower and reaches the level of 1.1 g/l of milk and for "weak" alleles (F, G) it is close to 0.45 g/l of milk. In the case of "null" alleles (0^1, 0^2, N) this protein is not synthetized. This fact is explained by morphological observations at a cellular level of mammary tissue performed by Martin et al. (1999). It was reported that epithelial cells of homozygous goats with "weak" alleles of α_{s1}-casein (E, F, G and O) were characterized by a dramatic swelling of the rough endoplasmic reticulum mainly due to the accumulation of proteins, what strongly suggests a dysfunction in secretion mechanisms. In similar studies Barłowska et al. (2007a) indicated that a group of animals produced milk with the highest contents of casein (over 2.4 %) and protein (over 3.0 %) and individuals with "strong" α_{s1}-casein variants were predominant. They constituted approximately 70 % (as for protein) and over 85 % in a group of goats which milk was associated with the highest casein level. A contrary tendency was observed in a group of goats producing milk with low protein (≤2.4 %) and casein (≤2.0 %) contents. There, the individuals with "medium" α_{s1}-casein genotypes (57 – 59 %) predominated. According to

Pierre et al. (1996) goat milk with AA genotype of α_{s1}-casein indicated a higher contents of total nitrogen and fat compared to milk obtained from goats with 00 genotype. Cheeses of this type of milk were also more firm, had a higher yield and contained less volatile aromatic compounds. In the research of Clark & Sherbon (2000) it was presented that the milk obtained from goats with at least 1 "strong" (B[1], B[2], B[3] or C) genetic variant of α_{s1}-casein contained more dry matter, SNF, protein and α_{s1}-casein compared to the milk of goats with "weak" variants of α_{s1}-casein (F or D) or homozygotes with "null" variants of α_{s1}-casein (00). However, genetic variants of α_{s1}-casein were not closely correlated with milk coagulation properties. According to Pop et al. (2008) formation of the characteristic "goat flavour" is also associated with the genotype of α_{s1}-casein (CSN1S1). Cheese manufactured from milk with AA genotype has a weaker "goat flavour" compared to the one produced from milk with FF genotype. This is explained by the fact that goat milk with FF genotype of CSN1S1 has a higher lipase activity in comparison to milk obtained from goats with AA genotype. There is relatively little research that clarifies the relationship of genetic variants of κ-casein with the technological parameters of goat milk. According to Chiatti et al. (2007) genetic variants of κ-casein may have an association with the contents of protein in milk (including casein) in the following trend: BB>AB>AA.

3.2.3. Sheep

There is not much research explaining the relationship of genetic variants of ovine milk proteins with the technological parameters of milk. Only in a few studies the correlation between β-lactoglobulin variants and base chemical composition was demonstrated. Çelük & Zdemür (2006) evaluating the Awassi and Morkaraman breeds demonstrated that genetic variants of β-lactoglobulin were only associated with the contents of protein and fat in the milk of Awassi sheep, but had no relation to the contents of dry matter, acidity and milk coagulation time in 2 breeds of sheep.

4. The modification of milk composition through genetic engineering

The measures to modify the chemical composition of milk in order to achieve the desired health benefits or processing properties are of increasing importance in dairy biotechnology. The mammary gland is a bioreactor which allows manufacturing of proteins of foreign species. One of the possible changes in the milk composition is the introduction of new or the development of existing milk protein genes which may increase the nutritional value of the product or improve its properties as a raw material for processing (for example by genetic modification of milk it is possible to increase the heat resistance). It is also possible to use the process of "humanization" of cow's milk by the partial replacement of the cattle proteins with those of a human. An important modification of milk is a reduction of lactose contents which adversely affects the quality of cheeses and other dairy products and is not well tolerated by many people as a food ingredient. The decrease of lactose contents in milk may be achieved by inactivating or reducing the expression of α-lactalbumin gene or by introducing active in the mammary gland lactase gene or bacterial β-galactosidase gene (*lacZ* gene). Studies are also conducted towards reducing or eliminating the contents of the main allergen of cow milk, which is β-lactoglobulin, through the inhibition of expression of BLG gene (*knock-out*). This protein is not present in human milk (Charon & Świtoński, 2000). The progress in the identification of genetic engineering methods utilized in dairy biotechnology will depend, in the near future, not only on the development of molecular biology but mainly on the social acceptance of the research in this area.

5. Conclusion

The extensive use of the achievements of biological and zootechnical sciences in the second half of twentieth century made it possible to attain a significant increase in the productivity of animals. What is more, the wide and easy access to molecular technologies has influenced the evaluation of animals in a way that the practical use of polymorphism of genes as markers of functional traits is commonly utilized. One of such examples is the polymorphism of milk proteins. The available bibliography suggests that these issues were analysed in all four main species of animals determining the global production of milk, i.e. in cattle, buffaloes, goats and sheep, though the cattle have been examined to the greatest extent.

The milk protein of greatest interest as a genetic marker in cattle is the κ-casein, whereas in goats it is αs₁-casein. Based on the results of many studies it is assumed that AA genotype of κ-casein is associated with higher milk production, while the BB genotype with higher contents of base chemical composition (proteins, including casein and fat). This genotype is also related to higher stability of casein micelles, shorter flocculation, firmer clot formation and consequently with a better performance of cheese. In the case of goats, a positive association between variants of "strong" (A, B, C) alleles of αs₁-casein with the composition of milk was demonstrated. Homozygous goats with "strong" alleles of CSN1S1 produced milk with higher contents of proteins, calcium, fat and with a smaller diameter of casein micelles. From the milk of these goats it is possible to obtain more cheese and the curd is more firm in comparison to homozygous goats with "medium" and "weak" alleles of αs₁-casein. Similar dependences were not observed in sheep and buffaloes, perhaps because of the lack of research in that area.

Author details

Barłowska Joanna, Wolanciuk Anna, Litwińczuk Zygmunt and Król Jolanta
University of Life Sciences in Lublin, Poland

6. References

Arora, R., Bhatia, S., Mishra, B. P., Sharma, R., Pandey, A. K., Prakash, B. & Jain, A. (2010). Genetic Polymorphism of the β-Lactoglobulin Gene in Native Sheep from India. *Biochem. Genet.*, Vol. 48, pp. 304–311

Azevedo, A. L. S., Nascimento, C. S., Steinberg, R. S., Carvalho, M. R. S., Peixoto, M. G. C. D., Teodoro, R. L., Verneque, R. S., Guimarães, S. E. F. & Machado, M. A. (2008). Genetic polymorphism of the kappa-casein gene in Brazilian cattle. *Genet. Mol. Res.*, Vol. 7, No. 3, pp. 623-630

Bai, W. L., Yin, R. H., Zhao, S. J. Y., Zheng, C., Zhong, J. C. & Zhao, Z. H. (2008). Short Communication: Characterization of a κ-Casein Genetic Variant in the Chinese Yak, *Bos grunniens. J. Dairy Sci.*, Vol. 91, pp. 1204–1208

Barłowska, J. (2007). Nutritional value and technological usability of milk from cows of 7 breeds maintained in Poland. Post-DSc dissertation. Lublin, Poland: Agriculture Academy in Lublin. Available from Univ. of Life Sciences in Lublin.

Barłowska, J., Litwińczuk, Z., Florek, M. & Kędzierska-Matysek, M. (2007a). Milk yield and its composition of 4 Polish goat breeds with different genotypes of αs_1-casein (in Polish). *Vet. Med.*, Vol. 63, No. 12, pp. 1600-1603

Barłowska, J., Litwińczuk, Z., Kędzierska-Matysek, M. & Litwińczuk, A. (2007c). Polymorphism of caprine milk αs_1-casein in relation to performance of four Polish goat breeds. *Pol. J Vet Sci.*, Vol. 10, No. 3, pp. 159-163

Barłowska, J., Litwińczuk, Z., Król, J. & Kędzierska-Matysek, M. (2007b). Relationship of β-lactoglobulin and κ-casein genetical variants with chosen indexes of milk technological usefulness of Red Polish and Whitebacks cows. *Ann. Anim. Sci.*, suppl. 1, pp. 43-47

Bemji, M. N., Ibeagha-Awemu, E. M., Osinowo, O. A. & Erhard, G. (2006). Casein (CSN3) variability of the Nigerian Red Sokoto goat. *Nigerian J. Genet.* Vol. 20, pp. 1-6

Beuzen, N. D., Stear, M. J. & Chang, K. C. (2000). Molecular markers and their use in animal breeding. *Vet. J.*, Vol.160, pp. 42–52

Bogdzińska, M. (2011). The utilization of the latest achievements of genetics in the improvement of functional characteristics of the animals (in Polish). *Anim. Prod. Rev.*, No. 6, pp. 2-4

Boichard, D., Grohs, C., Bourgeois, F., Cerqueira, F., Faugeras, R., Neau, A., Rupp, R., Amigues, Y., Boscher, M. Y. & Levéziel, H. (2003). Detection of genes in sequencing economic traits in three French dairy cattle breeds. *Genet. Sel. Evol.*, 35, pp. 77-101

Bonfatti, V., Di Martino, G., Cecchinato, A., Vicario, D. & Carnier, P. (2010). Effect of β-κ-casein (CSN2-CSN3) haplotypes and β-lactoglobulin (BLG) genotypes on milk production traits and detailed protein composition of individual milk of Simmental cows. *J. Dairy Sci.*, Vol. 93, pp. 3797-3808

Bonvillani, A. G., Di Renzo, M. A. & Tiranti, I. N. (2010). Genetic polymorphism of milk protein loci in Argentinian Holstein cattle. *Genet. Mol. Biol.*, Vol. 23, No. 4, pp. 819-823

Caravaca, F., Amills, M., Jordana, J., Angiolillo, A., Agüera, P., Aranda, C., Menéndez-Buxadera, Sánchez, A., Carrizosa, J., Urrutia, B., Sànchez, A. & Serradilla, J. M. (2008). Effect of αs_1-casein (CSN1S1) genotype on milk CSN1S1 content in Malagueña and Murciano-Granadina goats. *J. Dairy Res.*, Vo. 75, pp. 481–484

Çardak, A. D. (2005). Effects of genetic variants in milk protein on yield and composition of milk from Holstein-Friesian and Simmentaler cows. *S. Afr. J. Anim. Sci.*, Vol. 35, No. 1, pp. 41-47

Caroli, A. M., Chessa, S. & Erhardt, G. J. (2009). Invited review: Milk protein polymorphisms in cattle: Effect on animal breeding and human nutrition.*J. Dairy Sci.*, Vol. 92, pp. 5335 – 5352

Caroli, A., Chessa, S., Bolla, P., Budelli, E. & Gandini, G. C. (2004). Genetic structure of milk protein polymorphisms and effects on milk production traits in a local dairy cattle. *J. Anim. Breed. Gen.*, Vol. 121, No. 2, pp. 119-127

Çelük, Ş. & Zdemür, S. (2006). β-Lactoglobulin variants in Awassi and Morkaraman sheep and their association with the composition and rennet clotting time of the milk.*Turk. J. Vet. Anim. Sci.*, Vol. 30, pp. 539-544

Charon, K. & Świtoński, M. (2000). Genetyka zwierząt. Warszawa. Wydawnictwo Naukowe PWN.

Chen, S. Y., Costa, V., Azevedo, M., Baig, M., Malmakov, N., Luikart, G., Erhardt, G. & Beja-Pereira, A. (2008). Short Communication: New Alleles of the Bovine κ-Casein Gene Revealed by Resequencing and Haplotype Inference Analysis. *J. Dairy Sci.*, Vol. 91, pp. 3682–3686

Chessa, S., Ceriotti, G., Dario, C., Erhardt, G. & Caroli, A. (2003). Genetic polymorphisms of αs1-, αs2- and κ-casein in Maltese goat breed.*Ital. J. Anim. Sci.*, Vol. 2, Suppl. 1, pp. 58-60

Chianesse, L., Garro, G., Mauriello, R., Laezza, P., Ferranti, P. & Addeo, F. (1996). Occurrence of five αs1-casein variants in ovine milk. *J. Dairy Res.*, Vol. 63, pp. 49-59

Chiatti, F., Chessa, S., Bolla, P., Cigalino, G., Caroli, A. & Pagnacco, G. (2007). Effect of κ-casein polymorphism on milk composition in the Orobica goat. *J. Dairy Sci.*, Vol. 90, No. 4, pp. 1962-1966

Citek, J., Panicke, L., Rehout, V. & Prochazkova, H. (2006). Study of genetic distances between cattle breeds of Central Europe. *Czech J. Anim. Sci.*, Vol. 51, No. 10, pp. 429-436

Clark, S. & Sherbon, J. W. (2000). Genetic variants of alpha s1–CN in goat milk: breed distribution and associations with milk composition and coagulation properties. *Small Rum. Res.*, Vol. 38, pp. 135-143

Comin, A., Cassandro, M., Chessa, S., Ojala, M., Dal Zotto, R., De Marchi, M., Carnier, P., Gallo, L., Pagnacco, G. & Bittante, G. (2008). Effects of composite β- and κ-casein genotypes on milk coagulation, quality, and yield traits in Italian Holstein cows. *J. Dairy Sci.*, Vol. 91, pp. 4022 – 4027

Cosenza, G., Pauciullo, A., Coletta, A., Di Francia, A., Feligini, M., Gallo, D., Di Berardino, D. & Ramunno, L. (2011). Short communication: Translational efficiency of casein transcripts in Mediterranean river buffalo. *J. Dairy Sci.*, Vol. 94, pp. 5691-5694

Creamer, L. K. & Harris, D. P. (1997). Association between milk protein polymorphism and milk production traits. Proc. IDF Seminar "Milk Protein Polymorphism II" North Palmerston, New Zeland, pp. 22-37

Da Salva, I. T. & Del Lama, M. A. (1997). Milk protein polymorphisms in Brazilian Zebu cattle. *Braz. J. Genet.*, Vol. 20, No. 4, pp. 625-630

Devold, T. G., Nordbø, R., Langsrud, T., Svenning, C., Jansen Brovold, M., Sørensen, E. S., Christensen, B., Ådnøy, T. & Vegarud, G. E. (2010). Extreme frequencies of the αs1-casein "null" variant in milk from Norwegian dairy goats – Implications for milk composition, micellar size and renneting properties. *D. Sci. Tech.*, DOI: 10.1051/dst/2010033

Di Luccia, A., Mauriello, R., Chianesse, L., Moio, L. & Addeo, F. (1990). κ-casein polymorphism in caprine milk. *Sci. Tech. Latt. Cesearia*, Vol. 41, pp. 305-314

Ehrmann, S., Bartenschlager, H. & Geldermann, H. (1997). Quantification of gene effects on single milk proteins in selected groups of dairy cows. *J. Anim. Breed. Genet.*, Vol. 114, pp. 121-132

El Nahas, S. M., de Hondt, H. A. & Womack, J. E. (2001). Current status of the River Buffalo *(Bubalus bubalis L.)* gene map. *J. Hered.*, Vol. 92, No. 3, pp. 221-225

El-Hanafy, A. A., El-Saadani, M. A., Eissa, M., Maharem, G. M. & Khalifa, Z. A. (2010). Polymorphism of β-lactoglobulin gene in barki and damascus and their cross bred goats in relation to milk yield. *Biotech. Anim. Husb.*, Vol. 26, No. 1-2, pp. 1-12

Ellegren, H. (2004). Microsatellites: simple sequences with complex evolution. *Nat. Rev. Genet.*, Vol. 5, pp. 435-445

El-Shazly, S. A., Mahfouz, M. E., Al-Otaibi1, S.A. & Ahmed, M. M. (2012). Genetic polymorphism in β-lactoglobulin gene of some sheep breeds in the Kingdom of Saudi Arabia (KSA) and its influence on milk composition. *Afr. J. Biotechnol.*, Vol. 11, No. 19, pp. 4330-4337

Erhardt, G., Juszczak, J., Panicke, L. & Krick-Saleck, H. (1998). Genetic polymorphism of milk proteins in Polish Red Cattle: a new genetic variant of β-lactoglobulin. *J. Anim. Breed. Genet.*, Vol. 115, pp. 63-71

Farrell, H. M., Jimenez-Flores, R., Bleck, G. T., Brown, E.M., Butler, J. E. & Creamer, L. K. (2004). Nomenclature of the proteins of cows' milk – sixth revision. *J. Dairy Sci.*, Vol. 87, pp. 1641-1674

Feleńczak, A., Fertig, A., Gardzina, E. & Jezowit-Jurek, M. (2004). Technological characteristics of milk obtained from cows of Red-White and Simmental breeds and their relationship with the polymorphism of κ-casein (in Polish). *Ann. Anim. Sci.*, Suppl. 19, pp. 55-58

Feleńczak, A., Gil, Z., Adamczyk, K., Zapletal, P. & Frelich, J. (2008). Polymorphism of milk κ-casein with regard to milk yield and reproductive traits of Simmental cows. *J. Agrob.*, Vol. 25, No. 2, pp. 201-207

FitzGerald, R. J. (1996). Exploitation of casein variants. Cab International "Milk composition, production and biotechnology", Hamilton, New Zealand, pp. 153-172

Giambra, I. J., Chianese, L., Ferranti P. & Erhardt, G. (2010). Genomics and proteomics of deleted ovine CSN1S1*I. *Int. Dairy J.*, Vol. 20, pp. 195–202

Hallen, E. A., Wedholm, A. A. & Lunden, A. (2008). Effect of β-casein, κ-casein and β-lactoglobulin gynotypes on concentration of milk protein variants. *J. Anim. Breed. Genet.*, Vol. 17, pp. 791-799

Hanusová, E., Huba, J., Oravcová, M., Polák, P. & Vrtková, I. (2010). Genetic Variants of Beta-Casein in Holstein Dairy Cattle in Slovakia. *Slovak J. Anim. Sci.*, Vol. 43, No. 2, pp. 63 – 66

Heck, J. M. L., Schennink, A., van Valenberg, H. J. F., Bovenhuis, H., Visker, M. H. P. W., van Arendonk, J. A. M. & van Hooijdonk, A. C. M. (2009). Effect of milk protein variants on the protein composition of bovine milk. *J. Dairy Sci.*, Vol. 92, 1192-1202

Heidari, M., Ahani Azari, M., Hasani, S., Khanahmadi, A. & Zerehdaran, S. (2009). Association of genetic variants of β-lactoglobulin gene with milk production in a herd and a superior family of Holstein cattle. *Iran. J. Biotech.*, Vol. 7, No. 4, pp. 254 – 257

Henderson, D. A. & Marshall, D. M. (1996). Kappa-casein genotype effects in a multiple breed beef cattle population. *J. Anim. Sci.*, Supp.1, Vol.74, pp. 121

Huang, W., Penagaricano, F., Ahmad, K. R., Lucey J. A., Weigel, K. A. & Khatib, H. (2012). Association between milk protein gene variants and protein composition traits in dairy cattle. *J. Dairy Sci.*, DOI: 10.3168/jds.2011-4757

Iannuzzi, L., Di Meo, G. P., Perucatti, A., Schibler, L., Incarnato, D., Gallagher, D., Eggen, A., Ferretti, L., Cribiu, E. P. & Womack, J.(2003). The river buffalo (*Bubalus bubalis*, 2n = 50) cytogenetic map: Assignment of 64 loci by fluorescence in situ hybridization and R-banding. *Cytogenet. Genome Res.* Vol. 102, pp. 65-75.

Ibeagha-Awemu, E. M., Bemji, M. N., Osinowo, O. A., Chiatti, F., Chessa, S. & Erhardt, G. (2005). Caprine milk protein polymorphisms: Possible applications for African goat breeding and preliminary data in Red Sokoto. In: The role of biotechnology in animal agriculture to address poverty in Africa: opportunities and challenges. Proceedings of 4th All Africa Conference on Animal Agriculture and the 31st Annual Meeting of the Tanzania Society for Animal Production (TSAP), Arusha, Tanzania, 20-24 September, pp. 323-358

Ibeagha-Awemu, E. M., Prinzenberg, E. M., Jann, O. C., Luhken, G., Ibeagha, A. E., Zhao, X. & Erhardt, G. (2007). Molecular Characterization of Bovine CSN1S2*B and Extensive Distribution of Zebu-Specific Milk Protein Alleles in European Cattle. *J. Dairy Sci.*, Vol. 90, pp. 3522–3529

Ikonen, T., Ojala, M., Ruottinen, O. & Georges, M. (2001). Association between casein hyplotypes and first lactation milk production traits in Finish Ayrshire cows. *J. Dairy Sci.*, Vol. 84, pp. 507-514

Ilie, D., Sălăjeanu, A., Magdin, A., Stanca, C., Vintilă, C., Vintilă, I. & Gócza, E. (2007). Genetic polymorphism at the κ-casein locus in a dairy herd of Romanian Spotted and Brown of Maramures breeds.*Lucrări ştiinţifice Zootehnie şi Biotehnologii*, Vol. 40, No. 1, pp. 101-106

Imafidon, G. I. & Ng-Kwai-Hang, K. F. (1991). Effect of genetic polymorphism on the thermal stability of β-lactoglobulin and κ-casein mixture. *J. Dairy Sci.*, Vol. 74, pp. 1791-1802

Jann, O., Ceriotti, G., Caroli, A. & Erhardt, G. (2002). A new variant in exon VII of bovine b-casein gene (CSN2) and its distribution among European cattle breeds. *J. Anim. Breed. Genet.*, Vol. 119, pp. 65-68

Jordana, J., Amills, M., Díaz, E., Angulo, C., Serradilla, J. M. & Sánchez, A. (1996). Gene frequencies of caprine αs1-casein polymorphism in Spanish goat breeds. *Small Rum. Res.*, 20, 215-221

Jõuru, I., Henno, M., Värv, S., Kaart, T. & Kärt, O. (2007). Milk protein genotypes and milk coagulation properties of Estonian Native cattle. *Agric. Food Sci.*, Vol. 16, pp. 222-231

Kamiński, S. (2001). Polymorphism of bovine milk proteins. Dissertations and monographs (in Polish). Published by Warmińsko-Mazurski University, Olsztyn.

Karima, K. Gh. M., Nawito, M. F. & Abdel Dayem, A. M. H. (2010). Sire selection for milk production traits with special emphasis on kappa casein (CSN3) gene. *Global J. Molecular Sci.*, Vol. 5, No. 2, pp. 68-73

Khatkar, M. S., Thomson, P. C., Tammen, I. & Raadsma, H. W. (2004). Quantitative trait loci mapping in dairy cattle: review and meta-analysis. *Genet. Sel. Evol.*, No. 36, pp. 163-190

Król, J. (2003). The association between genetic variants of milk proteins with milk yield of beef cattle and the results of rearing of their offspring. Part I – Hereford breed (in Polish). *Annales UMCS*, Sec. EE, Vol. XXI, No. 1, pp. 81-89

Krukovics, S., Daroczi, L., Kovacs, P., Molnar, A., Anton, I., Zsolnai, A., Fesus, L. & Abraham, M. (1998). The effect of β-lactoglobulin genotype on cheese yield. *EAAP Publ.*, Vol. 95, pp. 524-527

Krzyżewski, J., Ryniewicz, Z., Strzałkowska, N. & Bagnicka, E. (2002). The concentration of selected macro- and microelements in goat milk depending on the polymorphic variants of αs1-casein (in Polish). *Prace i Mat. Zoot.*, Vol. 14, pp. 93-101

Krzyżewski, J., Strzałkowska, N., Ryniewicz, Z., Bagnicka, E. & Oprządek, A. (2000). The relationship between polymorphic variants of alfa-s1-CN and also the daily yield and chemical composition of goat milk during lactation (in Polish). *Zesz. Nauk. AR Wrocław*, Vol. 399, pp. 189-198

Kučerova, J., Matějiček, A., Jandurova, O. M., Sorensen, P., Němcova, E., Štipkova, M., Kott, T., Bouška, J. & Frelich, J. (2006). Milk protein genes CSN1S1, CSN2, CSN3, LGB and their relation to genetic values of milk production parameters in Czech Fleckvieh. *Czech J. Anim. Sci.*, Vol. 51, No. 6, pp. 241–247

Li, Y. C., Korol, A. B., Fatima, T., Beiles, A. & Nevo, E. (2002). Microsatellite: genomic distribution, putative functions and mutational mechanisms: a review. *Molec. Ecol.*, No. 11, pp. 2453-2465

Li, Y. C., Korol, A. B., Fatima, T. & Nevo, E. (2004). Microsatellite within genes: structure, function and evolution. *Mol. Biol. Evol.*, No. 21, pp. 991-1007

Litwińczuk, A., Barłowska, J., Król, J. & Litwińczuk, Z. (2006). Milk protein polymorphism as a marker of functional traits of dairy and beef cattle (in Polish). *Vet. Med.*, Vol. 62, No. 1, pp. 6-10

Litwińczuk, A., Kędzierska-Matysek, M. & Barłowska, J. (2007). Performance and quality of milk of various genotypes of αs1-casein obtained from goats from the region of Wielkopolska and Podkarpacie (in Polish). *Vet. Med.*, Vol. 63, pp. 192-195

Litwińczuk, A., Kędzierska-Matysek, M., Król, J. & Barłowska, J. (2004). Productivity of white improved and non-improved goat breeds characterized with different genetic variants of αs1-casein (in Polish). *Zesz. Nauk. Przeg. Hod.*, Vol. 72, No. 3, pp. 133-139

Litwińczuk, Z. & Król, J. (2002). Polymorphism of main milk proteins in beef cattle maintained in East-Central Poland. *Anim. Sci. Pap. Rep.*, Vol. 20, No. 1, pp. 33-40

Liu, Z. J. & Cordes, J. F. (2004). DNA marker technologies and their applications in aquaculture genetics. *Aquaculture*, No. 238, pp. 1-37

Liu, Z. J., Li, P., Kocabas, A., Ju, Z., Karsi, A., Cao, D. & Patterson, A. (2001). Microsatellite-containing genes from the channel catfish brain: evidence of trinucleotide repeat expansion in the coding region of nucleotide excision repair gene RAD23B. *Biochem. Biophys. Res.Commun.*, Vol. 289, pp. 317-324

Lodes, A., Buchberger, J., Krause, I., Aumann, J. & Klostermeyer, H. (1997). The influence of genetic variants of milk protein on the compositional and technological properties of milk. Content of protein, whey protein and casein number. *Milchwissenschaft*, Vol. 52, pp. 3-8

Luhken, G., Caroli, A., Ibeagha-Awemu, E. M. & Erhardt, G. (2009). Characterization and genetic analysis of bovine as1-casein/ variant. *Anim. Gen.*, Vol. 40, pp. 479–485

Lunden, A., Nilsson, M. & Janson, L. (1997). Marked effect of β-lactoglobulin polymorphism on the ratio of casein to total protein in milk. *J. Dairy Sci.*, Vol. 80, pp. 2996-3005

Mahé, M. F., Manfredi, E., Ricordeau, G., Piacere, A. & Grosclaude, F. (1994). Effects of α-s1 casein polymorphism on the dairy performances of Alpine goat breed. *Genet. Sel. Evol.*, Vol.26, pp. 151-157

Marini, A. B. A., Rashid, B. A., Musaddin, K. & Zawawi, I. (2011). αs1-Casein gene polymorphism in Katjang, Jamnapari, Boer and Boer-feral goats in Malaysia. *J. Trop. Agric. and Fd. Sc.*, Vol. 39, No. 1, pp. 1-5

Marle-Köster, E. & Nel, L. H. (2003). Genetic markers and their application in livestock breeding in South Africa: A review. *South Afr. J. Anim. Sci.*, Vol. 33, No. 1, pp. 1-10

Martin, P., Michale, O. B. & Grosclaude, F. (1999). Genetic polymorphism of caseins: A tool to investigate casein micelle organization. *Intern. Dairy J.*, Vol. 9, pp. 163-171

Matějíček, A., Matějíčková, J., Němcová, E., Jandurová, O.M., Štípková, M., Bouška, J. & Frelich, J. (2007). Joint effects of CSN3 and LGB genotypes and their relation to breeding values of milk production parameters in Czech Fleckvieh. *Czech J. Anim. Sci.*, Vol. 52, No. 4, pp. 83–87

McLean, D. M., Graham, E. R. B., Ponzoni, R. W. & McKenzie, H. A. (1984). Effect of milk protein genetic variants on milk yield and composition. *J. Dairy Res.*, Vol. 51, pp. 531-546

Mercier, J. C., Grosclaude, F. & Ribadeau-Dumas, B. (1971). Structure primaire de la caseine αs1 bovine. Sequence complete. *Eur. J. Biochem.*, Vol. 23, pp. 41-51

Moatsou G., Vamvakaki A. N., Mollé D., Anifantakis E. & LéoniL J. (2006). Protein composition and polymorphism in the milk of Skopelos goats. *Lait*, Vol. 86, pp. 345–357

Moatsou, G., Moschopoulou, E., Molle, D., Gagnaire, V., Kandarakis, I. & Leonil, J. (2008). Comparative study of the protein fraction of goat milk from the Indigenous Greek breed and from international breeds, *Food Chem.*, Vol. 106, pp. 509-520

Mohammadi, A., Nassiry, M. R., Elyasi, G. & Shodja, J. (2006). Genetic polymorphism of β-lactoglobulin in certain Iranian and Russian sheep breeds. *Irani. J. Biotech.*, Vol. 4, No. 4, pp. 265-268

Mroczkowski, S., Korman, K., Erhard, G., Piwczyński, D. & Borys, B. (2004). Sheep milk protein polymorphism and its effect on milk performance of Polish Merino. *Arch. Tierz.*, Vol. 47, pp. 114-121

Mroczkowski, S., Korman, K., Piwczyński, D. & Erhard, G. (2002). The influence of sheep genotype of αS1-casein on milk productivity of Polish Merino in the first three lactations (in Polish). *Zesz. Nauk. Przeg. Hod.*, Vol. 63, pp. 139-144

Naqvi, A. N. (2007). Application of Molecular Genetic Technologies in Livestock Production: Potentials for Developing Countries. *Adv. Biol. Res.*, Vol. 1, No. 3-4, pp. 72-84

Ng-Kwai-Hang, K. F. (1997). A review of the relationship between milk protein polymorphism and milk composition/milk production. Proc. IDF Seminar "Milk Protein Polymorphism II" North Palmerston, New Zeland, 22-37

Ng-Kwai-Hang, K. F., Otter, D. E., Lowe, E., Boland, M. J. & Auldist, M. J. (2002). Influence of genetic variants of β-lactoglobulin on milk composition and size of casein micelles. *Milchwissenschaft*, Vol. 57, pp. 303-306

Nilsen, H., Olsen, H. G., Hayes, B., Sehested, E., Svendson, M., Nome, T., Meuvissen, T. & Lien, S. (2009). Casein hyplotypes and their association with milk production traits in Norwegian Red cattle. *Genet. Sel. Evol.*, Vol. 41, No. 24, pp. 1-12

Nudda, A., Feligini, M., Battacone, G., Campus, R. & Pulina, G. (2000). Effects of β-lactoglobulin genotypes and parity on milk production and coagulation properties in Sarda dairy ewes. *Zootecnica E Nutrizone Animale*, Vol. 26, pp. 137-143

Olenski, K., Kamiński, S., Szyda, J. & Cieslinska, A. (2010). Polymorphism of the beta-casein gene and its associations with breeding value for production traits of Holstein-Friesian bulls. *Livestock Sci.*, Vol. 131, pp. 137-140

Pierre, A., le Quere, J. L., Famelart, M. H. & Rousseau, F. (1996). Cheeses from goat milks with or without αs1-CN. In: Production and Utilization of Ewe and Goat Milk. Proc. Of the IDF/Greek National Committee of IDF/CIRVAL Seminar, Crete, Greece, pp.322

Pisarchik, A. V. & Kartel, N. A. (2000). Simple repetitive sequences and gene expression. *Mol. Biol.*, Vol. 34, pp. 303-307

Pop, F. D., Balteanu, V. A. & Vlaic, A. (2008). A comparative analysis of goat αs1-casein locus at protein and DNA levels in carpathian goat breed. *Bulletin UASVM Anim. Sci. Biotech.*, Vol. 65, No. 1-2, pp. 1843-5262

Ramunno, L., Longobardi, E., Pappalardo, M., Rando, A., Di Gregorio, P., Cosenza, G., Mariani, P., Pastore, N. & Masina, P. (2001). An allele associated with a non-detectable amount of αs2-casein in goat milk. *Anim. Gen.*, Vol. 32, pp. 19-26

Remeuf, F. (1993). Influence du polymorphisme génétique de la caséine alfa S1 caprine sur les caractéristiques physico-chimiques et technologiques du lait. *Lait*, Vol. 73, pp. 549-557

Ren, D. X., Miao, S. Y., Chen, Y. L., Zou, C. X., Liang, X. W. & Liu, J. X. (2011). Genotyping of the k-casein and β-lactoglobulin genes in Chinese Holstein, Jersey and water buffalo by PCR-RFLP. *Journal of Genetics*, Vol. 90, No. 1, e1–e5.

Rincón, G., Armstrong, E. & Postiglioni, A. (2006). Analysis of the population structure of Uruguayan Creole cattle as inferred from milk major gene polymorphisms. *Genet. Mol. Biol.*, Vol. 29, No. 3, pp. 491-495

Robitaille, G., Britten, M. & Petitclerc, D. (2001). Effect of a differential allelic expression of kappa-casein gene on ethanol stability of bovine milk. *J. Dairy Res.*, Vol. 68, pp. 145–149

Ryniewicz, Z., Krzyżewski, J. & Jasińska, L. (1998). The relationship between polymorphic variants of goat's CSN1S1 and cow's LGB and the susceptibility of total proteins to enzymatic hydrolysis of milk (in Polish). *Prace i Mat. Zoot.*, Vol. 53, pp. 75-82

Ryniewicz, Z., Reklewska, B., Krzyżewski, J., Strzałkowska, N. & Gałka, E. (2000). The possibilities of improving the national population of goats and their milk properties and the correlation to recent research results obtained in IGiHZ PAN in Jastrzębiec (in Polish). Annals of Warsaw Agricultural University – SGGW, *Anim. Sci.*, Vol. 37, pp. 21-29

Sacchi, P., Chessa, S. & Budelli, E. (2005). Casein Haplotype Structure in Five Italian Goat Breeds. *J. Dairy Sci.*, Vol. 88, No. 4, pp. 1561- 1568

Schmidely, P., Meschy, F., Tessier, J. & Sauvant, D. (2002). Lactation response and nitrogen, calcium, and phosphorus utilization of dairy goats differing by the genotype for αs1-caseine in milk, and fed diets varying in crude protein concetration. *J. Dairy Sci.*, Vol. 85, pp. 2299-2307

Shende, T. C., Sawane, M. P. & Pawar, V.D. (2009). Genotyping of Pandharpuri buffalo for κ-casein using PCR-RFLP. *Tamilnadu J. Vet. Anim. Sci.*, Vol. 5, No. 5, pp. 174-178

Stoneking, M. (2001). Single nucleotide polymorphisms: From the evolutionary past. *Nature*, Vol. 409, pp. 821-822

Strzałkowska, N., Bagnicka, E., Jóżwik, A., Krzyżewski, J. & Ryniewicz, Z. (2004). Chemical composition and some technological milk parametres of Polish White Improved Goats. *Arch. Tierz.*, Vol. 47, pp. 122-128

Strzelec, E. & Niżnikowski, R. (2009). Single nucleotide polymorphism (SNP) of selected genes in representatives of the family Bovidae with particular emphasis on domestic goats. *Anim. Prod. Rev.*, No. 7, pp. 7-14

Supakorn, Ch. (2009). The Important Candidate Genes in Goats - A Review. *Walailak J. Sci. Tech.*, Vol. 6, No. 1, pp. 17-36

Sztankóová, Z., Senese, C., Czerneková, V., Dudková, G., Kott, T., Mátlová, V. & Soldát, J. (2005). Genomic analysis of the CSN2 and CSN3 loci in two Czech goat breeds. *Anim. Sci. Pap. Rep.*, Vol. l, No. 23, pp. 67-70

Teneva, A. & Petrović, M. P. (2010). Application of molecular markers in livestock improvement. *Biotech. Anim. Husb.*, Vol. 26, No. 3-4, pp. 135-15

Teneva, A. (2009). Molecular markers in animal genome analysis. *Biotech. Anim. Husb.*, Vol. 25, No. 5-6, pp. 1267-1284

Torres-Vázquez, J. A., Vázquez Flores, F., Montaldo, H. H., Ulloa-Arvizu, R., Posadas, M. V., Vázquez, A. G. & Morales, R. A. A. (2008). Genetic polymorphism of the αs1-casein locus in five populations of goats from Mexico. *Electronic J. Biotech.*, Vol.1, No.3, pp. 2-11

Tsiaras, A. M., Bargouli, G. G., Banos, G. & Boscos, C. M. (2005). Effect of kappa-casein and βeta-lactoglobulin loci on milk production traits and reproductive performance of Holstein cows. *J. Dairy Sci.*, Vol. 88, pp. 327–334

Vassal, L. & Manfredi, E. (1994). Des lait plus riches. *Chevre*, No. 201, pp. 33-36

Vătăşescu-Balcan, R. A., Georgescu, S. E., Manea, M. A., Dinischiotu, A. & Costache, M. (2007). Analysis of beta-lactoglobulin and kappa-casein genotypes in cattle. Conference paper *Archiva Zootechnica*, Vol. 10, pp. 103-106

Veress, G., Kusza Sz., Bősze, Z., Kukovics, S. & Jávor, A. (2004). Polymorphism of the αs1-casein, κ-casein and ß-lactoglobulin genes in the Hungarian Milk Goat. *S. Afr. J. Anim. Sci.*, Vol. 34, Supp. 1, pp. 20-23

Vlaic, A., Balteanu, V. A. & Carsai, T. C. (2011). Milk Protein Polymorphism Study in Some Romanian Sheep Breeds. *Bulletin UASVM Anim. Sci. Biotech.*, Vol. 68, No. 1-2, pp. 27-31

Walawski, K., Sowiński, G., Czarnik, U. & Zabolewicz, T. (1994). Beta-laktoglobulin and kappa-casein polymorphism in relation to production traits and technological properties of milk in the herd of Polish Black-and-White cows. *Genet. Pol.*, Vol. 35, No. 1/2, pp. 93-108

Winkelman, A. M. & Wickham, B. W. (1997). Associations between milk protein genetic variants and production traits in New Zealand dairy cattle. In: Milk protein polymorphism. Proceedings of the IDF Seminar held in Palmerston North, New Zealand. Int Dairy Fed, 38-46.

Yahyaoui, M. H., Pena, R. N., Sanchez, A. & Folch, J. M. (2000). Rapid communication: Polymorphism in the goat β-lactoglobulin proximal promoter region1. *J. Anim. Sci.*, Vol. 78, pp. 1100-1101

Relationship Between Kappa Casein Genes (CSN3) and Industrial Yield in Holstein Cows in Nariño-Colombia

Gema Lucia Zambrano-Burbano, Yohanna Melissa Eraso-Cabrera, Carlos Eugenio Solarte-Portilla and Carol Yovanna Rosero-Galindo

Additional information is available at the end of the chapter

1. Introduction

Most dairy farms in Colombia are grouped into four geographical areas, termed "competitive regions", according to Resolution 000017 of 2012, issued by the Ministry of Agriculture and Rural [1] Development. The majority of the high-tropics herds, which are mostly specialized dairy farms, are located in Antioquia, Cundinamarca and Nariño provinces, while dual-purpose dairy farms are more abundant in the basins of the low tropics [2].

One of the most limiting factors affecting the competitiveness of specialized dairy farms in Nariño is the low quality of the product, in terms of milk composition. According to [3], the low protein content in milk negatively affects the efficiency of industrial processes to produce curd and cheese. Therefore, the improvement of the milk composition in this area is an important matter to take into account when planning nutrition and the genetic variables for optimal animal production [4]. The compositional quality of milk depends not only on environmental factors, but also on genetic traits such as the Kappa Casein gene (CSN3), which has been widely studied in order to establish relationships between its polymorphisms with the percentage of total milk protein and industrial yield [5].

Research conducted in various regions has reported mixed results regarding the associations between the AA, AB and BB genotypes of CSN3 and milk yield [6 - 8] However, given the complexity of gene expression, it is necessary to compare these results with those obtained under environmental conditions typical of this area in Colombia.

Therefore, in this study we established the relationships between the CSN3 genotypes and curd yield and the percentage of milk protein in Holstein cows, which are the predominant breed in the Andean region of Nariño.

2. Genetic improvement of the dairy cattle in colombia

2.1. General review

The main purpose of animal breeding is to obtain, by selection, future offspring with superior performance in relation to their predecessors. To achieve this objective, the frequency of desirable genes should be increased from one generation to another, which leads to an increase of genotypes with higher productivity. The most important tool with the greatest impact on animal breeding has been the control of production and the reliable permanent recording of pedigrees as well as of all the productive and reproductive events of the herds. Having this information, it has been possible to quantify genetic variability and identify the superior genetic merits of animals through mathematical models, through more and more refined means. This procedure corresponds to the so-called classical breeding schemes, which have greatly contributed to increased production and productivity in virtually all domestic species.

The classical methods of animal breeding have experienced a rapid development since the last decade of the twentieth century thanks to molecular techniques, which have allowed, among other things, the more precise quantification of genetic variability, the better understanding the phenomena of genes and the diagnosis of a great number of diseases [9]. DNA molecular markers are expressed in various forms and can always be located in the same site of the genome from generation to generation, which is an advantage of genetic analysis [10].

Today, the use of molecular markers is almost routine, following the development of PCR [11]. This is a technique for the direct amplification of DNA from tissue samples, such as blood, skin, hair and - in general - any tissue. PCR is based on the use of polymerizing enzymes and the initiation of oligonucleotides, known as primers.

Nowadays, studies of genetic polymorphism in cattle and other domesticated species are closely related to production traits and health and, for this purpose, techniques capable of detecting small variations in the DNA molecule are used [12]. One of the biggest advantages of genetic studies using PCR and DNA molecular markers is to reduce the generation interval, as can be identified in very early life stages, including embryonic individuals with superior genotypes, which increased efficiency in breeding programs.

2.2. Overview of milk production in Colombia

Colombian dairy has excelled over the past 30 years in terms of its dynamics, reflected in high rates of output growth. In the 1970s, milk production grew at a rate of 4.7% on average per year. Over the next decade, it accelerated its expansion, reaching an annual rate of 6.5%, while in the 1990s growth decreased, but succeeded in satisfactory rates of 3.8% per year, reaching a figure of 5.877 billion litres of milk fluid in 2001 [13]. The FAO estimate for 2005 indicates that the country produced 6.77 billion metric tons of milk, with the Atlantic region contributing 40% of total production, the Occidental adding 17%, and with Central Pacifica contributing 34% and 9%. FEDEGAN [14] (2006) estimated that the percentage growth between 2004 and 2005 was 3.2%, which for the same period was lower than that achieved in countries with little experience in cattle, such as Costa Rica, Peru, Ecuador and Bolivia. However, Colombia ranked as the third largest producer in Latin America with 6.02 million tons in 2003, after Brazil and Argentina with 23 million and 7.7 million tons respectively.

The Agrocadenas observatory [2] stands out in that milk production has had a tendency to grow and which seem likely to be sustained over the next five years. This fact requires the search for new domestic and international markets and, therefore, the improvement of all links in the dairy chain, so as to increase the productivity and profitability of milk production systems.

The milk produced in Colombia comes from a specialized and dual purpose. It is estimated that the country has 25 million cattle, of which 11 million are engaged in milk production, 10 million in dual purpose and one million in the specialized system [14]. 88% of production is assigned for the dairy industry and 12% for the marketing of raw milk and the feeding of calves [16].

According to data from DANE and FEDEGAN [17, 14] in the living areas of the High Tropics 80% of the production system consists of specialized dairy breeds, with Holstein, Normando Brown Swiss, Jersey and Creole crossbreeds. In the dairy of the Tropic High of Nariño, also predominantly Holstein breed, which was introduced in the region for over a century, primarily considering its high production capacity? However, nowadays a need to consider different traits - especially those related to the compositional quality of milk, fertility and longevity - has been detected [18].

According to Agronet [19] and the Monitoring Unit Price of milk, from the Ministry of Agriculture and Rural Development-MARD in Colombia [1], in the region where the Department of Nariño is included, the average of fat is 3.63%, the protein is 3.10%, the total solids is 12.00% and the colony forming units vary within a range between 25,000 and 175,000. These figures are calculated on a permanent basis in the Animal Breeding Program at the University of Nariño - Meg@lac - for their importance in the selection process, which has been taking place since late 2006. Meanwhile, Agronet [19] indicates that in this region the protein content is below the national average. This weakness in compositional quality negatively affects the competitiveness of the livestock in Nariño, because the protein content is directly related to the biological value of milk, and their capacity for industrialization, making it mandatory to design and implement strategies leading to its improvement.

2.3. Milk composition

Milk is the normal product of the secretion of the mammary gland of female mammals. Physically, it is a complex fluid with more than one hundred substances in different stages, such as suspension, emulsion and true solution, which are the determinants of the nutritional quality and properties that mark it as a feedstock for the production of a large number of products [20]. As to the average composition of bovine milk among different breeds, Table 1 shows the composition of milk from the main dairy breeds in Colombia.

Breed	Fat(%)	Protein (%)	Lactose (%)	Ash(%)	Total solids (%)
Holstein	3.54	3.29	4.68	0.72	12.16
Brown Swiss	3.99	3.64	4.94	0.74	13.08
Ayrshire	3.95	3.48	4.60	0.72	12.77
Guernsey	4.72	3.75	4.71	0.76	14.04
Jersey	5.13	3.98	4.83	0.77	14.42
Shorthom	4.00	3.32	4.89	0.73	12.9

Table 1. Composition of milk from different breeds of cattle [21]

The variation in the constituents of milk is due to several factors that alter their composition, structure and properties. Among the most influential factors, the species, breed, lactation number, age, pathological conditions, and environmental factors such as diet, climate and the milking system, may be mentioned [22].

Such variations are observed more easily in components such as fat, lactose, ash and - more importantly - in proteins [23]. Usually, milk proteins have been divided into two groups, depending on their behaviour by acidification to pH 4.6 and isoelectric point at room temperature. Such groups refer to a soluble fraction containing whey proteins, namely alpha-lactalbumin (α-LA) and Beta-Lactoglobulin (β-LG), which represent 20% of the total proteins. The other fraction is insoluble, which is where we can find the caseins alpha s1 (αs1-Cs), alpha s2 (αs2-Cs), Beta (β-Cs) and Kappa (κ-Cs), which represent 80% of total milk proteins. The protein content in milk is especially affected by the influence of nutritional and genetic factors, considering the latter to be the most important, because they are the genes that allow the animal to produce milk of a certain quality, in an environment that provides the conditions for this to happen [24].

The average protein content in cow milk is 3.5% and an effective way to increase it is the selection of animals with the best genotypes, which, once identified, can be spread extensively using reproductive biotechnologies such as artificial insemination, multiovulación and embryo transfer. In species with a long generation interval, such as cattle, the genetic progress is relatively slow compared with that achieved in other species. This is one reason as to why in the last two decades use has been made of molecular markers as an option to increase genetic progress, as it facilitates the identification of desirable genotypes at an early age and - even in embryos - decreases the generation interval thereby increasing the genetic gain each year [25].

2.4. Genetic bases of kappa-casein (K-Cs)

2.4.1. Overview of K-Cs

Proteins are the major milk components though they differ in function and biological value, so it is necessary to study them individually in order to determine their physico-chemical and functional differences. In bovine milk, proteins represent approximately 3.5% of its total components. Caseins (Cs) constitute 80% of the protein fraction while whey proteins account for 20% [26].

Among the casein fractions are included αs1-Cs, αs2-Cs, β-Cs and κ- Cs casein. In the whey proteins, which are synthesized in higher concentrations, are α-LA and β-LG and lower concentrations of lactoferrin (LFE) and defensins (DFS), among others [26].

Several studies have confirmed that casein and whey proteins have an influence on the properties of milk for industrial processing, especially in terms of coagulation time, curd firmness and performance [27 to 30]. The κ-Cs has a clearly different structure from other caseins, since it is smaller, being comprised of 169 amino acids, and it is phosphorylated by only one phosphate group, which causes fewer interactions with calcium ions over other caseins. However, it shares with β-Cs the property of having predominantly hydrophilic and hydrophobic areas clearly marked and separated.

A peculiarity of this casein is the presence of a net positive charge zone between amino acids 20 and 115. This area allows the interaction of casein with polysaccharides that are negatively charged. The chain also has two groups of cysteine and is the only casein which has partly glycosylated molecules. The carbohydrate group is comprised of a trisaccharide or a tetrasaccharide or else is linked to a threonine residue.

Besides the presence of hydrophilic and hydrophobic areas, caseins interact together to form a colloidal dispersion which consists of spherical particles called micelles. The casein micelle is a very stable colloidal system in milk. This fact has important practical implications related to both the formation of casein gels and the stability of dairy products during thermal treatment, concentration and storage. For these reasons, the microstructure of the casein micelle has been intensively studied during the past five decades, since their knowledge is crucial in cheese making [31].

The formation of the casein micelle may be affected by changes in pH, salt concentration, temperature and hydrophilic regions. Under normal conditions of pH and salt concentration, casein micelles are well hydrated, having about 3.7 grams of water connected per gram of protein. When these conditions change, the casein micelles are destabilized by the acidity and proteolysis of the κ-Cs, which is also related to the allelic variant for this protein fraction.

As to the acid, there are two effects: first, a decrease in pH generates the breaking of the bonds between the phosphate and calcium ion by reducing the ionisation of the phosphates. Second, the repulsion between the micelles is reduced when the pH approaches the isoelectric point of casein. At a pH of about 4.5 and a temperature above 20 °C, casein curds are added forming a slightly mineralized.

A clear example of how changes in the hydrophilic regions and temperature affect the stability of the micelle is its treatment with chymosin. By adding chymosin, the κ-Cs loosens, by proteolysis, its hydrophilic region thereby facilitating aggregation. At low temperatures of refrigeration, the hydrophobic forces which hold together the molecules of the β-Cs are weakened, causing them to expose their hydrophilic region to the outside, increasing micelle hydration and volume. As a consequence, at refrigeration temperatures, there is no aggregation of casein, neither by the action of the acidity nor by that of chymosin.

Other factors which greatly affect micelle aggregation are the calcium and salt content. The loss of Ca++ leads to the dissociation of β-Cs without the disintegration of the micelle. Furthermore, the salt content affects the activity of serum calcium and the amount of calcium phosphate of micelles.

2.5. Molecular basis of polymorphisms

The milk proteins are synthesized in Lactocytes, and their expression is co-dominant. Moreover, not all of the alleles encode exactly the same protein, due to the alteration or substitution of one or more amino acids in the polypeptide chain; some alleles are even null, i.e. they do not code for any protein. Individuals heterozygously synthesize a protein, one for each allele that they possess, in the homologous chromosomes [32].

The caseins, meanwhile, are encoded by four characteristic genes called polymorphic autosomal CSN1S1, CSN1S2, CSN2 and CSN3, physically located in a 250 Kb region of chromosome 6, very close together, particularly at position 6q31-33 [33]. Genes αs1-Cs, αs2-Cs and β-Cs are within the 6q31 locus in a region of 140 Kb, while the gene of the κ-Cs is located in a region between 95 and 120 Kb.

Figure 1 shows the primary structures of the caseins. The αs1-Cs consists of 199 amino acids and has a molecular weight of 23,000 Dalton [34]; of these amino acids, 17 correspond to proline, with two hydrophobic regions separated by a polar region which contains the phosphate groups (Figure 1a). For its part, has the αs2-Cs 207 amino acids in its conformation (Figure 1b) and has a molecular weight of 25,000 Daltons and 10 amino acids of a proline type. Finally, the κ-Cs has 169 amino acids in conformation (Figure 1c) and its molecular weight ranges from 19,006 to 19,037 Daltons.

The primary structure of the β-Cs is formed by a chain of 209 amino acids with five phosphate groups, which provides for the property of strongly binding calcium ions. The molecular weight ranges from 23,983 to 24,000 Daltons (Fig. 1d), depending on the allelic present variation t [35]. As with the κ-Cs, Cs has a β-polar and a hydrophobic area. In the polar end, the phosphates' attached groups are concentrated on the amino acids serine and are more hydrophilic. However, the β-Cs are the most hydrophobic casein fractions, containing a larger number of proline compared to any other type of casein [36]. The amino acid sequence of the β-Cs was established in 1972 by Ribadeau & Dumas and a review of the same part by Yan and Wold (1984) found four differences in the original sequence, where three of them corresponded to a changing of glycine by glutamic acid at positions 117, 175 and 195, and a fourth in a reversal of the amino acids proline to leucine at positions 137 and 138 respectively [37].

Figure 2 describes the sequences and nucleotide positions for the proteins αs1-Cs, Cs αs2-, β and κ-Cs-Cs available on the NCBI database [41] through the Genbank.

a)

```
1                         10                              20
H-Arg-Pro-Lys-His-Pro-Ile-Lys-His-Gln-Gly-Leu-Pro-Gln-Glu-Val-Leu-Asn-Glu-Asn-Leu-

21                        30                              40
Leu-Arg-Phe-Phe-Val-Ala-Pro-Phe-Pro-Glu-Val-Phe-Gly-Lys-Glu-Lys-Val-Asn-Glu-Leu
------------------- (αs1-CN A)

41                        50                              60
Ser* Lys-Asp-Ile-Gly-Ser-Glu-Ser-Thr-Glu-Asp-Gln-Ala-Met-Glu-Asp-Ile-Lys-Gln-Met-
            (αs1-CN D) –Thr          (αs1-CN E) –Glu-

61                        70                              80
Glu-Ala-Glu-Ser- Ile-Ser- Ser- Ser-Glu-Glu-Ile-Val-Pro-Asn-Ser-Val-Glu-Gln-Lys-His-

81                        90                              100
Ile-Gln-Lys-Glu-Asp-Val-Pro-Ser-Glu-Arg-Tyr-Leu-Gly-Tyr-Leu-Glu-Gln-Leu-Leu-Arg-

101                       110                             120
Leu-Lys-Lys-Tyr-Lys-Val-Pro-Gln-Leu-Glu-Ile-Val-Pro-Asn-Ser-Ala-Glu-Glu-Arg-Leu-

121                       130                             140
His-Ser-Met-Lys-Glu-Gly-Ile-His-Ala-Gln-Gln-Lys-Glu-Pro-Met-Ile-Gly-Val-Asn-Gln-

141                       150                             160
Glu-Leu-Ala-Tyr-Phe-Tyr-Pro-Glu-Leu-Phe-Arg-Gln-Phe-Tyr-Gln-Leu-Asp-Ala-Tyr-Pro-

161                       170                             180
Ser-Gly-Ala-Trp-Tyr-Tyr-Val-Pro-Leu-Gly-Thr-Gln-Tyr-Thr-Asp-Ala-Pro-Ser-Phe-Ser

181                       190                             199
Asp-Ile-Pro-Asn-Pro-Ile-Gly-Ser-Glu-Asn-Ser-Glu-Lys-Thr-Thr-Met-Pro-Leu-Trp-OH

          (αs1-CN C; αs1-CN E) –Gly-
```

b)

```
1                         10                              20
H-Lys-Asn-Thr*-Met-Glu-His-Val-Ser-Ser-Ser-Ser-Glu-Glu-Ser-Ile-Ile-Ser-Gln-Glu-Thr-Tyr-

21                        30                              40
Lys-Gln-Glu-Lys-Asn-Met-Ala-Ile-Asn-Pro-Ser-Lys-Glu-Asn-Leu-Cys-Ser-Thr-Phe-Cys-
                                                     -Gly- (αs1-CN C)

41                        50                              60
Lys-Glu-Val-Val-Arg-Asn-Ala-Asn-Glu-Glu-Glu-Tyr-Ser-Ile-Gly-Ser-Ser-Ser-Glu-Glu-
                                            -Thr*- (αs1-CN C)
                                            (αs1-CN D) -------------

61                        70                              80
Ser-Ala-Glu-Val-Ala-Thr-Glu-Glu-Val-Lys-Ile-Thr-Val-Asp-Asp-Lys-His-Tyr-Gln-Lys-

81                        90                              100
Ala-Leu-Asn-Glu-Ile-Asn-Gln-Phe-Tyr-Gln-Lys-Phe-Pro-Gln-Tyr-Leu-Gln-Tyr-Leu-Tyr-

101                       110                             120
Gln-Gly-Pro-Ile-Val-Leu-Asn-Pro-Trp-Asp-Gln-Val-Lys-Arg-Asn-Ala-Val-Pro-Ile-Thr

121                       130                             140
Pro-Thr-Leu-Asn-Arg-Glu-Gln-Leu-Ser-Thr*-Ser-Glu-Glu-Asn-Ser-Lys-Lys-Thr-Val-Asp-
                                                        -Ile- (αs1-CN C)

141                       150                             160
Met-Glu-Ser-Thr-Glu-Val-Phe-Thr-Lys-Lys-Thr-Lys-Leu-Thr*-Glu-Glu-Glu-Lys-Asn-Arg-

161                       170                             180
Leu-Asn-Phe-Leu-Lys-Lys-Ile-Ser-Gln-Arg-Tyr-Gln-Lys-Phe-Ala-Leu-Pro-Gln-Tyr-Leu-

181                       190                             200
Lys-Thr-Val-Tyr-Gln-His-Gln-Lys-Ala-Met-Lys-Pro-Trp-Ile-Pro-Lys-Thr-Lys-Val-

201                       207
Ile-Pro-Tyr-Val-Arg-Tyr-Leu-OH
```

c)

```
1                         10                              20
H-Arg-Glu-Leu-Glu-Glu-Leu-Asn-Val-Pro-Gly-Glu-Ile-Val-Glu-Ser-Leu-Ser-Ser-Ser-Glu-
                                                            Lys(Variant D)

                    ⇓ → Y1-caseins
21          30            Lys(Variant E)  40
Glu-Ser-Ile-Thr-Arg-Ile-Asn-Lys-Lys-Ile-Glu-Lys-Phe-Gln-Ser-Glu-Glu-Gln-Gln-Gln-
                                                       Ser   Lys
                                                     (Variant C)

41                        50                              60
Thr-Glu-Asp-Glu-Leu-Gln-Asp-Lys-Ile-His-Pro-Phe-Ala-Gln-Thr-Gln-Ser-Leu-Val-Tyr-

61                        70                              80
Pro-Phe-Pro-Gly-Pro-Ile-Pro-Asn-Ser-Leu-Pro-Gln-Asn-Ile-Pro-Pro-Leu-Thr-Gln-Tyr-
                                                  His (Variant A¹, B, C)

81                        90                              100
Pro-Val-Val-Val-Pro-Pro-Phe-Leu-Gln-Pro-Glu-Val-Met-Gly-Val-Ser-Lys-Val-Lys-Glu-

                    ⇓ → Y2-caseins
101         110            ⇓                    120
Ala-Met-Ala-Pro-Lys-His-Lys-Glu-Met-Pro-Phe-Pro-Lys-Tyr-Pro-Val-Glu-Pro-Phe-Thr-
                    ⇑ Gln (Variant A²)
                    ⇓ → Y3-caseins

121                       130                             140
Glu-Ser-Gln-Ser-Leu-Thr-Leu-Thr-Asp-Val-Glu-Asn-Leu-His-Leu-Pro-leu-Pro-leu-Leu
Arg (Variant B)

141                       150                             160
Gln-Ser-Trp-Met-His-Gln-Pro-His-Gln-Pro-Leu-Pro-Pro-Thr-Val-Met-Phe-Pro-Pro-Gln-
                                                            Leu (Variant F)

161                       170                             180
Ser-Val-Leu-Ser-Leu-Ser-Gln-Ser-Lys-Val-Leu-Pro-Val-Pro-Gln-Lys-Ala-Val-Pro-Tyr

181                       190                             200
Pro-Gln-Arg-Asp-Met-Pro-Ile-Gln-Ala-Phe-Leu-Leu-Tyr-Gln-Glu-Pro-Val-Leu-Gly-Pro-

201         209
Val-Arg-Gly-Pro-Phe-Pro-Ile-Ile-Val-OH
```

d)

```
1                         10                              20
pyrGlu-Glu-Gln-Asn-Gln-Glu-Gln-Pro-Ile-Arg-Cys-Glu-Lys-Asp-Glu-Arg-Phe-Phe-Ser-Asp-
                                                            His (Variant F¹)

21                        30                              40
Lys-Ile-Ala-Lys-Tyr-Ile-Pro-Ile-Gln-Tyr-Val-Leu-Ser-Arg-Tyr-Pro-Ser-Tyr-Gly-leu-

41                        50                              60
Asn-Tyr-Tyr-Gln-Gln-Lys-Pro-Val-Ala-Leu-Ile-Asn-Asn-Gln-Phe-Leu-Pro-Tyr-Pro-Tyr-

61                        70                              80
Tyr-Ala-Lys-Pro-Ala-Ala-Val-Arg-Ser-Pro-Ala-Gln-Ile-Leu-Gln-Trp-Gln-Val-Leu-Ser-

81                        90                              100
Asn-Thr-Val-Pro-Ala-Lys-Ser-Cys-Gln-Ala-Gln-Pro-Thr-Thr-met-Ala-Arg-His-Pro-His-
                                                            His (Variant C)
                                                            Cys (Variant G)

                    ⇑Chymosin
101         110                                 120
Pro-His-Leu-Ser-Phe-Met-Ala-Ile-Pro-Pro-Lys-Lys-Asn-Gln-Asp-Lys-Thr-Glu-Ile-Pro-

121                       130                             140
Thr-Ile-Asn-Thr-Ile-Ala-Ser-Gly-Glu-Pro-Thr-Ser-Thr-Pro-Thr-Ile-Glu-Ala-Val-Glu-
                                                            Thr (variant A)

141         150            160
Ser-Thr-Val-Ala-Thr-Leu-Glu-Ala-Ser-Pro-Glu-Val-Ile-Glu-Ser-pro-pro-Glu-Ile-Asn-
                    Asp (Variant A)       Gly (Variant E)
                    Val (Variants F and F¹)

161         169
Thr-Val-Gln-Val-Thr-Ser-Thr-Ala-Val-OH
```

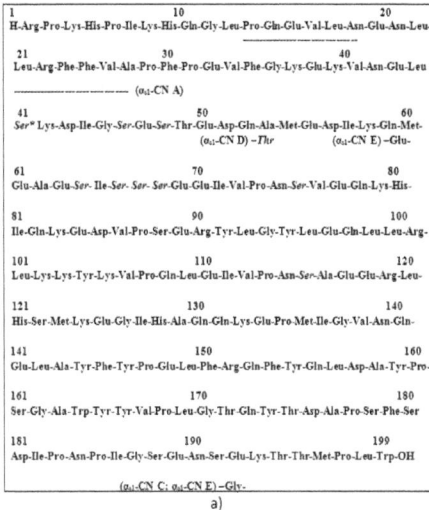

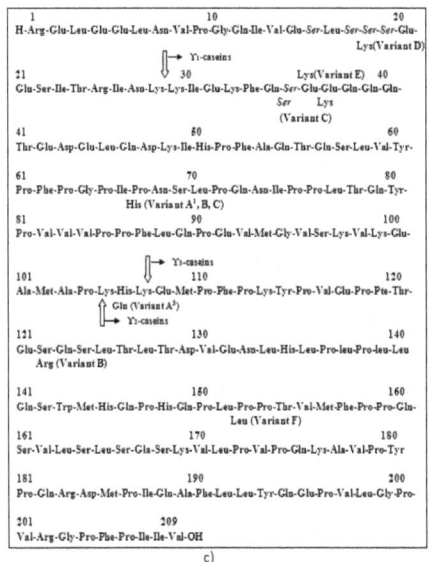

Figure 1. Primary structure of the casein from milk; a) αs1-Cs [34], b) αs2-Cs [38], c) β-Cs [39] y d) κ-Cs [40].

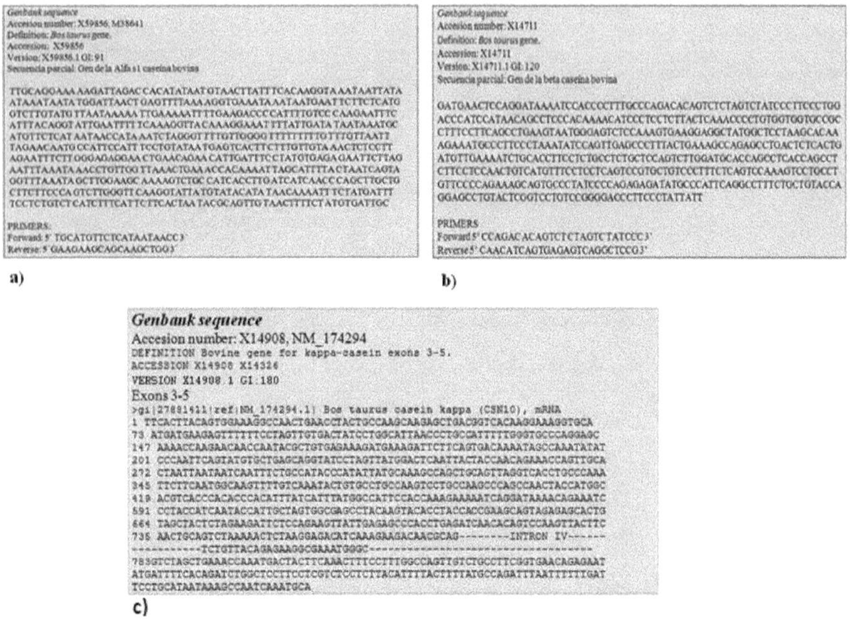

a)

Genbank sequence
Accession number: X59856, M33641
Definition: Bos taurus gene.
Accession: X59856
Version: X59856.1 GI: 91
Secuencia parcial: Gen de la Alfa s1 caseína bovina

TTGCAGGAAA AAGATTAGAC CACATATAAT GTAACTTATT TCACAAGGTA AATAATTATA
ATAAATAATA TGGATTAACT GAGTTTTAAA AGGTGAAATA AATAATGAAT TCTTCTCATG
GTCTTGTATG TTAATAAAAA TTGAAAAATT TTGAAGACCC CATTTTGTCC CAAGAATTTC
ATTTACAGGT ATTGAATTTT TCAAAGGTTA CAAAGGAAAT TTTATTGATA TAATAAAATGC
ATGTTCTCAT AATAACCATA AATCTAGGGT TTTGTTGGGG TTTTTTTGTT TGTTGTTAATT
TAGAACAATG CCATTCCATT TCCTGTATAA TGAGTCACTT CTTTGTTGTA AACTCTCCTT
AGAATTTCTT GGGAOAGGAA CTGAACAGAA CATTGATTTC CTATGTGAGA GAATTCTTAG
AATTTAAATA AACCTGTTGG TTAAACTGAA ACCACAAAAT TAGCATTTTA CTAATCAGTA
GGTTTAAATA GCTTGGAAGC AAAAGTCTGC CATCACCTTG ATCATCAACC CAGCTTGCTG
CTTCTTCCCA GTCTTGGGTT CAAGGTATTA TGTATCACATA TAACAAAATT TCTATGATTT
TCCTCTGTCT CATCTTTCAT TCTTCACTAA TACGCAGTTG TAACTTTTCT ATGTGATTGC

PRIMERS
Forward 5' TGCATGTTCTCATAATAACC 3'
Reverse 5' GAAGAAGCAGCAAGCTGG 3'

b)

Genbank sequence
Accession number: X14711
Definition: Bos taurus gene.
Accession: X14711
Version: X14711.1 GI: 120
Secuencia parcial: Gen de la beta caseína bovina

GATGAACTCCAGGATAAAATCCACCCCTTTGCCCAGACACAGTCTCTAGTCTATCCCTTCCCTGG
ACCCATCCATAACAGCCTCCCACAAAACATCCCTCCTCTTACTCAAACCCCTGTGGTGGTGCCGC
CTTTCCTTCAGCCTGAAGTAATGGGAGTCTCCAAAGTGAAGGAGGCTATGGCTCCTAAGCACAA
AGAAATGCCCTTCCCTAAATATCCAGTTGAGCCCTTTACTGAAAGCCAGAGCCTGACTCTCACTG
ATGTTGAAAATCTGCACCTTCCTCTGCCTCTCGCTCGCTTGGATGCACCAGCCTCACCAGCCT
CTTCCTCCAACTGTCATGTTTCCTCCTCAGTCCGTGCTGTCCCTTCTCAGTGCCAAAGTCCTGCCT
GTTCCCCAGGAAAGCAGTGCCCTATCCCCAGAGAGATATGCCCATTCAGGCCTTTCTGCTGTACCA
GGAGCCTGTACTCGGTCCTGTCCCGGGGGACCCTTCCCTATTATT

PRIMERS
Forward 5' CCAGACACAGTCTCTAGTCTATCCC 3'
Reverse 5' CAACATCAGTGAGAGTCAGGCTCCG 3'

c)

Genbank sequence
Accesion number: X14908, NM_174294
DEFINITION Bovine gene for kappa-casein exons 3-5.
ACCESSION X14908 X14326
VERSION X14908.1 GI:180

Exons 3-5
>gi|27881611|ref|NM_174294.1| Bos taurus casein kappa (CSN10), mRNA
    1 TTCACTTACAGTGGAAAGGGCCAACTGAACCTACTGCCAAGCAAGAGCTGACGGTCACAAGGAAAGGTGCA
   73 ATGATGAAGAGTTTTTCCCTAGTGTTGGACTATGCTGGGCATTAAACCCTGCGATTTTTGGGTGCCCAGGAGGC
  147 AAAACCAASAACAACCTCATCAGCTGTGAGAAAGATGAAAGATTCTTCAGTGACAAAATAGCCAAATATAT
  201 CCCAATTCAGTATGTGCTGAGCAGGTATGCTAGTTATGGACTCAATTACTACCAACAGAAACCAGTTGCA
  272 CTAATTAATAATCAATTTCTGCCATACCCATATTATGCAAAGCCAGCTGCAGTTAGGTCACCTGGCCAAA
  345 TTCTTCAATGGCAAGTTTTGTCAAATACTGTGCCTGCCAAGTCCTGCCAAGCCCAGCCAACTACCATGGC
  419 ACGTCACCCACACCCACATTTATCATTTATGGCCATTCCACCAAAGAAAAATCAGGATAAAACAGAAATC
  591 CCTACCATCAATACCATTGCTAGTGGCGAGCCTACAAGTACACCTACCACCGAAGCAGTAGAGAGCACTG
  664 TAGCTACTCTAGAAGATTCTCCAGAAGTTATTGAGAGCCCACCTGAGATGAACACAGTCCAAGTTACTTC
  735 AACTGCAGTCTAAAAACTCTAAGGAGACATCAAAGAAGACAACGCAG--------INTRON IV-----
-----------TCTGTTACAGAGAAGGCGAAATGGGC-------------------------------
  783GGTCTAGCTGAAACCAAATGACTACTTCAAACTTTCCTTTGGCCAGTTGTCTCGCCTTCGGTGAACAGAGAAT
ATGATTTTCACAGATCTGGCTCCTTCCTCCGTCTCCTCTTACATTTTACTTTTATGCCAGATTTAATTTTTTGAT
TCCTGCATAATAAAGCCAATCAAATGCA

Figure 2. Figure 2 Nucleotide sequence of the genes of caseins present in bovine milk: a) Sequence αs2 αs1-Cs-Cs and, b) Sequence of the β-Cs c) Sequence of the κ-Cs.

2.6. Allelic variants of κ-Cs

The gene for κ-Cs has been studied extensively in order to establish the relationship between their polymorphisms or allelic variants and the industrialization of milk, especially in relation to cheese processing. Several findings emerged from the study of the genetic polymorphism of milk proteins which have been conducted with the purpose of improving the different sectors of the dairy industry [42]. Animal breeding contributes to the genetic basis of knowledge on which the production characters are based [43]. In research aimed at determining the relationship between various DNA markers and production traits, a special emphasis has been placed on the association between the genetic polymorphism of proteins and the physicochemical properties of milk [44].

Milk protein polymorphism is caused by mutation, which produces the genetic variants detected by electrophoresis [42]. These variants are identified by the letters A, B and C and in bovine species it is observed that the frequency of each allelic form varies between breeds [20].

The polymorphism of milk proteins in cattle was first described more than 50 years ago [45]. The first studies were based on protein polymorphism from milk samples, but thanks to technical advances in the field of molecular genetics, until 2005 DNA polymorphisms had already been reported, of which 8 corresponded to variants different for αs1-Cs [46 - 48], 4 for αs2-Cs, 12 for β-Cs, 11 for κ-Cs, 11 for β-LG, and 3 for α-LA in Bos taurus [49].

From 1983 to 2007, 11 allelic variants have been reported for κ-Cs (A, B, C, E, F, G, H, I, A1, A2, A3), but in *Bos taurus* variants A and B are the most frequent alleles and not all of them appear in all races [33, 50, 51, 47, 52, 53].

Both in cattle and in goats, associations between certain protein polymorphisms and production traits have been shown, especially those related to the composition of protein and fat, although the results of several studies do not coincide because, in some cases, associations have been found but not in others [54, 55]. Therefore, it is important to confirm these findings under the conditions of the Tropical High Nariño.

Other studies have found lower industrial yields of cheese for the A allele when compared with allele B. The results for this variable have not been matched in all the investigations and, also, it is necessary to confirm them under our conditions. In some studies, it has been established that there is a significant effect on the conversion of milk into cheese, according to the specific genotype κ-Cs, given that in most cases the milk from animals with the BB genotype is homozygous and increased cheese yield; however, the milk from animals with the AA genotype has lower yields. This has been explained by a lower percentage of casein genotypes AA and as a result a greater proportion of large micelles [56].

In the case of the Tropic High Nariño, research conducted by the Genetic Improvement Program of the University of Nariño in southern Colombia, South America [18], shows gene frequencies close to fixation for allele A of the αs2-Cs, making it impossible to establish a direct association between the genotypes of the protein fraction with productive and reproductive performance variables [57].

These sample studies for the αs1-Cs show the absence of statistically significant differences on the above characteristics. These results do not agree with those obtained by Haenlein [58] and Graml [59] in relation to the stained brown bavárico and Guernsey, who claim that cows carrying the AB genotype reach a higher milk production compared to those with the homozygous genotype AA. On the contrary, [6, 60, 61] they reported the highly significant effect of genotype BB on this feature in the Holstein breed.

For the β-Cs protein fraction it was observed that the genotypes AA and AB have a significant effect on milk production. These results are similar to those described by several investigators on the same race, where the A allele is closely linked to an increased volume of milk per lactation [42, 62, 56, 63, 64].

Finally, our studies on the average figures for the production of a kilogram of curd for κ-Cs indicate that milk from animals with the genotype BB produce better yields of curd, requiring only 5.46 litres of milk [28]. This is explained by the formation of more stable micelles and smaller in the milk from these animals. Similar results are reported by [65], indicating that the B allele has better heat resistance, a shorter clotting time and produces a more consistent curd.

3. Materials and methods

3.1. Type of study

The present study was conducted in the municipality of Pasto, in south-western Colombia, located at north latitude 1°13′22″, west longitude 77°16′22″, at a height of 2690 m, with an average temperature of 12 °C and a relative humidity of 82% [66]. A total of 348 cows were sampled to determine their CSN3 genotype using PCR-SSCP. Once identified by this criterion, lactating cows were selected and classified according to their lactation stage, i.e., initial (first), mid (second), and final (third).

3.2. Identification of genotypes

The molecular identification of AA and BB homozygous and AB heterozygous genotypes for the CSN3 gene was conducted at the Animal Breeding laboratory of the University of Nariño, following the methodology described by [12] and modified [4].

3.3. Collection of the milk samples

Milk samples were obtained from animals whose output was adjusted to 305 days and an adult equivalent (PL) using the factors indicated by [24] for Colombian Holstein cattle. The experimental unit was each animal in its corresponding lactating stage. Four litres of milk were obtained from each cow during the morning milking and immediately transported under refrigeration (4 °C) to the processing plant. Upon receipt of the sample, three litres of milk were used for the assessment of the curd yield and one litre for physicochemical, microbiological and compositional analysis. An EKOMILK (KAM Milkama 98-2A) industrial milk analyser was used to assess milk composition. The values obtained for acidity and fat using the analyser were confirmed by manual testing. Acidity was determined by a qualitative change in colour after mixing one volume of milk with an alkaline solution of sodium hydroxide, adding phenolphthalein as an indicator (0.1 N). Fat content was determined using [67]. The number of colony-forming units (CFU) and the degree of environmental contamination of the milk were included in the microbiological analysis.

3.4. Curd

A protocol was established at the processing plant for the evaluation of curd yield. The equipment and implements used to obtain the curd were a centrifuge (Funke Gerber, Berlin Germany), autoclave (All American, Wisconsin, USA), stainless steel containers, electric stoves, burettes (Schott, Mainz, Germany), pipettes (Schott , Mainz, Germany), thermo-hydrometer (Funke Gerber, Berlin Germany), analytical balance (Mettler Toledo, Mexico DC), thermometers (B & S, Germany), curd knives to cut 1 x 1 cm, industrial moulds (500 g), and bags for packaging. The curd obtained was classified as "peasant type", according to the processing plant. This product is characterized by 60% moisture content. The fat and protein contents were not standardized prior to milk processing.

3.5. Evaluation of the curd yield

After 10 hours of cooling, the curd yield was calculated taking into account the volume of
processed milk and the final weight of the curd. The calculations were based on the
following formula:

$$RC = \frac{VL}{WC}$$

Where:

- RC = curd yield, expressed in litres of milk required to produce one kg of curd;
- VL = volume of milk;
- WC = final weight of curd.

Statistical analysis.

Analyses were performed using a linear model which included the fixed effects of the
lactation stage (lactation was divided into thirds), genotype, and the interaction effect of the
lactation stage by genotype. The percentage of fat in milk was included as a covariate. The
statistical model is expressed as:

$$y_{ijk} = \mu + \tau_j + \alpha_k + (\tau,\alpha)_{jk} + \beta_1(x_{1i} - \bar{x}_1) + \varepsilon_{ijk}$$

Where:

$$y_{ijk}$$

= curd yield, associated with the jth genotype, the kth lactation stage, and the interaction of
the jth genotype with the kth lactation stage, taking into account the percentage of fat in the
milk.

- μ= Media comon to all observations;
- τ_j = Effect of the jth genotype. J= 1, 2, 3;
- αk = Effect of the kth lactation stage. k = 1, 2, 3;
- $\tau\alpha$ = Interaction effect of the jth genotype with the kth lactation stage;
- $\beta_1(x_{1i} - \bar{x}_1)$ = Lineal effect of the covariate "percentage of fat in the milk";
- ε_{ijk} = Experimental error associated with the jth genotype in the kth lactation stage and
  the interaction between the jth genotype with the kth lactation stage.

Each genotype in each lactation stage was regarded as a treatment. Each treatment had three
replicates and thus, in total, 27 experimental units were evaluated. The age was not included
in the model as a covariate because it was not statistically significant. The analysis was
performed with the GLM procedure of SAS statistical software version 9.20 and Enterprise
SAS Guide version 4.2. (2009).

4. Results

Kappa - Casein (CSN3) genotypes: According to [12], the PCR-SSCP molecular technique identifies more than four allelic variants for the CSN3 gene. In our studied population, only two allelic variants were found (A and B), identifying the homozygous genotypes AA, BB and heterozygous AB, in accordance with the electrophoretic pattern described by [12] (Figure 3). These results are consistent with those reported by other researchers, in which the A and B alleles have the highest frequency for the CSN3 gene in dairy breeds [8, 33, 50, 47].

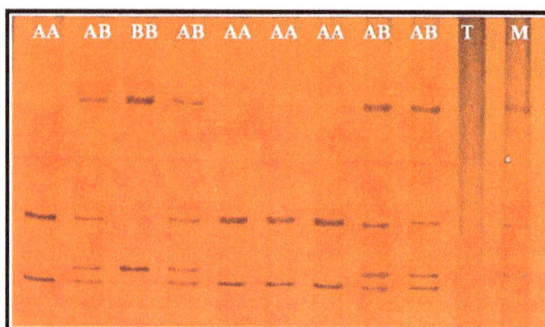

Figure 3. Electrophoretic bands generated by PCR-SSCP for the CSN3 gene in Holstein cattle of the High Tropics in Nariño.

Table 2 presents the percentages of protein (PP), body fat (BF), total solids (PST), and litres required to produce one kg of curd (L / Kg), separated for each genotype.

Genotype	Variable	Average	Standard deviation	Minimum	Maximum
AA	L/Kg	6.722	0.712	5.500	7.700
	PP	3.076	0.105	3.000	3.320
	PG	3.117	0.582	2.310	4.120
	PST	11.699	0.756	10.670	13.290
AB	L/Kg	6.111	0.653	5.000	7.000
	PP	3.062	0.114	2.900	3.260
	PG	3.718	0.659	2.510	4.470
	PST	12.269	0.774	11.140	13.470
BB	L/Kg	5.456	0.615	4.300	6.400
	PP	3.188	0.080	3.080	3.340
	PG	3.924	0.367	3.330	4.500
	PST	12.730	0.477	11.950	13.330

PP: protein (%); PG: fat (%); PST: total solids (%); L/Kg: litters of milk required to make one kilogram of curd

Table 2. Descriptive statistics of the variables evaluated according to the CSN3 genotype.

4.1. Analysis of variance for curd yield

The genotype, lactation stage and the linear effect of the fat percentage of the milk (included as a covariate) were statistically significant (p <0.05) in the ANOVA. The coefficient of determination was 0.624, indicating that the effects included in the model explain curd yield by 62.4%.

The least square means for curd yield were compared using the Tukey – Kramer test, concluding that the BB genotype had the highest yield compared to genotypes AA and AB, and no significant differences were found between AA and BB, as shown in Figure 4.

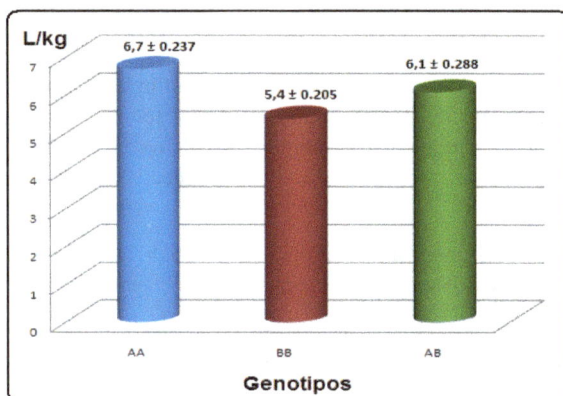

Figure 4. Number of milk litres required to produce one kilogram of curd for three κ-Cs genotypes of the Holstein breed in the Nariño High Tropic.

Finally, according to the Tukey – Kramer multiple comparison test, the only differences found for protein percentage were between the first and the second third of lactation. This is in agreement with reports by [68]. However, no differences were observed among the different genotypes.

The results of this study are consistent with those reported by [6, 7], who concluded that the BB genotype for κ-Cs determines the best milk properties for cheese production because of the greater firmness in the curd and smaller time required for the formation of small micelles. With regard to the homozygous AA genotype, these animals had lower casein contents and, as a consequence, a higher proportion of large micelles, which reduces the curd yield efficiency. Furthermore, according to [12], there is a positive relation between the κ-Cs genotype and the milk protein content, and this protein content influences the clotting time required by rennet as well as the firmness and the cheese yield, showing higher values in milk from cows with genotype BB with respect to the homozygous AA [69, 64], reaching differences that, in some cases, amount to up to 3% [70].

The study of [71] using Limonero cattle indicated that variant B of the κ-Cs could be used to improve the efficiency for transforming milk into cheese. A similar conclusion was arrived at in [5] for the same variant of the gene in the Harton del Valle cattle breed. In summary, as stated by [4], these studies found that bovine milk from cows with the BB genotype for κ-Cs has greater stability in relation to heating and freezing, requiring less clotting time, producing a more consistent curd, and increasing cheese yield. Those results are in agreement with the present study. These results confirm the need to reorient the selection process in the Nariño High Tropics, where κ-Cs genotype should be considered as an important criterion for defining selection objectives. This trait should be included along with other factors that are relevant to the region, such as the functional type, the longevity and the somatic cell count, consistent with current selection trends for the Holstein breed in many countries [72].

5. Conclusions

The use of molecular techniques - an unprecedented event in the region - to determine the genotypes of the κ-Cs and its implementation is useful for the correct identification of individuals according to their genetic constitution and allows for the development of other studies for measures, such as population structure and genetic diversity for the gene CSN3 in the High Tropic of Nariño.

In the Holstein breed and under the conditions of the High Tropic of Nariño, the highest yield was obtained with curdled milk from animals with the genotype BB.

The highest content of total protein was obtained in animals with the genotype BB and in the last third of lactation, factors that did not produce an interaction effect.

The results of this research provide are a good guide to reviewing and developing breeding programs in the Tropic Alto de Nariño for Holstein cattle, since in order to improve the compositional quality of milk and in order to increase industrial performance, it is necessary to include the feature of the genotype for κ-Cs in a comprehensive genetic evaluation which includes, along with other important factors, the increasing of the gene frequency of the B allele for kappa casein.

Author details

Gema Lucia Zambrano-Burbano[*], Yohanna Melissa Eraso-Cabrera,
Carlos Eugenio Solarte-Portilla and Carol Yovanna Rosero-Galindo
University of Nariño, Animal Genetic Improvement Program, Pasto, Colombia

6. Acknowledgement

We would like to acknowledge the Colombian Ministry of Agriculture and Rural Development, the Breeding Program at the University of Nariño, and the Nariño Dairy Cooperative Dairy - Colácteos, for providing the resources necessary for the completion of this study.

---

* Corresponding Author

7. References

[1] Ministerio de Agricultura y Desarrollo Rural Colombia, resolución número 000017 del 2012 "por la cual se establece el sistema de pago de la leche cruda al productor".

[2] Observatorio Agrocadenas. Segundo informe de coyunturas de leche. (2006). Disponible en internet www.agrocadenas.gov.co

[3] González M.F, Barrales V y Valenzuela A.S (2004). Desarrollo de un modelo para evaluar la factibilidad productiva y económica de la progenie resultante del cruzamiento de vacas holstein friesian con toros de las vacas francesa montbeliarde y normando. http://www.uc.cl/agronomia/e_publicaciones/Documentosderabajo/doc

[4] Solarte C, Rosero C, Eraso Y, Zambrano G, Cárdenas H, Burgos W (2009) Frecuencias alélicas del gen Kappa caseína en la raza Holstein del trópico alto de Nariño–Colombia. LRRD, Vol 21(1).

[5] Naranjo J, Posso A, Cárdenas H y Muñoz J.E (2007) Detección de la variantes alélicas de la kappa caseína en bovinos hartón del valle. Revista Acta Agronómica, Vol 56(1): 43-47.

[6] Ng-Kwai-Hang KF, Hayes JF, Moxley JE, Monardes HG (1984) Association of genetic variants of casein and milk serum proteins in milk fat and protein production by dairy cattle. Journal Dairy Science, Vol 67: 835.

[7] Mclean D.M, Graham H.A, Mckenzie (1982) Estimation of casein composition by gel electrophoresis. Int Dairy Congr Moscow 1(2): 221.

[8] Requena F.D, Agüera E.I, Requena F (2007) Genética de la caseína de la leche en el bovino Frisón. Redvet, Vol 8(1).

[9] Uffo O y Martinez S (2002) Amplificación por PCR de los genes que codifican para lactoalbúmina, la lactoblobulina y la K-caseína de una vaca alta productora de leche y dos de sus descendientes e identificación de las variantes alélicas por RFLP. Revista Salud Animal, Vol 24:22-26.

[10] Luque I y Herráez A (2001) Biología molecular e ingeniería genética Texto ilustrado. Ediciones Harcourt S.A. Madrid España.

[11] Mullis KB y Faloona F (1987) Specific synthesis of DNA in vitro via a polymerase-catalyzed chain reaction. Methods enzymol, Vol 155:335-350.

[12] Barroso A, Dunner S, Cañon J (1998) Detection of Bovine Kappa-casein Variants A, B, C y E by Means of Polymerase Chain Reaction-single strand conformation polymorphism (PCR-SSCP). Technical note. Journal Animal Science, Vol 76: 1535-1538.

[13] Belalcazar, A (2002) La ganadería bovina en Colombia 1970-1991. Coyuntura Agropecuaria. Vol 9 (2): 113-118.

[14] FEDEGAN. Federación Colombiana de Ganaderos (2006) En: www.fedegan.org.co/

[15] Holmann, F; L. Rivas, J; Carulla, B; Rivera, L; Giraldo, M; Martínez, A. Medina y A. Farrow (2006) Producción de leche y su relación con los mercados: caso colombiano. In: Rony Tejos (ed.). X Seminario Manejo y Utilización de Pastos y Forrajes en Sistemas de Producción Animal. Fundapastos. Guanare. Venezuela. 149-156.

[16] FAO, FAOSTAT (2005) Base de datos estadísticos. www.faostat.fao.org.

[17] Departamento Administrativo de Nacional de Estadística Colombia. DANE (2006).

[18] Programa de Mejoramiento Genético (2009) Caracterización y Evaluación Genética de la Población Bovina Lechera del Trópico Alto de Nariño para la Conformación de Núcleos de Selección, Universidad de Nariño, Pasto – Colombia.

http://promegalac.udenar.edu.co/wp-
content/uploads/2010/05/Informe_Final_Proyecto_48-1.pdf

[19] Agronet y la unidad de seguimiento de precios de la leche del Ministerio de Agricultura y Desarrollo Rural de Colombia-MADR (2012).
.://www.agronet.gov.co/www/docs_agronet/2006103010565_INFORME_Leche_JULIO.pdf

[20] Alais, Ch (1985) Ciencia de la leche, principios de la técnica lechera. Editorial Revertè España.

[21] Amiot, J (1994) Ciencia y tecnología de la leche. Universite laval Quebec. Editorial Acribia. España.

[22] Walstra, P; Geurts, T; Noomen, A; Jellema, A. y Van Boekel, M (2001) Ciencia de la leche y tecnología de los productos lácteos. Editorial Acribia S.A Zaragoza, España. 730.

[23] Latrille, L (1999) Calidad de la leche. En: Producción Animal Universidad Austral de Chile. Valdivia, Chile. 215– 236.

[24] Cerón M, Tonhati H, Costa C, Solarte C, Benavides C (2003) Factores de ajuste para producción de leche en bovinos Holstein colombiano. Revista Colombiana de Ciencias Pecuarias, Vol 16: 26-32

[25] Montaldo, H. y Barría, N (1998) Mejoramiento Genético de Animales. Ciencia al día, Vol 2:19.

[26] Gigli I, Riggio V, Monteleone G, Cacioppo D, Rosa A y Maizon D (2007) Relationship between beta lactoglobulin and subclinical mastitis in Valle del Belice sheep breed. Journal of Animal Science, Vol 6(1):140-142.

[27] Alipanah M, Klashnikova L y Rodionov G (2005) Kappa-casein genotypic frequencies in Russian breeds black and red pied cattle. Iranian Journal of Biotechnoly, Vol 3 (3):191-194.

[28] Zambrano G, Eraso Y, Solarte C, Rosero C (2010) Kappa casein genotypes and curd yield in holstein cows. Revista Colombiana de Ciencias Pecuarias, Vol 23(4):422-428.

[29] Martínez CG, González HF (2000) El ganado criollo Sanmartinero (SM) y su potencial productivo. Animal Genetic Resources, Vol 28: 7-18.

[30] Jakob T y Puhan Z (1992) Technological properties of milk as influenced by genetic poymorphism of milk proteins - A review. International Dairy Journal, Vol 2 (3):157-178.

[31] Ferrandini E, Castillo M, López MB, Laencina J (2006) Modelos estructurales de la micela de caseína. Departamento de tecnología de alimentos nutrición y bromatología. Universidad de Murcia.

[32] Hayes B, Hagesaether N, Ådnøy T, Pellerud G, Berg PR y Lien S (2006). Effects on production traits of haplotypes among casein genes in Norwegian goats evidence for a site of preferential recombination. Genetics, Vol 174:455-464.

[33] Chessa F, Chiatti G, Ceriotti A, Caroli C, Consolandi G, Pagnacco and Castiglioni B (2007) Development of a single nucleotide polymorphism genotyping microarray platform for the identification of bovine milk protein genetic polymorphisms. Journal of Dairy Science, Vol 90(1): 451-464.

[34] Koczan D, Hoborn G, Seyfert HM (1993). Characterization of the bovine αS1-casein gene C-allele based on a Mae III polymorphism. Animal Genetics, Vol 24(1):74.

[35] Harboe MK y Budtz P (1999) the production, action and application of rennet and coagulants. In technology of cheese making. Law, B.A Ed. Sheffield academic press, Sheffield p: 33-65.

[36] Marziali AS, NG-Kwai-Hang KF (1986) Relationships between milk protein polymorphisms and cheese yielding capacity. Journal of Dairy Science, Vol 86(5):1193-1201.

[37] Farrell HM, Jiménez-Flores R, Bleck GT, Brown EM, Butler JE, Creamer LK, Hicks CL;
     Hollar CM, Ng-Kwai-Hang KF, Swaisgood HE (2004) Nomenclature of the proteins of
     cows' milk-sixth Revision. Journal of Dairy Science, Vol 87(6):1641-1674.

[38] Brignon G, Ribadeau-Dumas B, Mercier JC, Pelissier JP, Das BC (1977). Complete amino
     acid sequence of bovine alphaS2-casein. FEBS Letters. Vol 76:274-279.

[39] Ribadeau-Dumas B, Brignon G, Grosclaude F, Mercier JC (1972). Structure primaire de la
     caseine β-bovine. Sequence complete. European Journal of Biochemestry, Vol 25:505-14.

[40] Medrano JF y Aguilar-Córdova E (1990) Genotyping of bovine kappa-casein loci
     following DNA sequence amplifition. Boi/ technology, Vol 8:144-146.

[41] National Center for Biotechnology Information - NCBI. Nucleic Acids Research (2007)
     Vol 35, Database issue: D21-D25.

[42] Ng-Kwai-Hang KF (1997) A review of the relationship between milk protein
     polymorphism and milk composition - milk production. International Dairy Federation,
     Vol 9702:22-3.

[43] San Primitivo Tirados F (2001) La mejora genética animal en la segunda mitad del siglo
     XX. En: Archivos de Zootecnia Vol 50:517-546.

[44] Litwinczuk Z, Król J (2002) The yield and composition of beef cow milk and the results
     of calf rearing. Anim. Sci. Papers and Reports Vol 20(1): 199-204.

[45] Flower DR, North AC y Sansom CE (2000) The lipocalin protein family: Structural and
     sequence overview. Biochimica Biophysica Acta, Vol 148(2):9-24.

[46] Aschaffenburg R y Thymann M (1965) Simultaneous phenotyping procedure for the
     principal proteins of cow's milk. Journal of Dairy Science, Vol 48: 1524-1526.

[47] Eigel WN, Butler JE, Erstrom CA, Farrel HM, Harwalker VR, Jenness R and Whitney R
     (1984) Nomenclature of protein of cow's milk: fifth revision. Journal of Dairy Science,
     Vol 67: 1599-1631.

[48] David VA y Deutch AH (1992) Detection of bovine alpha s1-casein genomic variants
     using the allele-specific polymerase chain reaction. Animal Genetics, Vol 23(5): 425-429.

[49] Prinzenberg EM, Gutscher K, Chessa S, Caroli A y Erhardt G (2005) Caprine kappa-
     casein (CSN3) polymorphism: new developments of the molecular knowledge. Journal
     of Dairy Science, Vol 88(4):1490-1498.

[50] Soria LA, Iglesias GM, Huguet MJ, Mirande SL (2003) A PCR-RFLP test to detect allelic
     variants of the bovine kappa-casein gene. Anim Biotechnol, Vol 14(1):1-5.

[51] Atherton HV, Newlander JA (1977) Chemistry and Testing of Dairy Products. 4th
     Edition AVI publishing Co, Westport, Connecticut, USA.

[52] Addeo F, Anelli G, Chianese L, Di Luccia A, Mauriello R, Petrilli P (1983) Identification of
     bovine casein variants by gel isoelectric foucosing. Milchwissenschaft, Vol 38(10): 586-588.

[53] Di Luccia A, Addeo F, Corradini C, Mariani P (1988) Bovine kappa Casein C: an
     electrophoretic study. Scienza e tecnica lattiero-casearia, Vol 39(6): 413 – 422.

[54] Fitzgerald R (1997) The relationship between milk protein polymorphism and
     manufacture and functionality of dairy products. In "Milk protein polymorphism" Ed.
     Creamer, LK International Dairy Federation, Brussels, pp 355-371.

[55] Trujillo E, Camargo M, Norieda D (2000) Genotipificación de kappa-caseína bovina y
     evaluación de frecuencias genotípicas y alélicas de sus polimorfismos en cuatro razas.
     Act Biol, Vol 22:145-152.

[56] Bovenhuis H, Johan AM, Arendonk V, Korver S (1992) Associations between milk protein polymorphisms and milk production traits. Journal of Dairy Science, Vol 75:2549-2559.

[57] Solarte Portilla CE, Rosero Galindo CY, Eraso Cabrera YM, Zambrano Burbano GL, Barrera DC, Martínez OA, Guerrón ML y Cháves Galeano FP (2011): Polimorfismo de las fracciones caseínicas de la leche en bovinos holstein del Trópico Alto de Nariño. Livestock Research for Rural Development, Vol 23(136).

[58]Haenlein G, Gonyon D, Mather R, y Hines H (1987). Associations of bovine blood and milk polymorphisms with lactation traits: Guernseys. Journal Dairy Science Vol, 70:2599-2609

[59] Graml R, Buchberger J, Klostermeyer H y Pirchner F (1985) Pleiotrope wirkungen von β-lactoglobulin - und casein-genotypen auf milchinhaltsstoffe des bayerischen fleckviehs und braunviehs. Journal Animal Breeding and Genetics, Vol 102(1-5): 355-370.

[60] Boettcher PJ, Caroli A, Stella A, Chessa S, Budelli E, Canavesi F, Ghiroldi S, Pagnacco G (2004) Effects of casein haplotypes on milk production traits in Italian Holstein and Brown Swiss cattle. Journal Dairy Science, Vol 87(12), 4311–4317.

[61] Gurcan EK (2011) Association between milk protein polymorphism and milk production traits in Black and White dairy cattle in Turkey. African Journal of Biotechnology, Vol 10(6):1044-1048.

[62] Çardak AD (2005) Effects of genetic variants in milk protein on yield and composition of milk from holstein-Friesian and simmentaler cows. South African Journal of Animal Science, Vol 35(1):41-47.

[63] IGENITY® (2008) Marcando el future http//www.ignity.com Argentina: 0800 444 2582 | Uruguay: 0800 2222 | Chile: 3676997

[64] Van Eenennaam A y Medrano JF (1991) Differences in allelic protein expression in the milk of heterozygous Kappa casein cows. Journal Dairy Science, Vol 74:1491-1496.

[65] Rohallah A, Mohammadreza MA, Shahin MB (2007) Kappa-casein gene study in Iranian Sistani cattle breed (Bos indicus) using PCR-RFLP. Pakistan Journal of Biological Science, Vol 10(23): 4291 – 4294.

[66] IGAC (2006) Instituto Geográfico Agustín Codazzi. Colombia. URL: www.igac.gov.co/

[67] British Standards Instituttion (1955) Gerber Method for determination of fat in milk and milk products. Londres.

[68] Comerón EA, Aronna MS, Roggero M, Brizi N, Romero LA, Cuatrín A (2003) Sistema de pastoreo líderes-seguidores con vacas lecheras a dos niveles de suplementación. Rev Arg Prd Anm, Vol 23(1):103-104.

[69] Schlieben S, Erhardt G, Senft B (1991) Genotyping of bovine kappa-casein (kappa-CNA, kappa-CNB, kappa-CNE) following DNA sequence amplification and direct sequencing of kappa-CNE PCR product. Animal Genetics, Vol 22(4):333-342.

[70] Aleandri R, Buttazzoni LG, Schneider JC (1990) The effects of milk protein polymorphisms on milk components and cheese-production ability. Journal Dairy Science, Vol 73:241-255.

[71] Rojas I, Aranguren J, Portillo M, Villasmil Y, Valbuena E, Rincón X, Contreras G, Yañez L (2009) Polimorfismo genético de la kappa-caseína en ganado criollo limonero. Revista Científica Venezuela, Vol 19(6):645-649.

[72] Heins BJ, Hansen LB, Seykora AJ (2006) Production of Pure Holsteins versus crossbreds of Holstein with Normande, Montbeliarde, and Scandinavian Red. Journal Dairy Science, Vol 89:2799-2804.

Regulation of Milk Protein

The Effect of the Photoperiod and Exogenous Melatonin on the Protein Content in Sheep Milk

Edyta Molik, Genowefa Bonczar, Tomasz Misztal,
Aneta Żebrowska and Dorota Zięba

Additional information is available at the end of the chapter

1. Introduction

The first product of mammary gland in mammals is colostrum, which differs from milk with respect to density, colour and reaction, and is mildly acidic (pH of 6.4). Protein substances in colostrum account for approximately 60% of the dry matter, including more than 80% whey protein and almost 20% casein protein. Colostrum contains 19 exogenous aminoacids. Specifically, colostrum casein and phosphates form a complex of macromolecular micelles that contains a number of exogenous aminoacids (valine, leucine, proline, and lysine), and whey proteins found in the colostrum in the form of a colloid have a high content of cystine, cysteine and lysine [1]. Milk, the product of the mammary gland of mammals, is the first food of a young mammal, and its composition fully reflects the nutritional needs of the organism. The diverse composition of milk in different animal species results from the specific nutritional and physiological needs of the young. Thus, great importance is attached to the level of the components, such as protein, fat, lactose, and mineral salts, which are likely to change due to a number of factors [2,3].

2. Characterisation of nitrogen compounds in ruminant milk

Protein is one of the more important components of sheep milk and represents 95% of all nitrogen compounds (total nitrogen). The remaining 5% of nitrogen compounds form non protein nitrogen compounds [2,4]. Milk proteins represent a heterogeneous group of compounds, which differ with respect to their composition and properties. The protein content in sheep milk is diverse and varies from 4.75% to 7.2%, which is attributable to a variety factors, including the breed, lactation period, health of the udder, diet, age, number of fed lambs and climate conditions [2,4,5,6]. Numerous studies have shown that the diverse protein content of milk is connected with the breed of the sheep that the milk was obtained.

Another factor affecting the content of the milk is the lactation period. The level of protein in colostrum generated within the first hours after lambing can be as high as 19.5% [4]. The protein content in sheep milk in subsequent lactation weeks gradually increases and reaches a maximum level in the last week [3]. Ramos and Juarez [4] report that the protein content in sheep milk at the end of lactation can be higher than 6.8-8.9%. The studies by Bonczar [7] reveal that subclinical mastitis causes an increase in the protein content. However, Novotna et al. [8] concluded that the protein level in milk containing more than 200,000 somatic cells per 1 ml is significantly lower compared to milk with a low number of cells. Proteins in sheep milk are diverse with respect to the molecular weight, amino-acid composition, structure and properties, which constitutes a base to distinguish two groups of proteins: caseins and whey proteins.

3. Characterisation of casein and whey proteins in ruminant milk

Sheep milk, like milk from other ruminants, belongs to casein milk, which is milk with a high casein content compared to the level of total nitrogen compounds, as opposed to mare milk or human milk, which contains less casein and more whey protein (shown in Table 1).

Species	Protein (%)	Casein (%)	Whey proteins (%)	Share of casein in the total protein content (%)
Human	1.1	0.4	0.7	36.3
Cattle	3.4	2.8	0.6	78.1
Buffalo	4.8	4.2	0.6	87.5
Goat	2.9	2.5	0.4	80.6
Sheep	5.5	4.6	0.9	83.6
Horse	2.0	1.3	0.7	65.0
Reindeer	10.2	8.5	1.7	83.3
Camel	3.3	2.9	0.4	87.9

Table 1. The content of protein, casein and whey proteins in human milk and milk of various animal species adapted from [39,40]

From a chemical point of view, casein, which is the most important milk protein, is a phosphoprotein that accounts for 75–83.8% of all nitrogen compounds in sheep milk [4]. Casein is not a homogeneous protein and is composed of four fractions: α_s–casein (48.5%), β–casein (38.1%), κ–casein (13.4%) and γ-casein (5%). The casein content in ruminants, which is higher than cow milk, determines the high production efficiency of the material. The major fraction of casein in milk from small ruminants is β-casein, which contains approximately 55% of the total nitrogen content, whereas casein represents only 36% of the nitrogen content in cow milk. In casein from small ruminants, it is possible to observe a high concentration of the α_{s2} fraction (10-30%), whereas in cow casein, it only accounts for 12%. The content of κ-casein in goat and sheep milk is 20% on average, and this value is greater than cow milk, which is only 14% (Table 2.) [9,10].

Components	Milk						
	Cow		Sheep		Goat		Human
	$\bar{x}$	S(E)	$\bar{x}$	S(E)	$\bar{x}$	S(E)	n
Water	87.3	85.8-88.8	80.6	75.7-86.1	86.6	79.9-89.3	87.2
Dry matter	12.7	11.2-14.5	19.4	13.9-24.3	13.4	10.7-20.1	12.8
Proteins, including:	3.2	3.05-3.85	5.6	4.47-6.83	3.9	2.8-5.0	1.2
Casein	2.5	2.0-3.0	4.2	4.17-4.3	2.7	2.3-3.2	0.5
Whey proteins	0.6	0.5-0.8	1.1	0.9-1.3	0.6	0.4-0.7	0.7

Table 2. Comparison of the chemical composition of cow's milk, sheep milk, goat milk and human milk (g/ 100 g) adapted from [41]

As in proteins, the casein content is variable, which is caused by different factors, including the sheep breed, lactation period, and health of the udder [5,11]. In the milk of different sheep breeds, the level of casein may vary from 4.4% to 5.3%. The lowest level of casein is found in the colostrum (21-50% of the total protein content). It was also reported that the level of casein in the milk from udders with mastitis was lower than in milk from healthy udders. Caseins form spherical micelles with a diameter that ranges from 80 nm to 190 nm and forms a colloidal solution [2,5]. Caseins are produced in the milk-producing cells of the udder. Bonczar and Paciorek [2] reported that casein micelles in sheep milk are richer in Ca compared to micelles in cow milk, and no $CaCl_2$ additive is required in the production of sheep cheese. Casein micelles are composed of the following fractions: α_{s1}-casein with 5 genetic variants (A, B, C, D, and E), α_{s2}-casein with 3 genetic variants (A, B, and F), β-casein with 3 genetic variants (A, B, and C) and κ-casein, which has no known genetic variants [4,12]. The share of the α_s-casein fraction compared to the total casein content ranges from 47.2–56.5%, and the share of β-casein ranges from 28.1-36.0% while the share of κ-casein ranges from 10.6-12.1% [5]. The amino-acid composition of casein is diverse, and glutamic acid is the most dominant aminoacid; however, some authors have varying opinions on the contents of proline and sulphur-containing aminoacids. Some studies have suggested that sheep milk does not contain these aminoacids at all, whereas other studies have shown that their level is relatively high [5,13]. Casein micelles are particularly abundant in calcium and colloidal phosphorus and contains approximately 53.14 (Ca) and 22.99 (P) mM/l in sheep milk, respectively, which is why there is no need to add calcium to the milk after pasteurisation in cheese production [5]. Casein undergoes a coagulation process induced by rennet, which is used in the production of cheeses made with rennet. The process can also cause the acidification of milk to its isoelectric point (pH of 4.6 at the temperature of 20ºC).

Whey proteins are the second basic group of proteins, which includes mostly albumins represented in milk by three major fractions: β–lactoglobulin, α–lactalbumin and serum albumin. The major fractions in whey proteins are β–lactoglobulins (67%), immunoglobulins (20%), and the remaining fractions contain α–lactalbumins and serum albumin. The proteins are found in milk, and they are difficult to isolate as their coagulum is fine and easily dispersed. Unlike casein, the proteins do not contain phosphorus; however, they include a considerable amount of cystine, cysteine and lysine [14]. The major fraction in whey proteins is β-lactoglobulin, followed by immunoglobulins, α-lactalbumin and serum albumin. Sheep milk has a high level of aminoacids compared to cow milk, particularly lysine, valine and serine, which have a favourable effect on the human organism [5,15]. The content of whey protein in milk is approximately 0.6%, which accounts for 20% of the nitrogen content. The proteins do not include phosphate residues; however, they contain a high content of sulpher-containing aminoacids. Unlike typical albumins, β-lactoglobulin, which represents approximately 50% of whey proteins, is not water-soluble; however, it is relatively easy to dissolve in dilute solutions of natural salts. α-Lactoalbumin and blood serum albumin are typical water-soluble albumin proteins. These proteins are more resistant to thermal denaturation compared to β-lactoglobulin. Compared to cow milk, whey proteins from sheep milk are more sensitive to heating. Pasteurisation of sheep milk at 65°C for 30 minutes causes denaturation of approximately 15% of the water-soluble proteins, whereas only 2.3% of cow milk proteins are denatured under these conditions. β-Lactoglobulin found in whey protein binds to heavy metals during heating, such as copper and iron, which prevents oxidation to the fat by the metals and significantly increases the shelf life of dairy products. A complex with iron is durable even after thermal denaturation. It has been shown that lactoferrin from sheep milk destroys *Micrococcus flavus* to a greater extent than cow lactoferrin [16]. Immunoglobulins, which account for approximately 10% of whey protein, are a mixture of high-molecular proteins with immunological properties and are composed of aminoacids and saccharides. We can distinguish three types corresponding to blood γ-globulins: G immunoglobulins (IgG), A immunoglobulins (IgA) and M immunoglobulins (IgM), which differ with respect to molecular weight and saccharide content. Despite the fact that the amino-acid composition and sequences of β-lactoglobulin and α-lactalbumin are known, not all of their biological functions have been determined, and α-lactalbumin, which is a regulatory protein, supports the binding of glucose to UDP-galactose [17].

4. Nutritional value of sheep milk proteins

Sheep milk is believed to have a high nutritional value compared to milk from other species of mammals, and it has a richer chemical composition. Clare and Swaisgood [18] stated that the compounds function in a number of biological functions in the human body, such as digestion, hormonal activity, immunity, neurology and nutrition. The most important peptides with strong biological properties include casokinines, lactorphines, casein phosphopeptides, casoxines and lactoferroxines. Due to the biological activity of these peptides, they can be used in the pharmaceutical industry in drug production and in the food industry in the production of dietary supplements. The antibacterial properties of peptides, such as those of casecidin, casocidin, isracidin, lactoferricin and lactoferrin, make it possible to use peptides as food additives to ensure its health safety [19]. The nutritional value of sheep milk results from the high content of components that are valuable to the human body as exogenous aminoacids, such as short-chain and medium-chain fatty acids, which may play an important role in the treatment of patients with metabolic disorders and in people suffering from hypercholesterolaemia [5,15]. Table 3.

The Effect of the Photoperiod and Exogenous Melatonin on the Protein Content in Sheep Milk   289

Amino-acids	In g/ 100 g of sheep milk	In g/100 g of casein
Tryptophan	0.084	1.3
Threonine	0.268	3.6
Isoleucine	0.338	5.1
Leucine	0.587	9.0
Lysine	0.513	7.3
Methionine	0.155	2.1
Cystine	0.035	0.8
Phenyloalanine	0.284	5.2
Tyrosine	0.281	5.6
Valine	0.448	6.7
Arginine	0.198	3.3
Histidine	0.167	3.3
Alanine	0.269	3.2
Aspartic acid	0.328	7.7
Glutamic acid	1.019	21.1
Glycine	0.041	1.7
Proline	-	10
Serine	0.492	5.0

Table 3. Amino-acid composition of sheep milk proteins adapted from [5, 15]

Specification	Recommended daily allowances for an adult individual	Cow's milk		Sheep milk	
		Content in a glass of milk	Satisfaction of recommended daily allowances (%)	Content in a glass of milk	Satisfaction of recommended daily allowances (%)
Energy [kJ]	10870	643	5.9	1128	10.4
Protein [g]	75	8.2	11.0	15	20
Fat [g]	100	8.8	8.8	17.5	17.5
Carbohydrates [g]	400	11.3	2.8	11.3	2.8
Calcium [mg]	800	300	37.5	483	60.3
Phosphorus [mg]	800	233	29.1	395	49.4
Iron [mg]	12	0.13	1.1	0.25	2.1
Magnesium [mg]	350	32.5	9.3	45	12.9
Vitamin C [mg]	75	2.4	3.2	10.4	13.9
Vitamin B_1	1.4	0.10	7.1	0.16	11.4
Witamin B_2	1.4	0.41	29.3	0.89	63.6

Table 4. Relative contributions of cow and sheep milks to recommended daily allowances for nutrients and energy, adapted from [20]

The amino-acid content in sheep milk is proportionally higher than in cow milk because sheep milk contains a higher level of serine, alanine, histidine, valine and lysine, whereas the content of cystine and glycine is lower [13]. The high nutritional value of sheep milk is also related to the proline content, which affects the production of haemoglobin, and the high cytological quality is determined by a relatively low content of somatic cells in the milk, which is approximately 200 cells/cm^3. The nutritional value of milk is appropriately illustrated by the degree to which the recommended daily allowance of an individual adult for nutrients and energy is met by drinking a glass of the milk. As it was reported by Haenlein [15], drinking two glasses of sheep milk (1/2 litre), or consumption of an equivalent portion of sheep yogurt or cheese, satisfies 162% of the recommended dietary allowance (RDA) of protein for an adult, 121% of the RDA for calcium and 200% of the RDA for riboflavin. Although two glasses of sheep milk is not enough to satisfy the allowances for all substances and energy, it meets the requirements better than cow milk (Table 4) [20]. One litre of sheep milk contains 900–1050 kcal, whereas the same amount of cow milk delivers only 700 kcal [21]. Table 4

5. Modulation of the protein content in the milk of seasonally breeding sheep

There is increased interest in products made from sheep milk because of the high content of components that are valuable for humans, and there is a need to deliver this product to the market throughout the year [5,22]. However, due to specific characteristics of the reproduction cycle in seasonally breeding sheep, the lactation period is from the time of lengthening days to the end of September, which does not allow for obtaining sheep milk and its products throughout the entirety of the year. Therefore, studies have been performed over the past several years with the goal of determining the role of day length and exogenous melatonin in the modulation of milk yields and chemical composition of milk in seasonally breeding sheep. It is well known that light influences the behaviour of animals. As early as 1925, Rowan [23] showed that lighting conditions have an effect on gonad activity in birds, and the processes connected with them are controlled by intraorganic signals. In farm animals, changes in day length play an important role in determining their yields. Introduction of a long photoperiod in cows produced an increase in the prolactin content and overall milk yield [24].

The photoperiod is of special importance to short-day animals (sheep) where the day length is related with changes in melatonin concentrations [25]. In sheep, changes in day length are perceived as a biological marker with respect to reproductive function and impact the processes of milk secretion and lactation. The major gland that acts as a messenger to transmit information from the external environment to the central regulating activities of the organism, which informs the organism of season changes, is the pineal gland, which is where melatonin is synthesised [26]. The biosynthesis of melatonin in most animals is dependent on the lighting conditions. The highest concentration of the hormone known as the *"hormone of darkness"* is at night [27-29]. It is typical for the circadian rhythm of melatonin secretion to increase its concentration in the dark phase and drop during the day, and this process occurs in all animals. Seasonal changes in melatonin secretion result from the activity of the biological clock, which sends a signal in the annual reproduction cycle in animals characterised by seasonal reproduction.

The endocrine mechanism of entering and maintaining lactation is not fully understood. However, it is clear that this process involves a number of hormones, which shows that the process relies on the activity of the hypothalamus and pituitary gland [30]. Currently, it is believed that milk yields in mammals are determined by genetic and environmental factors [31]. However, in recent years, more attention has been paid to the role of light, which is a factor that modulates the concentration of prolactin and depends on the annual rhythm of melatonin concentrations. In natural conditions, the maximum concentration of prolactin in the blood is identified in the long-day period, and at this time, the level of melatonin decreases. The lowest concentration of prolactin is observed during the short-day period, which is when the level of melatonin is the highest [32,33]. Experiments performed by Molik et al. [34,35] demonstrated that day length likely determines the milk yield of sheep. Sheep entering lactation in the short-day period had a milk yield of 50% compared to sheep milked in the long-day period. Changes in the secretion of prolactin during the lactation period of sheep have an effect on the amount of milk produced and on the synthesis of proteins, fat and immunoglobulins, which indicates that they determine its composition, quality and technological usability. The knowledge of physiological and endocrinological mechanisms that determine the lactation level may contribute to a better understanding of the relationship between the photoperiod and milk yields in sheep [36]. If the season determines the amount of milk produced, it can be presumed that it also modulates the chemical composition and nutritional value of products made from the milk.

6. The effect of a diverse photoperiod and exogenous melatonin on the protein and casein contents in sheep milk produced during the long-day period

A change in the day length is a vital factor that affects the yields of seasonally breeding farm animals. Therefore, the purpose of these experiments was to determine the effects of day length and exogenous melatonin on the protein and casein contents in sheep entering lactation in the long-day period. The experiments were performed on 60 Polish long-wool sheep mothers that were bred with a seasonal reproduction cycle, aged 4–5 years and weighed 60 ± 5 kg. Synchronisation of the sheep estrous cycle was performed with gestagens using the Chronogest® method. Polyurethane foams saturated with 40 mg of the cronolone preparation (Intervet; Holland) were inserted into the sheep vaginally for the period of 14 days. When the foam pads were removed, the ewes were administered 500 IU of PMSG (Serogonadotropin; Biowet, Drwalew). Estrous ensued 48-72 hours after administering the preparation, and its duration was controlled with a teaser ram. The breeding season for all groups of sheep occurred from the 15th to the 30thof September, whereas lambing ensued in the second half of February. Lambs were raised with mothers before being separated, and the mothers were milked. After separation of lambs on the 57th day of lactation, the sheep were divided into three groups: Group 1 (n=20 sheep, sheep from the control group maintained in natural day-length conditions), Group 2 (n=20 sheep, sheep with subcutaneous melatonin implants that released 18 mg of exogenous melatonin; Ceva Animal France), and Group 3 (n=20 sheep, sheep maintained in artificial short-day conditions; 16D:8L). After separation of lambs on the 57th day, the sheep were milked twice a day using an Alfa-Laval milking machine. Milk yields were measured every 10 days, and at that time, milk samples were collected from each group of sheep to determine the protein and casein contents. Feeding the sheep during the experiment was standardised. During the experiment, the sheep from groups 1 and 2 were maintained indoors with an option to use runs, and group 3 stayed indoors throughout the period that they were used for dairy production in a 20-m² room with artificial short-day conditions (16D:8L).The experiments demonstrated that the highest protein level was on the 60th milking day (May) from the sheep maintained in the 16D:8L conditions (6.87±0.01% protein), whereas milk from the sheep with melatonin implants had a protein content of (5.86±0.02%), which was significantly lower (P ≤ 0.01) (Table 5). Table 5.

Months of milking used	Polish longwool sheep					
	Control group		Exogenous melatonin		16D:8L group	
	$\bar{x}$	S(E)	$\bar{x}$	S(E)	$\bar{x}$	S(E)
April	5.47	0.07	5.83	0.49	6.19	0.04
May	6.63[B]	0.01	5.86[A]	0.02	6.87[B]	0.01
June	7.10[b]	0.07	7.15[b]	0.01	7.5[a]	0.07
July	6.8[B]	0.04	7.8[A]	0.01	7.09[B]	0.01

A, B,- average levels marked with different letters vary significantly in drawn samples at P ≤ 0.01,
a, b, - average levels marked with different letters vary significantly in drawn samples at P ≤ 0.05

Table 5. Effects of the diverse photoperiod and exogenous melatonin on the protein content (%)

In June, the protein level in the milk of sheep in the 16D:8L conditions was markedly higher (P ≤ 0.05), which was7.5±0.07%, whereas milk from sheep in the control group and with melatonin implants contained a lower level of protein (7.1±0.07% and 7.15±0.01%, respectively). In the fourth month of milking, the milk from sheep with melatonin implants had a protein content of 7.8±0.01%, which was significantly higher (P ≤ 0.01)than sheep without the implant, whereas the milk from sheep in the 16D:8L and control groups contained less protein(7.09±0.01% and 6.8±0.04%, respectively). Based on the results of the experiments, it was determined that the photoperiod represents an important factor that impacts the productivity of the animals. The protein content in the sheep with melatonin implants and sheep maintained in 16D:8L conditions increased from the first day of milking until the end of lactation. Upon analysing the changes in the casein content, it was determined that in the first month of milking (April), the lowest level of casein was identified in the milk of sheep raised under the 16D:8L conditions (2.28 ±0.03%) and in the milk of sheep with melatonin implants (2.32±0.03%), which was determined to be significant (P ≤ 0.01).Table 6

Monthsof milking used	Polish longwool sheep					
	Control group		Exogenous melatonin		16D:8L group	
	$\overline{x}$	S(E)	$\overline{x}$	S(E)	$\overline{x}$	S(E)
April	3.42[A]	0.01	2.32[B]	0.03	2.28[B]	0.03
May	4.6[a]	0.05	4.22[b]	0.02	4.36[b]	0.01
June	4.39[B.a]	0.02	5.35[B.b]	0.05	2.85[A]	0.04
July	3.53[A]	0.03	5.28[B]	0.02	5.04[B]	0.01

A, B,- average levels marked with different letters vary significantly in drawn samples at P ≤ 0.01,
a, b, - average levels marked with different letters vary significantly in drawn samples at P ≤ 0.05

Table 6. The effect of the diverse photoperiod and exogenous melatonin on the casein content (%)

In subsequent months of lactation, the highest casein content was observed in the milk from sheep in the control group. However, in June, the casein content in the milk from sheep in the group with melatonin implants increased (5.35 ±0.05%), whereas the casein content became significantly lower (P ≤ 0.01) and approached 2.85 ±0.04% in sheep maintained in the 16D:8L conditions. In the last month of milking, the highest casein content ((P ≤ 0.01) was observed in the milk from sheep with melatonin implants (5.28±0.02%), and the lowest casein levels were observed in milk from the control sheep (3.53±0.03%). These results indicated that the protein content grew as lactation progressed in the control group, which was because of a change in the milk yields. Administration of exogenous melatonin and simulation of the short-day period exerted a significant influence on the protein and casein contents in sheep milk. Administration of exogenous melatonin in the period of lengthening days caused an increase in the protein content of sheep milk beginning in the 60th day of lactation. Similarly, artificial simulation of the short-day conditions during the period of lengthening days resulted in an increase of the protein and casein contents in sheep milk. Casein is an important factor that functions in cheese curd formation; therefore, the high casein content in the milk of sheep with melatonin implants and sheep maintained in the 16D:8L artificial short-day conditions resulted in a higher processing efficiency.

7. The effect of the day length and exogenous melatonin on the protein and casein contents in sheep milk produced at the time of the diverse photoperiod

Given the intensified production of food with animal origins observed over the last few years as well as the enhanced nutritional awareness of consumers, the quality and particularly the nutritional and pro-health value of food [37] is a major criterion for choosing a food product. Therefore, the purpose of the present study was to determine the effects of day length and exogenous melatonin on the protein and casein contents in seasonally lactating sheep.

The experiments were performed on 60 Polish long-wool sheep mothers, which is a breed with a seasonal breeding cycle, and the sheep were 4 to 5 years of age and weighed 60 ± 5 kg. Synchronisation of the sheep estrous cycle was performed with gastagens using the

Chronogest® method. Polyurethane foam saturated with 40 mg of a cronolone preparation (Intervet; Holland) was inserted into the sheep vaginally for 14 days. When the foams were removed from the ewes, the ewes were administered 500 IU of PMSG. Estrous for the first group began on the 15thof September, and for the second and third groups, on the 15thof January. Ewes from the first group lambed in mid-February, and the lambs from the second and third groups lambed in mid-June: Group 1 (n=20 sheep, sheep lambed in February that were maintained in natural day-length conditions and served as the control group), Group 2 (n=20 sheep, sheep lambed in June and maintained in natural day-length conditions), and Group 3 (n=20 sheep, sheep lambed in June with subcutaneous melatonin implants to introduce a short-day signal as early as possible and were maintained in long-day conditions). Six weeks before lambing, 18 mg of exogenous melatonin (Ceva Animal; France) was introduced. Lambs were raised with mothers until the 56th day of life before they were separated from their mothers, and the mothers were milked. After the 57th day of lactation, the ewes were milked twice-a-day using an Alfa-Laval milking machine. Milk yields were measured every 10 days and collective milk samples were drawn from each group of sheep to determine the protein content. Feeding the sheep during the experiment was standardised. During the experiment, the sheep were raised indoors with an option to use runs. The conducted experiments demonstrated that shifting lactation to the period of shortening days and administration of exogenous melatonin caused an increase in the protein content. In the first month of significant milking ($P \leq 0.05$), the highest level of protein was observed in the milk of the sheep that lambed in June (6.7±0.8%) and in the milk of mothers with subcutaneous melatonin implants (6.3±0.8%; shown in Table 7).

Months of milking used	Polish longwool sheep					
	Sheep milking used in long days		Sheep milking used in short days		Sheep with exogenous melatonin	
	$\bar{x}$	S(E)	$\bar{x}$	S(E)	$\bar{x}$	S(E)
April	5.42[a]	0.7	-	-	-	-
May	5.7[a]	0.5	-	-	-	-
June	5.49[A]	0.6	-	-	-	-
July	6.2	0.8	-	-	-	-
August	6.8	0.8	6.7[b]	0.8	6.3[b]	0.8
September	-	-	7.49[b]	0.8	7.19[b]	0.9
October	-	-	8.05[B]	0.9	7.6[B]	0.9

A, B,- average levels marked with different letters vary significantly in drawn samples at $P \leq 0.01$,
a, b, - average levels marked with different letters vary significantly in drawn samples at $P \leq 0.0$

Table 7. The effect of the day length and exogenous melatonin on the protein content (%)

In the second month of significant milking ($P \leq 0.05$), the highest level of protein was recorded in the milk of the sheep that lambed in June (7.49±0.9%) and in the milk of sheep with melatonin implants (7.19±0.9%). At the same time, the lowest protein content was identified in the milk of sheep entering lactation in the long-day period (5.7±0.5%). In the third sample (P≤0.01), the highest level of protein was detected in the milk of the sheep that lambed in June (8.05±0.9%) and in the milk of sheep exposed to exogenous melatonin (7.19%±0.9%). It is important to note that in sheep from the control group, the highest protein content was observed in the last month of lactation (6.8±0.8%), whereas a comparable level of protein in the group of sheep milked in the period of shortening days and with melatonin implants was identified in the first month of milking. The identified changes in the protein content suggest that the milk of sheep entering lactation in the period of shortening days and exposed to the effects of exogenous melatonin can differ with respect to the casein content compared to the milk from sheep in the control group. Table 8.

| Months of milking used | Polish longwool sheep | | | | | |
| | Sheep milking used in long days | | Sheep milking used in short days | | Sheep with exogenous melatonin | |
	$\bar{x}$	S(E)	$\bar{x}$	S(E)	$\bar{x}$	S(E)
April	3.9[a]	0.01	-	-	-	-
May	4.31[a]	0.02	-	-	-	-
June	4.10a	0.03	-	-	-	-
July	4.63	0.03	-	-	-	-
August	5.06	0.04	5.2[b]	0.04	4.87[b]	0.02
September	-	-	5.73[b]	0.03	5.63[b]	0.03
October	-	-	4.95	0.02	5.59[b]	0.02

a, b, - average levels marked with different letters vary significantly in drawn samples at P ≤ 0.05

Table 8. The effect of the day length and exogenous melatonin on the casein content (%)

In sheep milked during the long-day period, the casein content increased as lactation progressed and milk yields dropped. The lowest level of casein was observed in the first month of milking (3.9±0.1%), whereas the highest casein content was observed in milk obtained during the last month of milking (5.06±0.4%). The shift of lactation to the period of shortening days resulted in an increase of the casein content in the examined milk in the first month of milking (5.2±0.4%). Administration of exogenous melatonin 6 weeks before lambing also contributed to a change in the casein content of the milk. In the first month of milking, the milk of sheep exposed to exogenous melatonin contained 4.87±0.2% of casein, whereas on the 90th day of milking, the milk contained as much as 5.59±0.2% casein. Thus, the milk from sheep milked in the short-day period and exposed to the effects of exogenous melatonin contained significantly more casein than milk from sheep milked in the long-day period. It is important to note that the milk from sheep lambed in June (groups 2 and 3) exhibited a higher casein content that corresponded with the casein content observed in the final phase of lactation from sheep milked under the long-day period. While analysing the total protein and casein contents in sheep milk and taking into account the whole milking period, it was shown that the protein and casein contents in sheep milk grew significantly by applying exogenous melatonin (Table 9). Similarly, the milk of sheep milked in the short-day period had a higher level of protein and casein than milk from sheep that were milked in the long-day period. Table 9.

Parameters	Polish longwool sheep					
	Sheep milking used in long days		Sheep milking used in short days		Sheep with exogenous melatonin	
	$\bar{x}$	S(E)	$\bar{x}$	S(E)	$\bar{x}$	S(E)
Protein (%)	5.92[AB]	0.49	7.30[A]	0.64	7.43[B]	0.71
Casein (%)	1.15[ab]	0.04	0.85[ab]	0.13	1.72[b]	0.19

A, B,- average levels marked with different letters vary significantly in drawn samples at $P \leq 0.01$
a, b, - average levels marked with different letters vary significantly in drawn samples at $P \leq 0.05$

Table 9. The effect of the day length and exogenous melatonin on the total protein and casein content in the period in which sheep are used for dairy production

Casein is a protein that plays a fundamental role in cheese production; its high content in sheep milk signifies a high production efficiency of this material. In milk from the examined groups of sheep, a significantly higher level of casein was observed in the group of sheep that lambed in June and is correlated with an increase in total nitrogen compounds, which had a significantly higher concentration in the milk from this group of sheep.

8. Conclusions

The increase of interest in nutritional and pro-health values of food has resulted in more attention to components that are of paramount importance to the health of people. In general, consumers pay attention to the quality of consumable products while searching for products whose composition is as close as possible to the "natural" product (i.e. the least processed foods). The chemical composition of sheep milk indicates a high technological value of the milk from this animal breed and its fitness for processing into cheese. The studies performed by Ciuryk et al. (1999) and Molik et al. (2006 and 2007) showed that the milk yields of sheep with strong seasonal breeding characteristics depend on the lambing time of the ewes. It was also concluded that shifting the lambing time to the spring and summer months has a negative impact on the duration of lactation and amount of milk obtained as well as its chemical composition, which suggested that, aside from genetic and environmentally related factors, an important role in the process of entering and maintaining lactation is assigned to the day length, which is a factor that modulates the level of melatonin and prolactin. Thus, only prolactin secretion changes in lactating sheep influences the amount of milk produced and the synthesis of milk proteins. In natural conditions, the maximum concentration of prolactin in sheep blood is observed in the long-day period as opposed to the time when the level of melatonin drops. The lowest concentration of prolactin is recorded during the short days when the level of melatonin is the highest. Shortening of the day length or prolonged administration of exogenous melatonin during the period of physiologically increased concentrations of prolactin causes a reduction in the secretion of this hormone. Changes in the secretion of prolactin during the lactation period in sheep undoubtedly have an influence on the amount of milk produced and its chemical composition [38]. The conducted studies revealed that administering exogenous melatonin and modulation of the day length had a significant influence on the protein and casein contents in sheep milk. However, shifting the lactation time to the period of shortening days and applying exogenous melatonin caused an increase in the protein content within the first month of milking. The protein content in the milk of sheep milked during the period of shortening days and exposed to the effects of exogenous melatonin was adequate in the last stage of lactation compared to sheep milked during the long-day period. The performed studies demonstrate that day length and melatonin are factors that modulate the protein content in the milk of seasonally breeding sheep.

Author details

Edyta Molik* and Dorota Zięba
Department of Swine and Small Ruminant Breeding, Biotechnology and Genomic Laboratory, Agricultural University in Krakow, Poland,

Genowefa Bonczar and Aneta Żebrowska
Department of Animal Product Processing, Agricultural University in Krakow, Poland,

Tomasz Misztal
Department of Endocrinology, The Kielanowski Institute of Animal Physiology and Nutrition, Polish Academy of Sciences, Poland

Acknowledgement

This research was supported by projects MNiSZW NN311245033 and DS/KHiOK/3242/2010.

9. References

[1] Mubois JL, Leorii J (1989) Peptides du lait a activite biologique. Lait. 69: 245-835.

[2] Bonczar G, Paciorek A (1999) Właściwości mleka owczego. Zesz. Nauk. AR. Kraków. 360: 37-48.

[3] Ciuryk S, Molik E, Bonczar G (1999) Wydajność i skład chemiczny mleka plennej owcy olkuskiej, długowełnistej i mieszańców po trykach charollais, Zesz. Nauk. PTZ. 43: 73 – 79. In Polish.

[4] Ramos M, Juarez M (2003) Sheep milk, Encyclopedia of Dairy Sciences. In: editor Roginski H, Fuquay JW, Fox PF. pp 2539-2545.

---

* Corresponding Author

[5] Anifantakis E M (1986) Comparison of the physico-chemical properties of ewe's and cow's milk, FIL. Doc. 202: 42-53.

[6] Molik E, Ciuryk S, Misztal T, Romanowicz K, Wierzchos E (2004) Effect of climatic conditions on milk yield of polish longwool sheep lambing on different dates. Scien. Messeng. Lviv. 6: 58-62.

[7] Bonczar G (1989) Zmiany składu chemicznego i cech fizycznych mleka owczego w zależności od stanu zdrowotnego wymienia. Zesz. Nauk. AR Kraków. 133: 2-133. In Polish.

[8] Novotna L, Kuchtik J, Dobes I, Sustova K, Zajicova P (2007) Effect of somatic cell count on ewe's milk composition, its propierties and quality of rennet curd. Acta. Universita. Brunensis. 55: 59-64.

[9] Szczepanik A, Libudzisz Z (2001) Przydatność technologiczna mleka koziego Przem. Spoż. 2: 35 – 36. In Polish.

[10] Wszołek M, Tamime AY, Muir DD, Barclay MNI (2001) Properties of Kefir made in Scotland and Poland using bovine, caprine and ovine milk with different starter cultures. Lebensm.-Wiss. u.-Technol. 34: 251 – 261.

[11] Bonczar, G (1990) The content of nitric substances in ewe's milk in relation to age, period of lactation, and health of udder. Zbornik Bioteh. Univer. EK v Ljubljani. 15: 549-556.

[12] Park YW, Juarez M, Ramos M, Haenlein GFW (2007) Physico-chemical characteristics of goat and sheep milk. Small. Rumin. Res. 68: 88-113.

[13] Haenlein GFW (1997) Nutritional value of dairy products of ewes and goats milk. Sheep Dairy. News. 13 (1): 10-16.

[14] Kędzior W (2005) Owcze produkty spożywcze. Aspekty Towaroznawcze. Warszawa: PWE. 250 p. In Polish.

[15] Haenlein GFW (2001) Past, Present, and Future Perspectives of Small Ruminant Dairy Research. J. Dairy Sci. 84: 2097 – 2115.

[16] Recio I, Visser S (2000) Antibacterial and binding characteristics of bovine, ovine and caprine lactoferrins: a comparative study. Inter. Dairy Jour. 10: 597-605

[17] Ziajka S (2008) Mleczarstwo - Volume 1. Olsztyn. UWM.380 p. In Polish.

[18] Clare DA, Swaisgood HE (2000) Bioactive Milk Peptides -A Prospectus. J. Dairy. Sci. 83: 1187 - 1195.

[19] Lahov E, Regelson W (1996) Antibacterial and immunostimulating casein-derived substances from milk: casecidin, isracidin peptides. Food Chem. Toxicol. 34: 131 – 145.

[20] Bonczar G (2001) Znaczenie mleka owczego w żywieniu człowieka. Przeg. Mlecz. 3: 125-128. In Polish.

[21] Ważna E (1997) Wartość mleka owczego. Por. Gos. 12. 1- 26. In Polish.

[22] Borys B, Pisulewski PM (2001) Jakość oraz możliwości kształtowania prozdrowotnych właściwości spożywczych produktów owczarskich. Rocz. Nauk. Zoot. 11: 67-86. In Polish.

[23] Rowan W (1925). Relation of light to bird migration and developmental changes. Nature. 115: 494-495.

[24] Reksen O, Tverdal A, Landsverk K, Kommisrud E, Boe KE, Ropstad E (1999) Effects of photointensity and photoperiod on milk yield and reproductive performance of Norwegian Red Catlle. J. Dairy Sci. 82: 810-816.

[25] Misztal T, Romanowicz K, Barcikowski B (1996) Seasonal changes of melatonin secretion in relation to the reproductive cycle in sheep. J. Anim Feed Sci. 56: 35-48.

[26] Reiter RJ (1988) Neuroendocrinology of melatonin. In: Melatonin Clinical Perspectives. editors: Miles A, Philbrick DRS, Thomson C, Oxford: Oxford University Press. pp 1-42 .

[27] Reiter RJ (1991a) Melatonin: that ubiquitously acting pineal hormone. News. Physiol. Sci. 6: 223-227.

[28] Reiter RJ (1991b) Pineal melatonin; cell biology of its synthesis and of physiology interactions. Endocr. Rev. 12: 151-180

[29] Arendt J (1993) Biological rythms. The mammalian pineal gland and its control of hypohalamic activity. Prog. Brain Res. 41: 149-174.

[30] Peaker M, Neville, MC (1991) Hormons in milk: chemical signals to the offspring's'. J. Endocrinol. 131: 1-3

[31] Molik E, Murawski M, Bonczar G, Wierzchos E (2008) Effect of genotype on yield and chemical composition of sheep's milk. Anim. Sci. Paper. Rep. 26: 211-218.

[32] Robinson JJS, Wigzell RP, Aitken JM, Wallace S, Ireland, Robertson IS (1992) Daily oral administration of melatonin from March on wards advances by 4 months the breeding season of ewes maintained under the ambient photoperiod at 57^0 N'. Anim. Reprod. Sci. 27: 141-160

[33] Misztal T, Romanowicz K, Barcikowski B (1999) Melatonin modulation of the daily prolactin secretion in intact and ovariectomized ewes, relation to phase of the estrous cycle and to the presence of estradiol. Neuroendocrinology. 69: 105-112.

[34] Molik E, Misztal T, Romanowicz K, Wierzchoś E (2006) The Influence of length day on melatonin secretion during lactation in asesonal sheep. Arch. Tierz. 49: 359-364.

[35] Molik E, Misztal T, Romanowicz K, Wierzchoś E (2007) Dependence of the lactation duration and efficiency on the season of lambing in relation to the prolactin and melatonin secretion in ewes. Liv. Sci. 107: 220-226

[36] Molik E, Misztal T, Romanowicz K, Zięba D, Wierzchoś E (2009) Changes in growth hormone and prolactin secretion in ewes used for milk under different photoperiodic conditions. Bull. Vet. Inst. Pulawy. 53: 389-393

[37] Zebrowska A, Bonczar G, Molik E (2009) Właściwości prozdrowotne mlecznych napojów fermentowanych. Przeg. Hod. 8: 24-26.

[38] Molik E, Bonczar G, Żebrowska, Misztal T, Pustkowiak H, Zieba D (2011) Effect of day length and exogenous melatonin on chemical composition of sheep milk. Archiv. Tierz. 54: 177-187

[39] Fox PF (2003) Milk. Encyclopedia of Dairy Sciences. London: Elsevier Science. pp.100
[40] Jandal JM (1996) Comparative aspects of goat and sheep milk. Small. Rum. Res. 22: 177 – 185.
[41] Obrusiewicz T (1994) Mleczarstwo. Part 1. WSiP. Warszawa. pp 250. In Polish.

Chapter 12

Milk Protein Synthesis in the Lactating Mammary Gland: Insights from Transcriptomics Analyses

Massimo Bionaz, Walter Hurley and Juan Loor

Additional information is available at the end of the chapter

1. Introduction

In recent years, and with the advent of more advanced molecular techniques, a compelling case has been made for more in-depth studies of molecular regulation of lactation, particularly in livestock species [1-3]. One of the underlying premises is that advances in our knowledge of the key control points regulating milk component synthesis in the mammary gland, e.g. milk protein, could only be made through more mechanistic studies focusing not only on large-scale mRNA expression [4] but also on post-translational events, e.g. phosphorylation/dephosphorylation of key proteins. The combination of the accumulated knowledge of regulatory mechanisms of milk protein synthesis in recent years has opened new frontiers where intervention may be made in order to improve milk protein synthesis.

2. Nutrition and milk protein synthesis

The synthesis of proteins requires the constituents of the protein synthesis machinery, as well as the availability of amino acids and a large supply of energy. Protein synthesis and turnover has a high-energy requirement and after ion transport is one of the most energetically costly processes in the cell. This is evidenced by the reduction of protein synthesis and ion transport during anoxic intervals (reviewed in [5]) but also by the decrease of overall protein synthesis as consequence of caloric restriction (reviewed in [6]). In mammals, the basal energy expenditure for maintenance of ion transport and protein re-synthesis has been estimated to be 30-40% and 9-12%, respectively [7].

The need of energy and protein during lactation increases dramatically. In dairy cows there is more than a 5-fold increase in energy and protein requirement from late gestation to lactation [8]. The fractional synthesis rate in the mammary gland of goats increases ca. 7-fold from the dry period to lactation, with synthesis of milk protein accounting for ca. 60% of the overall proteins synthesized in that tissue [9]. Recently, the use of more precise measurements has estimated that daily tissue protein synthesis (i.e., non-milk protein in mammary tissue) can represent up to 88% of the total protein synthesized in the goat mammary gland, which uses half of the available ATP supply generated in the lactating udder [10]. In the bovine, there is a 4-fold increase in mRNA translation in lactating compared to non-lactating mammary tissue [11]. The efficiency to transform dietary nitrogen into milk proteins is low (25-30%) [12]. The estimated fractional protein synthesis rate in the goat mammary gland is 42-130% and the turnover of the protein contributes 42-72% of the total (milk included) protein synthesized in the mammary gland [12]. Thus, protein synthesis is a highly active and energetically costly process, with only a minor part of the synthetic machinery apparently being used for production of milk proteins [10].

Milk protein yield is of great significance for the dairy industry. The amount and composition of proteins in milk is largely determined by the genetics of the animal, and is difficult to change through nutrition. However, due to the high requirement of protein synthesis for energy, the milk protein yield can be affected by the energy content in the diet [13]. This has been observed in dairy cows and sheep where energy in the diet was increased or decreased by feed restriction, and both overall protein yield and percentage were affected [14-17]. In the cow the effect is quite consistent. However, in sheep the direct relationship between feed intake and milk protein is not as consistent; for instance, some reports found that feeding a higher energy level compared with a low energy diet resulted in lower milk protein [18, 19]. In rodents, feed restriction significantly decreased milk production, and milk protein yield, but not the percentage of milk protein [20]. Interestingly, feed restriction and, in a more acute way, starvation, decreased mRNA synthesis of genes coding for milk proteins in rats [21]. The mRNA synthesis of those proteins fully recovered by re-feeding, indicating that the regulation of milk protein synthesis is, at the least in part, driven by a transcriptomic adaptation rather than driven only by availability of energy. There is also an inverse relationship between dietary fat content and milk protein production, as observed in dairy cows and sheep (reviewed in [15, 17]). Interestingly, the treatment of lactating animals with *trans*10,*cis*12-conjugated linoleic acid (t10,c12-CLA), a long-chain fatty acid (LCFA) with a well-established negative effect on milk fat production [22], increases production of milk and milk protein. This has been observed in grazing dairy cows (i.e., that are considered to have a limited energy intake) [23] and fed-restricted ewes [24]. The increase of milk protein by t10,c12-CLA treatment appears to be associated with a greater metabolizable energy availability due to the decrease of fat synthesis in the mammary gland and other peripheral organs, such as adipose [24, 25]. Overall, all the above reported studies support the large dependence of mammary milk protein synthesis on the availability of energy.

Along with the availability of energy, the availability of amino acids (AA) is critical for mammary protein synthesis. Earlier studies carried out to evaluate the effect of increasing dietary protein of dairy cows have failed to demonstrate a consistent positive effect on milk protein (reviewed in [26]). The rumen contains a large population of active microbes that use dietary proteins for their own metabolism, changing the composition of the available AA reaching the intestine. Post-ruminal infusion of casein can preserve its high biological value for milk protein synthesis, and theoretically, should provide the optimal composition of AA for milk protein synthesis. Post-ruminal infusion of casein has been reported to result in a general increase in milk protein synthesis [13, 27], however this is accompanied by a decrease in efficiency as the amount of infused protein is increased and resulting in a lower effect compared to dietary energy [27]. In further support of the importance of energy compared to AA supply, abomasal infusion of casein alone failed to increase milk protein yield, while infusion of starch significantly increased milk protein yield [28]. *In vitro* studies evaluating protein synthesis in bovine mammary acini cultures also indicate an effect of available energy on protein synthesis [29]. All the above studies support a significant role of energy availability for bovine milk protein synthesis and a smaller role for overall AA availability. In contrast, increases in dietary protein results in greater milk and milk protein yield in lactating sows [30].

Nevertheless, the availability of specific AA can be a limiting factor for milk protein synthesis [12, 13]. Among the many AA essential for the lactating bovine mammary gland [13, 28, 31], methionine and lysine are considered the most important [12, 32]. The AA in mammary gland in dairy cows are not solely used for protein synthesis, but also for production of energy through the Krebs cycle [12, 33]. Our recent transcriptomics studies in dairy cattle from pregnancy to lactation support that conclusion [34].

The positive effect of dietary energy on milk protein synthesis is partly a result of the availability of energetic precursors to produce intracellular energy transfer molecules (e.g., ATP, GTP, NADH, and NADPH). Another important role of energy is through the increase in insulin secretion as a consequence of greater dietary energy. A positive role of insulin in milk protein synthesis has long been recognized [35]. A strong positive effect of insulin on yield of milk protein was demonstrated in dairy cattle with chronically elevated insulinemia through the use of hypeinsulinemic-euglycemic clamp experiments (reviewed in [36]). The role of insulin in increasing milk protein is considered problematic in dairy cows due to the dramatic decrease in blood concentration of this hormone at the onset of lactation which remains low for at least the first two months post-partum [37], a period which coincides with peak milk yield. However, the pattern of insulinemia during lactation [37] appears to follow the pattern of milk protein concentration [4]. Recently, insulin has been demonstrated to play a key role in milk protein synthesis both in mice [38] and dairy cows [39]. Our own studies also support a pivotal role of insulin in milk protein synthesis [4]. The potential mechanisms are discussed in more detail in Section 5.

In summary, milk protein synthesis can be fine-tuned by nutrition through increased availability of metabolized energy (e.g., by feeding t10,c12-CLA besides increasing energy content in the diet and/or food intake) and specific AA. Metabolizable energy appears to be the most important factor through provision of energetic precursors for the protein synthesis process, but also through increasing insulin (and perhaps also insulin-like growth factor 1, see below), which in turn positively affect the protein synthesis signaling network.

3. Transcriptomics: a novel approach to study biology

The classically-considered transcripts (i.e., mRNA; that reflected the classical transcriptome until the discovery of large transcription of non-coding RNAs) encode genetic information about proteins; thus, the changes in expression of the genes making up the transcriptome can exert major influences on physiological functions. The transcriptomics in biology has expanded our knowledge at an unprecedented pace, and essentially forced the scientific community to embrace the notion of holism as it relates to animal function. The term "functional genomics" is generally defined as the study of the functions of genes, related proteins, and activity (i.e., metabolomics) [40]. Most of the functional genomics studies have been carried out using transcriptomics with the intent, besides determining/confirming functions of genes, to infer biological adaptation, i.e., by comparing the transcriptome in different conditions it is possible to infer the biological adaptation of cells, tissue, or organs. The study of the transcriptome has classically being carried out via DNA microarray technology, and more recently, via Next Generation Sequencing. The latter, while relatively new, appears to be replacing microarrays as the tool of choice for functional genomics studies in livestock [41]. Both approaches allow the simultaneous monitoring of the expression of most or, particularly for the latter, the entire transcriptome in tissues or cells. Currently-available types of microarray platforms include cDNA microarrays (usually pure products from PCR amplification of cDNA and EST clones that are 100 – 2,000 nucleotides long), oligonucleotide microarrays (60-70 nucleotides per DNA synthesized from single-stranded probes on the basis of sequence information in databases), and commercial platforms such as the Affymetrix GeneChips (35-200 nucleotides per DNA) or the Agilent whole-transcriptome microarray.

The so-called "bottom-up" or "top-down" approaches have been used successfully in model organisms to study the metabolic behavior at the cellular level. Whereas, the bottom-up approach relies on developing automated tools based on a mathematical model, the top-down approach encompasses data processing from 'omics' studies to pathways and individual genes of an organism [42]. In essence, the cell can be approached from both bottom to top (universality) or from top to bottom (organism specificity) equally well, i.e., from molecules to the scale-free networks or modules, or moving from a network scale-free and hierarchical nature to organism-specific modules [43].

The top-down approach originates from the transcriptome experimental data and information used to reconstruct metabolic models. This approach can help to unravel biological behavior and underlying interactions using 'omics' data which can be obtained via DNA microarrays [44], RNA sequencing (RNA-seq) [45] or other genome-enabled technologies. The flow of information in the top-down approach occurs from transcriptome and proteome to flux-balanced metabolic pathways [46]. This approach aims to discover new functional aspects of cellular behavior from the 'omics' data sets through the standard top-down methodologies. By default, this approach covers the whole genome, thus, it is considered as a "potentially complete" approach, i.e., it deals with all the genome-wide transcriptomic information [42, 47]. This has been widely used in model organisms during the past 15 years, and resulted in major advancement of regulatory metabolic networks [48].

	Cow	Mouse	Rat	Human	Pig	Sheep	Goat	Kangaroo
Body Weight (kg)	650	0.04	0.25	60	170	70	65	50
MG$ weight (g)	25,000	2.7	17.8	1,000	6,000	1,400	1,200	--
% MG/BW (kg/kg)	3.8	7.5	7.1	2.5	3.5	2.0	1.8	--
Milk yield (g/d)	40,000	6	43	600	10,000	2,000	4,500	50
Dry matter (DM) (%)	12.4	29.3	22.1	12.4	20.1	18.2	12.0	23.5
g milk/kg BW	61	150	172	10	59	29	40	1
g milk/g MG	1.6	2.2	2.4	0.6	1.7	1.4	1.7	--
Day of lactation	305	20	20	365	45	200	305	340
g DM milk/kg BW	7.6	44.0	38.0	1.2	11.8	5.2	8.3	0.2
g DM milk/kg MG	198.4	651.1	533.9	74.4	335.0	260.0	450.0	--
Parturition/year	1	8	5	1	2	1	1	1
	g/kg milk							
Total Proteins	28.9	111.5	84.0	11.7	53.0	62.1	34.5	35.8
Caseins	23.8	90.3	64.0	4.0	29.0	51.6	29.4	16.1
α-S1	9.1	25.0	--	--	--	--	--	19.6
α-S2	2.4	--	--	--	--	--	--	--
β	8.5	26.0	--	--	--	--	--	11.5
κ	3.0	--	--	--	--	--	--	--
Whey proteins	3.0	21.2	20.0	7.6	20.0	8.1	5.1	8.1
α-Lactalbumin	1.1	--	1.0	6.1	2.6	--	1.7	2.1
β-Lactoglobulin	2.8	--	--	--	--	--	--	--
Albumin	4.0	15.0	--	0.3	--	--	--	--
Lactoferritin	<0.1	--	--	2.8	--	--	--	--
Casein/Whey protein	7.9	4.2	3.2	0.5	1.4	6.3	5.7	2.0

Protein content relative to several other parameters

	Cow	Mouse	Rat	Human	Pig	Sheep	Goat	Kangaroo
g/day	1156.0	0.7	3.6	11.7	530.0	124.2	155.2	1.8
g/day protein synth.**	2196.0	0.8	--	--	975.1	--	241.0	--
% Milk protein/synth.	52.6	87.5	--	--	54.3	--	64.0	--
kg/lactation	352.6	0.01	0.1	4.3	23.9	24.8	47.4	0.6
kg/lact/kg BW	0.5	0.4	0.3	0.07	0.1	0.4	0.7	0.01
g/day/kg MG	46.2	247.8	202.9	7.8	88.3	88.7	129.4	--
kg/lact/kg MG	14.1	5.0	4.1	2.8	4.0	17.7	39.5	--
kg x # Lact/year/kg BW	0.5	3.0	1.4	0.07	0.3	0.4	0.7	0.01

*Estimates were made knowing that there is large variation between breed of animals and among human populations; data are based on references [50-63]; **Measured protein synthesis rate (or averaged among different stages of lactation) as reported in reference [64] for pig, [10] for goat, and [65] for mouse. For cows and goats it has been reported that the protein synthesis in mammary gland is between 1.3 to 2.5 the rate of milk protein secretion [12, 66]; thus, for cow an average of 2.0 was used.

$Mammary Gland (MG). General references [67, 68]. Specific references for milk composition are: bovine [4, 67, 69], mouse and rat [70-74], human [75], pig [55], sheep and goat [76], and Kangaroo [77-79]

Table 1. Estimated* body weight, milk protein composition, and total protein relative to other parameters in several species. Reported are the total proteins and the major protein fractions.

The application of functional genomics and bioinformatics allows for a thorough exploration of the biological complexity of organisms' tissues [34]. Together, these approaches form part of the systems biology framework, i.e., a way to systematically study the complex interactions in biological systems using a method of integration instead of reduction. One of the goals of applying systems concepts is to discover new emergent properties that may arise from examining the interactions between all components of a system to arrive at an integrated view of how the animal functions. It has been argued previously [3, 49] that application of the systems approach might lead to the discovery of regulatory targets that could be tested further (i.e., model-directed discovery) or help address a broader spectrum of basic and practical applications including interpretation of phenotypic data, metabolic engineering, or interpretation of lactation phenotypes.

4. Expression of genes coding for main milk proteins in lactating mammary gland

4.1. Introduction

The main proteins in milk are caseins and whey proteins (i.e., alpha-lactalbumin, beta-lactoglobulin, whey acidic protein [WAP], albumin, and immunoglobulin; proteins highly enriched in the milk serum after removal of casein). The milk protein content and composition of the main milk proteins with abundance of each of the caseins and whey proteins in several species (when available) is summarized in Table 1.

The milk fat globule membrane (MFGM) is also highly enriched with proteins even though those account for only 1-4% of total protein in the milk [80]. Reinhardt and Lippolis identified 120 proteins in an proteomics analysis of the bovine MFGM [81]. More recently, the same authors have compared the MFGM proteome between colostrum and milk and found 138 proteins in the MFGM [82]. Analysis of two MFGM-enriched milk fractions, a whey protein concentrate (WPC) and a buttermilk protein concentrate (BMP), identified as the major proteins associated with the WPC immunoglobulin k, lactoperoxidase, serpin A1, immunoglobulin lambda light chain variable region (Vl1a protein), serum albumin, lactoferrin, and CD9; and the most abundant associated with the BMP, besides caseins, were adipophilin, tripartite motif containing 11, and folate receptor [83]. The fraction enriched in BMP known to be part of the MFGM [80, 81], besides adipophilin, were lactadherin (PAS 6/7 or milk fat globule-EGF factor 8 protein or MFGE8), xanthine dehydrogenase, butyrophilin, and fatty acid binding protein [83]. The major MFGM proteins are mostly involved in the milk secretion, particularly the secretion of milk fat globules [84]. There are, however, many proteins present in a minor amount. Among those, of relatively high abundance are CD36 and mucin proteins [80, 81]. While a specific biological role for all the proteins associated with the BMP and WPC has not been clearly defined, it appears that at the least some of the latter proteins play a role in the defense mechanisms in the newborn and some appear to be related to human/animal health, particularly lactoferrin [80, 83, 85-87].

We have used the data in Table 1 to estimate the milk protein synthesis rate for the respective species. It is evident that rodents, and particularly mice, have the richest amount of proteins in milk among the compared species. Compared with cows, mice and rats have more than twice as much milk production per body weight (BW), from 3 to 6-fold more milk production considering the milk solid relative to the BW or weight of mammary gland, and more than 5-fold more milk protein synthesis per weight of mammary gland (Table 1). Therefore, rodents have, compared with cows, an extraordinary milk protein synthesis rate (based on the rough estimate in Table 1, each mammary epithelial cell in the lactating mouse produces and secretes approximately a quarter of its weight in milk protein per day). The mouse has an estimated rate of protein synthesis that is 5-fold greater than that estimated for the cow, and ca. 100-fold greater than estimated for humans (see Table 1). Overall, the monogastrics appear to have on average a greater rate of milk protein synthesis compared with ruminants, although sheep and goats are estimated to have a synthesis rate equal or greater than pigs (Table 1). The estimates offered in Table 1 have to be considered indicative and do not account for the proportion of epithelial cells in the mammary weight and the amount of degraded proteins, thus, those are underestimates of the mammary epithelial milk protein synthesis rates.

Milk yield, milk composition, energy output by lactation, and mammary gland weight for several species were reviewed more than 25 years ago [62, 68]. Some of those same data have been used in the present chapter. In those reviews it was shown that there is a linear correlation between either maternal weight and milk energy yield (i.e., larger the size of the animal greater the milk energy output, with an increase of milk energy output with number of litters), or mammary gland weight with milk yield. Linzell [62] suggested that the milk yield per kg of mammary tissue does not differ between species. Our data reported in Table 1 appear to come to a different conclusion. For instance, for Linzell, the mammary gland weight over BW is slightly lower in the mouse than in the Holstein cow, but our calculation in Table 1 showed that the mammary gland of the mouse is ca. 7% of BW while in high producing dairy cows it is only ca. 4% of BW. The same is true for the productive capacity of the mammary gland. From Linzell, the amount of milk per mass of mammary gland is similar among all species, but, from our calculation (Table 1), rodents have a greater capacity for milk synthesis compared with other species, particularly the high producing dairy cow. Our data, however, are consistent with the known greater uptake of oxygen (i.e., metabolism) in the mammary gland of smaller relative to larger animals (cited in [62]).

4.2. mRNA abundance and expression patterns of transcripts coding for milk proteins

The abundance of the milk proteins (with the exception of albumin, as discussed below) is highly-dependent on the transcription level, with additional important regulatory roles for translational and post-translational modifications [88]. In Figure 1 are reported the temporal patterns of transcription of caseins, alpha-lactalbumin, whey acid protein, and albumin from pregnancy to lactation in two ruminant species (dairy cow [89] and dairy goat [90]), two monogastric species (mouse [91] and pig [92]), and a marsupial (kangaroo; Macropus eugenii [93]). Those were the only microarray datasets publicly available at the time of this writing. In Figure 2 is reported the expression pattern from pregnancy to lactation for other proteins found to be associated with the MFGM, BMP, and/or WPC [80, 81, 83].

The temporal expression pattern of all the casein genes was similar, and was consistent for the two ruminant species, with *CSN1S2* followed by the *CSN1S1* as the most up-regulated casein genes in both species and with an overall greater up-regulation of casein genes in the cow relative to the goat (Figure 1). Alpha-lactalbumin was similarly up-regulated in all species except kangaroo (Figure 1). Both monogastrics had a large increase in expression of *CSN1S2* during lactation, with the mouse having a particularly large increase of the Csn1s2b isoform (Figure 1).

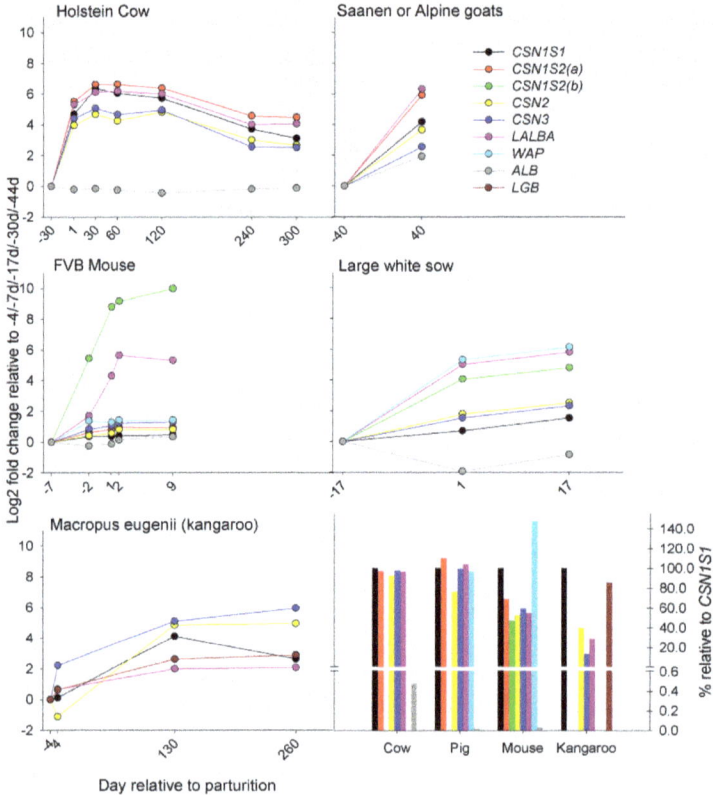

Figure 1. Temporal expression pattern of genes encoding casein alpha-S1, casein alpha-S2 (for mouse the genes are casein alpha s2-like A and casein alpha s2-like B), casein beta, casein kappa (*CSN1S1*, *CSN1S2*- (*a*) and (*b*) for mouse -, *CSN2*, and *CSN3*, respectively), alpha-lactalbumin (*LALBA*) and albumin (*ALB*) in dairy cows [89], mouse [91], pig [92], goat [90], and kangaroo (Macropus eugenii; [93]). Dataset, except for dairy cows and kangaroo, was obtained through NCBI Gene Expression Omnibus (GEO at http://www.ncbi.nlm.nih.gov/geo/): GSE14008 for goat, GSE4222 for mouse, and GSE30704 for pig. Data from the Series Matrix File were used. A statistical analysis using a false discovery rate for the overall time effect was run using GeneSpring GX7.. Data for the kangaroo are from [93]; only one isoform of alpha casein was reported. Important whey protein components such as immunoglobulin genes are not reported in the present figure. Expression of beta-lactoglobulin was only reported for the Kangaroo [93]. The relative mRNA abundance between the caseins, *LALBA*, and *ALB* is

also reported. Those data should be evaluated with caution because microarray platforms do not provide an absolute abundance value for each mRNA. This is due to technical reasons, such as the fact that the DNA oligos complementary for each cDNA used in the platforms can bind the cDNA with different efficiency and also because the highly expressed proteins can saturate the spot with consequent over-saturation and bleaching during microarray scanning. Despite those limitations, there is a relatively good agreement between the abundance of mRNA and microarray signal (i.e., number of cDNA with dyes bound to the oligos); thus, despite not being absolute, the comparison between microarray signals can be used to provide an initial idea of relative abundance between mRNA species. The high similarity in relative transcript abundance among caseins and between casein genes and alpha-lactalbumin in mammary gland of dairy cows is consistent with transcriptomics analysis of bovine milk somatic cells using the more accurate RNA sequencing technology [94]. In the figure are reported the relative percentage of the signal of the genes of interest compared to *CSN1S1* or Csn1s1. For the kangaroo, the percentage of beta-lactoglobulin mRNA compared to *CSN1S1* is also reported, and for the mouse, the relative abundance of Csn1s2a is reported (the Csn1s2b mRNA abundance is 47.5% of Csn1s1). The lack of data for some genes is a consequence of the absence of the respective oligos on the microarray. The gene symbols of the legend are in capital and italic as should be reported for cow, pig, and goat. Gene symbols for the mouse should be only first capital letter and not italic. All genes were deemed to be significantly affected by time with a false discovery rate < 0.10 [95] based on analysis using GeneSpring GX7 (Agilent). Exception was the *ALB* for the pig. For the kangaroo and the goat it was not possible to obtain the statistical data.

The other caseins, despite being significantly affected by time, were not highly up-regulated by lactation in those species. Also, the monogastric had a large increase in expression of WAP especially for pig, where the increase in expression was proportional to *LALBA* (Figure 1). The relative mRNA abundance of WAP appears to be as high as or higher than caseins in monogastric, particularly in the mouse (Figure 1).

In the kangaroo all transcripts coding for caseins had a greater increase in expression compared with *LALBA* which was up-regulated but only in a small magnitude during lactation in this species. The low importance of alpha-lactalbumin in this species has been previously established and its presence in milk does not change throughout lactation [78]. This is in line with the relatively low amount of lactose in the milk of the kangaroo [78, 96, 97].

It has been considered for a long time that the albumin present in bovine milk, which is probably identical to the serum albumin, is present as a consequence of permeability of the epithelial lining in the mammary gland. However, other evidence suggested that the goat mammary gland is able to synthesize a portion of the milk albumin [98]. Expression of albumin in the bovine mammary gland has been previously reported and studied in dairy cows [99]. Observations of increased albumin expression in mammary tissue explants challenged with lipopolysaccharides, and a higher expression observed in glands with mastitis, have led to the suggestion that albumin may have a role in the innate nonspecific defense system [99]. In all species where albumin mRNA was available (i.e., all except kangaroo, Figure 1) a level of expression was detected by the microarray platforms even though at very low levels compared to the highly abundant milk proteins. Interestingly, even when considering the limitations of the relative abundance between measured genes in Figure 1 (see Figure 1 caption), the dairy cow appears to have a relatively higher expression of *ALB* compared to the mouse and pig, where it is virtually absent (Figure 1). The expression of albumin decreased slightly, but significantly, in lactation compared to pregnancy in dairy cows (Figure 1). Milk content of albumin decreases rapidly during the first month of lactation, followed by a slight but steady increase up to 6 months of lactation, then rapidly decreasing reaching a nadir at around 200 day in milk [100]. The gene expression data support a basic, although low, expression of albumin in the mammary gland, however, the importance of this expression in the lactating gland is unclear.

The pattern in expression of the other major proteins in milk during pregnancy and lactation in several species is shown in Figure 2. Among the genes coding for proteins highly enriched in the WPC, mostly involved in the innate immune response [101], the data clearly shows that the mouse and pig have a large increase in expression of only lactoferrin (LTF), while the goat has a large increase in expression only for lactoperoxidase and a decrease in expression of LTF (Figure 2). Those data appear to be supported by an increase of milk lactoperoxidase content through lactation in goat [102]. Dairy cows have a moderate, although consistent, increase only for the genes coding for lactoperoxidase and lactoferrin (Figure 2). Collectively, those data might indicate an overall increase in the innate immune response of the mammary gland during lactation [103] and/or an increase in milk antimicrobial activity in all species during lactation. It appears that the goat, mouse and pig have a higher antimicrobial activity in the milk due to the spike of specific activities (i.e., lactoperoxidase in goat and lactoferrin in monogastric). However, despite the lack of change in expression of LTF from early to peak lactation (Figure 2), the amount of lactoferrin in milk of sows has been reported to peak during colostrogenesis and to decrease thereafter [104], while in the cow lactoferrin concentration in milk has been reported to both increase [105] or decrease [100] as lactation progress.

The expression in the mammary gland of milk proteins related to BMP and/or MFGM (Figure 2) indicates an overall higher increase in expression in the bovine mammary gland compared to the other species evaluated. In particular, there is a large expression of FABP3. The FABP3 is also called mammary-derived growth inhibitor because it has been known to be highly abundant in the bovine mammary gland [106] and has a high affinity for long-chain fatty acids (LCFA), particularly saturated LCFA. This protein may have a dual role in the mammary gland for the synthesis of milk fat, participating in the transfer of LCFA during triacylglycerol synthesis and carrying LCFA into the nucleus for activation of nuclear receptors such as peroxisome proliferator-activated gamma (PPARγ). The latter nuclear receptor has been suggested to be partly responsible for the transcription of milk fat synthesis-related genes [69]. Expression of FABP3 was significantly affected by stage of lactation in all species. Other BMP proteins for which the expression was generally increased in all species (expect the pig, due to their absence in the microarray platform) are butyrophilin and xanthine dehydrogenase, considered to have a pivotal role in milk fat secretion [69, 81]. Interestingly, the genes coding for the proteins reported to be among the most abundant in milk after casein and whey proteins [80, 83] all showed an increase in expression during lactation.

The above observations support the importance of transcriptional regulation of the milk proteins, as also previously suggested for dairy cows [34]. However, some of the data (e.g., patterns of expression and milk content of *LTF* in sows) might indicate that the abundance of some of the most enriched proteins in milk is regulated post-translationally, as previously shown [88].

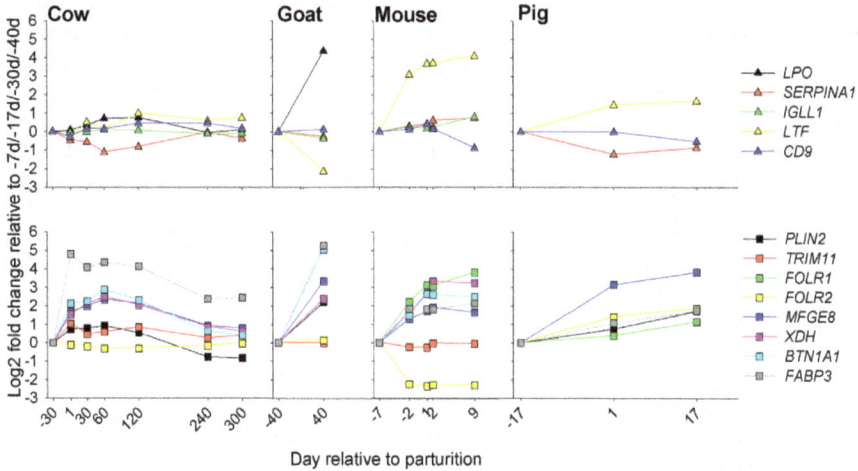

Figure 2. Temporal expression of genes coding for major proteins found in whey protein concentrate (WPC; upper panels) and in the buttermilk protein concentrate (BMP; lower panels) [83]. Data are for Holstein dairy cows [89], Saanen or Alpine goats [90], FVB mice [91] and Large white sows [92]. Data were recovered and treated as reported in the legend for Figure 1. Symbols of the genes denote *LPO* = lactoperoxidase; *SERPINA1* = serpin peptidase inhibitor, clade A (alpha-1 antiproteinase, antitrypsin), member 1; *IGLL1* = immunoglobulin lambda-like polypeptide 1; *LTF* = lactoferrin; and *CD9* = CD9 molecule for the genes coding for the WPC related-proteins. Symbol of the genes denote **PLIN2** = perilipin 2 or adipophilin; *TRIM11* = tripartite motif containing 11; *FOLR1* and *FOLR2* = folate receptor 1 and 2; **MFGE8** = milk fat globule-EGF factor 8 protein or lactadherin; **XDH** = xanthine dehydrogenase; **BTN1A1** = butyrophilin; and *FABP3* = fatty acid binding protein 3 for the genes coding for the BMP related-proteins. The ones in bold font are major MFGM proteins in the BMP fraction. With a FDR<0.10 as cut-off for overall time effect, all genes were significantly affected in mouse, while in the pig *CD9*, *FOLR2*, *IGLL1*, *SERPINA1* and *XDH* and in the cow *TRIM11* and *IGLL1* were not significantly affected by time. It was not possible to obtain the statistical data for the kangaroo and the goat.

4.3. Regulation of milk protein expression

It is well-established, at the least in rodents and ruminants, that hormones such as prolactin, growth hormone, insulin, insulin-like growth factor, thyroid hormone, parathyroid hormone, oxytocin, placental lactogen, and glucocorticoids are important, if not essential, in the regulation of lactation and some play an essential role in milk protein expression with noteworthy differences between species [107-109]. For instance in rats, prolactin concentration in plasma increases significantly just before parturition and remains high for the entire lactation [110], accompanied also by a concomitant increase in prolactin receptor [111] which appears similar to the pattern of prolactin receptor in mouse mammary during lactation [112]. In contrast, prolactin concentration in dairy cows increases dramatically at the onset of lactation and decreases afterwards to levels observed prior parturition, and prolactin receptor expression does not increase [4].

However, the prolactin concentration is acutely increased by suckling or milking in all species. One large difference between bovine and mouse is the frequency of milk removal. In general, in mouse pups suckle every 20 min [113] (ca. 72 times/day) while in suckling beef calves (which can be considered the "natural" milk removal in cattle) the frequency is between 3-5 sucklings/day with 5-10 min per suckling [114]. Dairy cows in conventional farms are milked 2-3 times a day with an average of ca. 5 min/milking [115]. Therefore, in species like the pig, mouse, and rat the higher prolactin concentration is likely due to short intervals pattern of nursing, while in ruminant the low prolactin concentration might be due to the longer nursing intervals. Interestingly, even though most of the lactogenic hormones are present in a relatively high concentration prior parturition, the drop in blood progesterone is essential in order to allow lactogenesis to proceed [108], highlighting the pivotal role of this hormone.

Several transcription factors control expression of the major milk proteins, particularly for caseins [107]. The caseins genes are all present on one chromosome (e.g., chromosome 6 for bovine and chromosome 5 for mouse) and in a cluster of single-copy genes [116]. The alpha- and beta-caseins have highly conserved regulatory motifs, while kappa-casein, even though the expression pattern is similar to the other caseins (especially in dairy cows, see Figure 1 as example), does not have similarly conserved regulatory motifs. Most of the investigation in regulation of milk protein expression have been concentrated on beta-casein and WAP in mice [107]. The regulatory motifs found to affect the expression of milk proteins have been reviewed more than a decade ago [107]. At that time it was well-established that the response element for signal transducer and activator of transcription 5 (STAT5), CCAAT/enhancer binding protein (C/EBP) beta (CEBPB), and glucocorticoid receptor (GR) were essential for inducing transcription of beta-casein and nuclear factor 1 (NF1) was essential for inducing transcription of WAP [107]. The Yin and Yang 1 protein (YY1) instead has a well-known negative effect on beta-casein expression [107]. More recently, octamer-binding transcription factor 1 (Oct1; official gene symbol is Pou2f1) was found to be involved in the expression of beta-casein in the mouse mammary gland [117]. Glucocorticoid receptor and STAT5 cooperate through a protein-protein interaction for the induction of beta-casein expression (reviewed in [118]). Interestingly, using a genetically modified mouse where it was possible to repress the Stat5a expression in the mammary gland during lactation, the repression of Stat5a expression negatively affected the transcription of Wap and Lalba, but not Csn2 [119].

An important role for the control of casein expression appears to be played by the extracellular matrix (ECM) and cell-to-cell adhesion. More than two decades ago it was clearly shown that ECM affects epithelial cell differentiation and function [120]. Cellular interaction with ECM is necessary in order for prolactin to activate STAT5 (reviewed in [107]). This role of ECM in regulation of CSN1S1 expression in rabbit mammary epithelial cells occurs by modifying chromatin structure through the laminin-integrins interaction and signaling [121]. The same work showed that, contrary to previously reported studies in mice, the ECM in rabbit epithelial cells activates expression of casein through a mechanism independent from STAT5 or CEBPB. In fact they showed an increase of histone acetylation (i.e., dissociation of histone from DNA allowing the latter to be accessible for transcription factors and polymerase) by the ECM. Very recently is was demonstrated that the RhoA- Rho kinase -myosin II pathway, involved in stress fiber formation, cellular contractility, cell migration, and polarity, has a negative effect on the activation of STAT5 by prolactin in mouse mammary epithelial cells [122]. The same study showed that mammary epithelial cells cultured on plastic or collagen I have a large activation of such pathways and, as a consequence, the stimulation of casein expression by prolactin is strongly impaired.

It is well-known that epithelial cells cultured on plastic do not express or produce caseins [123] but when they are cultured in 3D they regain the ability to express and secrete casein, as seen in porcine primary mammary epithelial cells [124]. Bovine mammary epithelial MAC-T cell line cultured on plastic expresses and releases a low amount of casein into the media [125, 126]; however, the amount released is often below the limit of detection [127]. When MAC-T cells are cultivated on a floating collagen gel the expression and production of casein is augmented dramatically [127]. Interestingly, the addition in the media of growth hormone (besides the presence of the insulin, prolactin, and hydrocortisone, the essential cocktail for milk protein synthesis) strongly induces the expression of casein and alpha-lactalbumin genes in MAC-T cells cultured on plastic with enhanced cell adherence capacity [126], partly contrasting with the findings supporting the importance of ECM (or 3D structure) for the expression of caseins. The dissimilarity also may be due to species differences. In the bovine, STAT5 responds to prolactin and other lactogenic growth factors and its activity increases during lactation mostly due to phosphorylation [128, 129]. However, in the bovine mammary gland when compared to the rodent, the role of STAT5 in controlling milk protein expression through the Jak2-Stat5 signaling pathway appears to be weak at best [130], highlighting a difference between the bovine and the mouse in protein synthesis regulation. It has been proposed that the transcription factor E74-like factor 5 (ELF5), a co-regulator of the STAT5 signaling and a mammary epithelial specific ELF gene, can compensate for the lack of prolactin signaling in inducing lactation and, particularly, milk protein synthesis in mice [131, 132]. Recent data in dairy cows showed a large increase in expression of *ELF5* during lactation supporting a major role of this transcription factor in controlling milk protein synthesis, considering that prolactin in this species is higher only at the onset of lactation [4].

In a study carried out in lactating rats, deficiency of prolactin and growth hormone was not as effective in shutting down the expression of casein genes as the removal of the suckling pups suggesting that other local factors are essential and more potent than endocrine in controlling milk synthesis in this species [133]. In mouse mammary epithelial cells the presence of laminin-rich basement membrane (i.e. ECM) was sufficient to induce expression of casein genes without the addition of any growth factor, highlighting the essential role of integrins [123, 134]. Interestingly, in goat mammary tissue *in vivo* and *in vitro* disruption of cell junctions strongly reduced milk yield and expression of casein and alpha-lactalbumin [135].

It also has been demonstrated that the expression of milk proteins is strongly under epigenetic regulation (reviewed in [136]). Among several epigenetic mechanisms known (reviewed in [137]) the studies in mammary gland have primarily focused on DNA methylation. The methylation of DNA represses the transcription of genes by two mechanisms, by a steric impediment for the transcription factors to bind to their binding sites or through recruiting methyl-CpG binding proteins that, by engaging additional proteins involved in compacting the DNA, renders the DNA inaccessible for the polymerase [137]. The promoter region of casein genes is hypo-methylated in the mammary gland but hyper-methylated in other tissues. Similarly, the *STAT5* promoter region is also hypo-methylated in the mammary gland [137]. Interestingly, the methylation status of the casein promoter region is affected by inflammation in the mammary gland. In regard to epigenetic, an active field of research now is the role of hereditary epigenetic patterns and fetal reprogramming in dairy cows and how those account for the performance of the animal, (reviewed in [137]) including expression of genes encoding major milk proteins.

Overall the findings reported above indicate that the regulation of milk protein synthesis is complex and multiple factors work in concert to bring about full activation of expression of genes coding for major milk proteins. The endocrine and mechanical cues independently stimulate expression of a plethora of genes involved in mammary differentiation and lactation in mice; but, apparently, the combination of those cues are more important in stimulating the expression of caseins, xanthine oxidoreductase, and several genes involved in milk protein synthesis, as well as genes coding for proteins involved in packaging and transport [138]. Finally, the recognized differences among species prompts a call for caution when extrapolating findings in one species to another [139].

Studies about regulation of other minor milk proteins are limited. For instance lactoferrin expression is known to be induced during mastitis (e.g., [140]) and lactoferrin expression increases dramatically after treatment of mammary epithelial cells with bacteria lipopolysaccharides or double stranded RNA through PKC, NF-kappaB, and MAPK pathways [141]. Transcript abundance of *LTF* is also under control of a specific miRNA (miR-214) [142]. Beta-lactoglobulin in sheep is regulated by the milk protein binding factor (MPBF/Stat5)[143]. The MPBF mediates the prolactin signal transduction in the lactating mammary gland by binding to gamma-interferon activation site DNA elements [144]. Finally, butyrophilin expression appears to be strongly under the control of Akt1 [145]. The Akt genes are serine-threonine protein kinase with a role in multiple pathways, including insulin signaling. The Akt proteins are essential for milk synthesis, as shown by the lack of production of all main milk components in Akt1 and Akt2-deficient mice [146]. The role of the three known Akt isoforms is different in the mammary gland, with Akt1 being the most important for lactogenesis [147].

In summary, the data presented in Figures 1 and 2, together with the understanding we have so far about the regulation in expression of milk protein, support a model were milk protein genes are coordinately expressed by concerted action of several hormones, cell-to-cell interaction, basement membrane, and ECM components. Those act ultimately through signaling pathways (e.g., Jak-STAT, insulin) that activate transcription factors (e.g., GR, STAT5, CEBPB, MPBF) that in turn drive increased expression of genes. There are likely complex interactions among those factors that still need to be studied in detail. A more complete picture may be obtained by investigating all the above factors simultaneously. This may be achieved using system biology approaches.

5. Networks encompassing insulin signaling and mTOR control milk protein synthesis: the missing link with nutrition

5.1. Introduction

As described above, milk protein synthesis is sensitive to energy level in the diet probably due to the increase in insulin and energy available for the costly process of assembling amino acids into proteins. A role for insulin in milk protein synthesis was suggested to be only indirect [108]. This suggestion made sense considering that blood insulin drops substantially during late pregnancy-early lactation. This hormonal pattern is relatively consistent in cattle [4, 148], sheep [149] and rats [150]. Despite the decrease of insulin, a role of this hormone in synthesis of the main milk constituents, particularly protein and lactose, in the bovine mammary gland was proposed more than 40 years ago [35]. Recent work has demonstrated a pivotal role of insulin in regulating milk protein synthesis both in the cow [39] and mouse [38].

5.2. How insulin affects milk protein synthesis

The mechanisms of insulin in regulating milk protein synthesis can be multiple, direct and indirect; however, the major activity of insulin occurs at two main steps of milk protein synthesis. The first step involves the control of gene expression of milk proteins and the second involves the regulation of translation. For the former, it is well-known that insulin has a strong positive role in activation of STAT5 through increase phosphorylation of the transcription factor [151]. In addition, expression of *ELF5*, the gene coding for a co-activator and amplifier of STAT5 signaling (see above), is induced by insulin in cattle and mouse mammary tissues [38, 39]. Those are direct or indirect effects of insulin on control of milk protein expression through STAT5-ELF5. The second role for insulin controlling milk protein synthesis is by regulating amount of translation via the mammalian target of rapamycin (mTOR) pathway.

5.3. Role of mTOR on regulation of protein synthesis: brief overview

A master role of mTOR (particularly mTORC1 among the two mTOR complexes) in the regulation of protein synthesis, particularly translation, in all tissues of mammals has been well defined [152]. A simplified model of the role of mTOR in protein synthesis is summarized in Figure 3.

Protein synthesis is basically inhibited by the association of the un-phosphorylated eukaryotic translation initiation factor 4E binding protein 1 (4EBP1) with the eukaryotic translation initiation factor 4E (eIF4E), preventing the formation of the translation initiation complex. The increase in insulin signaling increases specific phosphorylation of mTOR. This

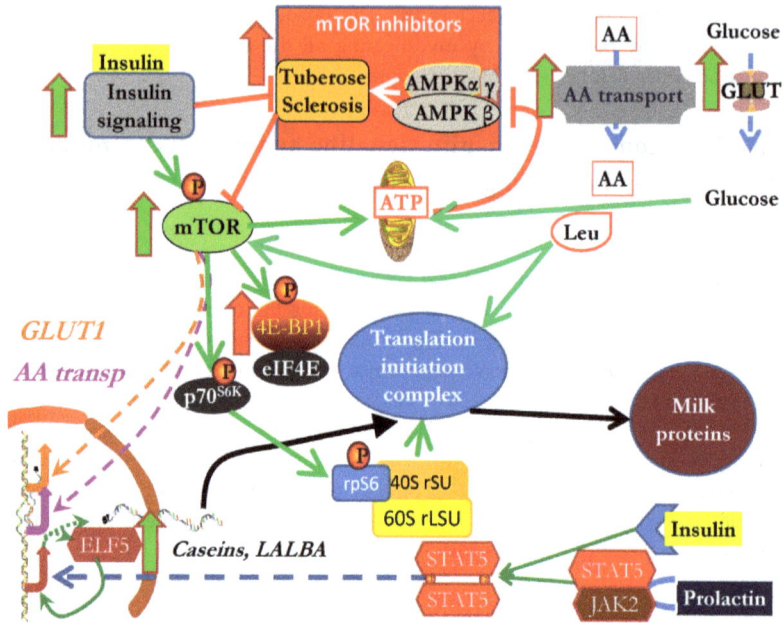

Figure 3. Simplified model of milk protein synthesis regulation by insulin-mTOR pathways (see detailed model in [4]). Green arrows denote activation or positive effect; dashed arrows denote either transport or indirect effect on transcription. Red cap flat lines denote inhibition. Phosphorylation is denoted by a round orange shape. When the mTOR complex is not activated the non-phosphorylated eukaryotic translation initiation factor 4E binding protein 1 (4E-BP1) binds eukaryotic translation initiation factor 4E (eiF4E) preventing protein synthesis. The mTOR complex is formed and activated by phosphorylation through the insulin signaling and enhanced by amino acids (particularly Leu). When the mTOR complex is activated it phosphorylates 4E-BP1 at multiple sites which results in release of the eiF4E. Once released, eiF4E forms a complex with other eukaryotic translation initiation factors that in turn bind the 40S ribosomal subunit initiating translation. The mTOR also induces translation by two additional mechanisms: indirectly activating the 40S ribosomal subunit through phosphorylation of RPS6-p70-protein kinase (p70^{S6K}) that in turn phosphorylates the RPS6 enhancing its activity (RPS6 is part of the 40S complex and the increase in RPS6 activity increase 40S complex activity as well); and indirectly preventing the inhibition of the translation elongation. The activity of mTOR complex is inhibited by the dimer formed by the two tuberous sclerosis (TSC1 and TSC2) proteins. The TSC complex is directly inhibited by insulin. Among others, the TSC complex is activated by AMP-activated protein kinase (AMPK). The latter is induced by low level of energy (i.e., higher AMP). The level of energy in mammary is mostly determined by glucose and AA availability to be metabolized into the TCA cycle. Thus, the increase in glucose import (availability of energy) can prevent inhibition of protein synthesis by inhibiting AMPK. At the same time, the activation of mTOR complex increases expression of glucose transporter GLUT1. In addition, mTOR appears to directly induce the mitochondrial ATP synthesis. The activation of transcription of milk protein synthesis by Jak2-Stat5 is also reported, where binding of prolactin and insulin to their receptors induces the phosphorylation of STAT5. Upon phosphorylation, STAT5 dimers translocate to the nucleus and induce transcription of several genes including *ELF5* and genes coding for milk proteins. ELF5 protein enhance the activity of STAT5

increasing transcription of milk protein genes. Vertical arrows denote increased gene expression of components of reported protein complexes during lactation in bovine as observed in [4]. Vertical green arrows denote protein complex with likely increase in abundance during lactation which activity increases protein synthesis. Vertical red arrows denote protein complex with likely increase in abundance during lactation which activity inhibits protein synthesis.

activates the formation of the mTOR complex that in turn phosphorylates 4EBP1 at multiple sites allowing release of eiF4E. Once released, eiF4E forms the translation initiation complex with other translation initiation factors. The complete translation initiation complex then binds the 40S ribosomal subunit forming the 43S pre-initiation complex and initiate translation of mRNA into protein [153]. Besides the above described mechanism, the activated mTOR complex enhances translation by additional mechanisms. A first mechanism is the increase activation of the 40S ribosomal subunit through phosphorylation of the ribosomal protein S6 kinase (S6K1) which in turn phosphorylates the ribosomal protein S6 (RPS6, a component of the 40S ribosomal subunit) enhancing its activity. A second mechanism is the inhibition through phosphorylation of the eukaryotic elongation factor-2 kinase (EEF2K). The active un-phosphorylated EEF2K inhibits translation by hindering the activity of the elongation factor 2 (eEF2); thus, inhibition of EEF2K frees eEF2 thereby boosting translation. A third mechanism is through an increase of cellular energy production by enhancing expression of glucose transporter 1 and some of the AA transporters, and by improving mitochondrial ATP synthesis [154]. This in turn provides energy in the form of ATP for the translation but also prevents the inhibition of mTOR through the AMPK (see below).

The mTOR complex is chiefly inhibited by the tuberous sclerosis (TSC) complex [152, 155]. The TSC complex is activated by several factors in coordination with AMPK [152, 156, 157]. The latter is activated by AMP level and the amount of AMP is determined by low intracellular energy level (i.e., higher AMP). The AMPK increases the TSC complex phosphorylation activating it and, consequently, inhibiting mTOR complex [158].

Interestingly, the mTOR complex is indirectly activated by AA, particularly leucine [153]. Besides activation of mTOR, leucine enhances overall protein synthesis via additional mechanisms [153, 159, 160].

5.4. How insulin and mTOR cross-talk control translation of milk protein

Several lines of evidence support a pivotal role of mTOR in milk protein synthesis in mice and cows (some of the studies carried out in cows are summarized in Table 2). Toerien and Cant measured phosphorylation status and abundance of several mTOR-related proteins in bovine mammary tissue and observed that lactating compared to non-lactating bovine mammary tissue is characterized by a higher abundance of eukaryotic translation initiation factor 2 alpha (eIF2α) protein, eIF4E protein, and the 4EBP1-eIF4E complex, percentage of eIF4E protein in the 4EBP1-eIF4E complex, and phosphorylation of RPS6 compared to non-lactating bovine mammary [161]. Bovine mammary acini cultured in lactogenic hormones and in a nutrient-rich medium (i.e., high glucose and 4-fold the concentration of amino acids

Molecular branch studied	Tissue or cell type	Main objective*	Main conclusions*	Reference
Protein phosphorylation	Mammary tissue biopsy	Role of dietary AA and starch fed to lactating cows on MPS	mTOR and RPSK6 phosphorylation is enhanced by starch and AA	[28]
Protein phosphorylation	MacT cells	Role of the level of essential AA availability and insulin on phosphorylation of several mTOR pathways proteins and MPS	Essential AA enhance MPS rate by enhancing phosphorylation of 4EBP1 and eEF2	[169]
Protein phosphorylation	MacT cells	Role of specific essential AA on phosphorylation and MPS	Phosphorylation of mTOR and RPS6K decreases in the absence of leucine and isoleucine, and leads to lower protein synthesis rate	[165]
Protein phosphorylation and mRNA expression	Mammary tissue	Role of mRNA translation on MPS regulation	Lactating mammary tissue is associated with greater expression of RPS6, RPS6K, and eIF isoforms, thus, they play a key role in MPS	[161]
Protein Phosphorylation	Mammary tissue biopsy	Role of mTOR signaling in nutritional regulation of MPS	Intravenous essential AA and glucose infusion enhance MPS via increased phosphorylation of mTOR	[164]
Protein phosphorylation	Primary mammary epithelial cells	Role of mTOR signaling in nutritional and hormonal regulation of MPS	Nutrients and hormones are capable of regulating MPS through phosphorylation of the mTOR signaling pathway	[29]
Protein phosphorylation	MacT cells	Role of IGF-1 on mTOR phosphorylation and regulation of MPS	Exogenous IGF-1 increased RPS6K and mTOR phosphorylation and stimulated global protein synthesis	[168]
mRNA expression	Mammary tissue biopsy	Expression profile of the gene networks associated with mammary protein synthesis during the lactation cycle	Regulation of mammary protein synthesis during lactation occurs via mTOR and involves adaptations in insulin signaling, glucose transport, and amino acid uptake.	[4]

*Amino acid(s) = AA; MPS = milk protein synthesis

Table 2. Published molecular studies of bovine mammary milk protein synthesis regulation.

found in plasma of well-fed cows) had an increase of S6K1 and 4EBP1 phosphorylation and dissociation between 4EBP1 and eIF2E [29]. Mixtures of all AA or Leu alone in medium strongly affects synthesis of beta-lactoglobulin in mammary epithelial cells of the mouse and the cow, with all AA being more potent than Leu alone [162]. The latter study demonstrated that synthesis of beta-lactoglobulin in bovine cells was associated with phosphorylation of 4EBP1 and S6K1 and that the major effect on protein synthesis of all AA or Leu was through mTOR. Media containing Leu increased milk protein synthesis, while Lys, His, Thr, or the combination of those AA decreased milk protein synthesis up to 65% via dephosphorylation of S6K1[163]. Infusion of His and Met+Lys did not depress milk protein synthesis in bovine mammary tissue and Leu infusion only had a numerical increase in protein synthesis, but Leu and Met+Lys increased phosphorylation of S6K1 and RPS6 [164]. In the latter study, the infusion of all 10 essential AA plus glucose or glucose alone enhanced milk protein synthesis in dairy cows to the same extent. Both infusions significantly decreased eIF2 phosphorylation but the AA plus glucose also significantly increased phosphorylation of S6K1 and RPS6. This last study highlighted a crucial role for glucose in milk protein synthesis, probably through insulin signaling and not involving mTOR. A positive relationship between mTOR phosphorylation and the fractional protein synthesis rates was also observed in MAC-T cells [165]. The same study showed that leucine and isoleucine are the only AA that significantly increased the phosphorylation of mTOR and S6K1.

The positive response in milk protein synthesis in dairy cows to increasing energy content of the diet is probably elicited through greater phosphorylation of RPS6 by insulin- mTOR signaling as recently demonstrated with abomasal starch infusions [28]. Besides insulin, insulin-like growth factor, growth hormone, leptin, and prolactin appear to increase protein synthesis in bovine mammary through the mTOR pathway [166-168].

5.5. How insulin and mTOR control bovine milk protein synthesis: an overall model suggested by transcriptomics analysis

As also described above, we previously proposed a model detailing the networks of factors involved in the regulation of milk protein synthesis in the bovine through the mTOR signaling pathway [4]. Figure 3 represents a simplified version of the proposed model plus main findings from an experiment performed where expression of genes coding for the main players in the insulin-mTOR signaling and other proteins involved in protein synthesis were measured from pregnancy to end of the subsequent lactation in bovine mammary gland [4]. The bovine mammary gland during lactation had a significant, but rather modest, increase in expression of mTOR (*FRAP1*), *EIF4E*, and *EEF2*, all genes coding for proteins involved in enhancing protein synthesis, but also several genes coding for proteins involved in the negative regulation of protein synthesis were up-regulated, such as one of the 4EBP isoforms (which bind eIF4E) and both TSC genes.

The above observations, together with data from other studies as described above, indicated that the bovine mammary gland during lactation increases slightly the abundance of the main players in the regulation of translation, however, due to the increase in expression of inhibitors, protein synthesis may be basically inhibited and mostly driven by phosphorylation that activate the proteins involved in the mTOR pathway and other translational-related proteins. Interestingly, we observed a significant increase in expression during lactation of genes coding for several proteins involved in the insulin signaling, such as the insulin receptor and insulin receptor substrate 1, as well genes coding for several of the insulin down-stream factors, such as *AKT3* and 3-phosphoinositide dependent protein kinase-1. Those data indicated that insulin signaling capacity is increased overall in mammary during lactation. In addition, we observed a large increase in expression during lactation of several AA transporters, particularly the one transporting branched-chain amino acids and the one related to mTOR activity (e.g., see leucine activity on mTOR above), as well as the three glucose transporters known to be important in bovine mammary [4]. The large increase in expression of those transporters strongly supports the known increase in mammary uptake of glucose and AA during lactation; however, this also suggests that the energy level inside the mammary gland increases, thereby preventing the indirect mTOR inhibition by AMPK (mostly through the use of AA, see [34]).

Collectively, the observations described above allow us to propose that mammary gland protein synthesis is basically inhibited, however the activity of insulin and the increased energy inside the mammary gland on the one hand induces the activity of the mTOR complex and on the other hand prevents its inhibition, with the end result of enhancing translation of genes coding for milk proteins. This model appears to explain many of the findings from other laboratories, as well as the consistent observed increase in milk protein content through energy in the diet (i.e., this augment energy availability and insulinemia). Considering that dairy cattle experience a sort of insulin insensitivity in tissues other than mammary gland during lactation [170], the low level of insulinaemia observed during lactation might be overridden in the mammary tissue by an increase in insulin signaling capacity occurring as a result of increasing abundance of key proteins in the insulin signaling. Besides insulin, as mentioned above, other hormones such as the insulin-like growth factor (IGF1, which is known to decrease during lactation compared to pregnancy in dairy cows) and growth hormone (GH) can enhance the activity of mTOR and, as consequence, milk protein synthesis. Overall the data suggest a system involving insulin (and likely IGF and GH) and mTOR that allows the mammary gland to fine-tune regulate the milk protein synthesis based on energy availability. The sensitivity of mTOR to AA can open up the possibility of increasing milk protein synthesis by providing a large amount of AA, particularly leucine and other branched-chain amino acid, in the diet with the aim of increasing mTOR activity. Studies discussed above provide support for such an approach but the exclusive use of AA in this case is not enough and might be detrimental if there is not also a concomitant increase in energy in the diet in the form of carbohydrates [28]. Contrary to observations in the bovine, there is not significant increase in expression of insulin receptor and FRAP during lactation in the porcine mammary gland [171]; however, phosphorylation studies are not available in this species to provide conclusions about the activity of mTOR during lactation.

One finding from the gene expression experiment mentioned above [4] was the lack of increase and even a decrease in expression of ribosomal proteins. This appears to contradict the overall increase in protein synthesis during lactation compared with pregnancy observed in the ruminant [9, 10, 164], as well the monogastric [65] mammary gland. The use of large transcriptome analyses might provide further clues and allow for development of testable hypotheses.

6. Unbiased transcriptome: novel findings on mammary protein synthesis by large transcriptomics analysis

The mammary gland is one of the tissues that changes the most morphologically and functionally during the lifetime of an active procreative female. The plethora of phenomena happening in this tissue during all the developmental phases make it suitable for study by 'omics' technologies; particularly useful has been application of microarray technology to study gene and miRNA expression. Several studies have been published about the use of microarray technologies to investigate transcriptomics adaptation of mammary gland during development, lactation, and involution (some references among others [34, 172-176]). Also the use of microarrays to study the miRNA in mammary gland development has seeing an increase in popularity [177-179].

Recently we have performed a large transcriptomics analysis of the bovine mammary gland from pregnancy to the end of subsequent lactation [34]. For the purpose of providing a functional interpretation of the data we have used the Dynamic Impact Approach (DIA) [89]. This method allows for an interpretation of the impact and the effect (i.e., increase or decrease) on biological functions inferred by the transcriptomics changes. The results of pathways analysis indicated that the most impacted and induced pathways during lactation were related to metabolism, with, among others, an overall increase in the synthesis of lactose, synthesis of lipids, and production of glycans [34]. Interestingly, and unexpectedly, the results indicated that the protein synthesis was not highly impacted and was overall inhibited during lactation. This was mostly due to an overall reduction in expression of ribosomal proteins [34]. However, the data also indicated a large increase of protein export and an increase of protein processing in the endoplasmic reticulum [34]. In order to evaluate if the reduction of overall protein synthesis capacity in mammary tissue during lactation is a unique feature of the bovine or it is also common to other species we have here performed a functional analysis of microarray data generated in mouse [91] and pig mammary tissue [92] during pregnancy and lactation using the DIA and compared with the results of the same analysis carried out in the bovine mammary tissue using data from the above mentioned experiment [34].

The interpretation of the transcriptomics data in the mouse and pig and the comparisons between species face several limitations. For instance the transcriptomics data from mouse and pig can be affected by the large change in proportion between parenchyma and mammary fat pad. In the mouse, the proportion of fat pad can be >60% of the entire mammary gland at the beginning of pregnancy but is <20% at peak lactation, while epithelial cells may make up 30% and 80% of the tissue, respectively, in the same timeframe [176, 180]. In primiparous gilts there is a decrease of percentage of mammary lipid weight from ca. 90% at the beginning of pregnancy [181] to ca. 45% at the end of lactation [182]. In multiparous dairy cows the change in proportion of fat pad and epithelial tissue is limited with small fluctuations around ca. 75% of epithelial [58, 183]. However, the change in percentage of epithelial tissue during the 2 last days of pregnancy compared to full lactation in the mouse is only about 10% (from ca. 70 to 80%), and in the pig the fat content of the mammary gland decreases only from ca. 60% to ca. 50% from the last 3 weeks prior parturition to peak lactation, indicating that the change in tissue composition has probably a minor effect on the gene expression data in that timeframe.

Figure 4 offers a comparison of the number of differentially expressed genes (DEG) and the DIA results of pathways associated with milk protein synthesis in mouse [91], pig [92], and dairy cow [34] from the end of pregnancy to mid- to late-lactation (peak lactation for pig and cow but midway during a consistent increase in milk yield in mouse). From that comparison the mouse and cow had a greater transcriptomics adaptation to lactation relative to the pig. In the mouse, most of the DEG were down-regulated and the application of a 2-fold cut-off clearly indicated that most of the genes highly affected by lactation were down-regulated. In the pig, most of the DEG were up-regulated while in the bovine there was an equivalent number of up- and down-regulated DEG; however, when a 2-fold cut-off was applied in the bovine we observed a larger proportion of DEG with a large change to be up-regulated. Those comparisons suggest that the mouse and bovine mammary glands rely heavily on a change in transcription of a large number of genes in order to initiate, increase, and maintain lactation. Surprisingly, expression of most of the genes in the mouse mammary gland was significantly down-regulated in order to "allow" for lactation, in spite of the relatively large milk production and protein synthesis rate (see above and Table 1). This concept was previously proposed by the analysis of a larger dataset in the mouse [175]. In contrast, the bovine mammary gland requires a large increase in expression of several genes in order to activate lactogenesis. Due to the large number of highly affected down-regulated genes in the mouse, most of the pathways were deemed to be inhibited by the DIA analysis, while in the pig and cow most of the pathways were estimated to be induced (results not shown).

Among the pathways selected for their putative importance in milk protein synthesis the DIA results suggested that the mouse mammary gland was featured by a large decrease of protein degradation, suggested by the 'Proteasome' pathway (being one of the 10 most impacted pathways and inhibited in mouse), but also by a decrease in abundance of protein synthesis machinery components as well capacity for charging tRNA with amino acids (see 'Ribosome' and 'Aminoacyl-tRNA biosynthesis' in Figure 4). Other selected pathways had no apparent induction, or a slight inhibition such as for 'Jak-STAT' and 'Insulin' signaling pathways, or an evident inhibition as for the 'SNARE interactions in vesicular transport' (Figure 4). In the pig, the components of the mammary protein synthesis machinery and those for protein degradation were not largely affected by lactation, particularly for the

'Ribosome', while the other pathways were highly impacted and induced the day after parturition, except for 'Jak-STAT' and 'Insulin' signaling pathways (Figure 4). As observed previously by analysis of the same dataset but with more time points and using a more strong statistical approach [34], for the bovine mammary gland the components of the protein synthesis machinery were induced just before parturition followed by a continuous reduction, while the charging of amino acids to tRNA appeared to be only slightly induced during lactation. The export of protein, and to a lesser magnitude the 'Protein processing in the endoplasmic reticulum', was largely impacted and induced during lactation in this species (Figure 4). The signaling pathways related to protein synthesis in the bovine had a greater impact during lactation compared to the monogastrics, with a clear induction of the 'Jak-STAT signaling' and slight inhibition of both 'mTOR signaling' and 'Insulin signaling' pathways (Figure 4).

A reduction of the protein synthesis machinery was a common feature between the mouse and cow. As discussed above the reduction of the components of the protein synthesis machinery in the cow, but even more importantly in the mouse with its tremendous level of milk protein synthesis per mass of mammary gland (see Table 1), is a curious paradox. In contrast, measurement of the ribosomal proteins in mammary tissue of rabbits and sheep (reviewed in [184]) indicated an increase in quantity and formation of polyribosomes in lactating vs. non-lactating mammary tissue.

The mouse produces a large amount of milk per body weight or per weight of mammary tissue, larger than cow or pig (Table 1). In mice the mammary RNA/DNA ratio increases significantly during lactation [184, 185] and the peak in the percentage of protein in milk is reached at ca. 10 days postpartum [186], but milk yield, and total milk protein yield, increases until the end of lactation (ca. 20 day in milk) [186]. In the mouse mammary gland the amount of protein doubles and the rate of protein synthesis triples from end of pregnancy (18 day of pregnancy or 2 days prior parturition) to the beginning of lactation (3 day in milk), and protein content increases 4-fold and protein synthesis rate increases 8-fold from end of pregnancy to 15 day in milk [65]. Similarly, rat acinar cells have a ca. 9-fold increase in synthesis of secreted proteins and ca. 5-fold increase of non-secreted proteins in early lactation [187]. Interestingly, the moment of maximum protein content of the mouse milk seems to coincide with the greatest inhibition of the 'Proteasome' and 'Ribosome' pathways (Figure 4).

In the cow the maximum percentage of milk protein occurs at the end of lactation but the milk protein yield reaches a plateau at around 30 day in milk and remains at that level for up to 6 months [4]. As for the mouse and other species (e.g., pig, see [188]) RNA increases around 2-fold at peak lactation relative to pregnancy [189]. In the dairy goat the rate of protein synthesis in mammary tissue increases ca. 7-fold from dry-off to lactation [9]. There are not equivalent data for the bovine, but it is likely that the protein synthesis rate increases during lactation by the same order of magnitude.

Summarizing the above observations, during lactation there is an increase in RNA, an increase in overall protein synthesis rate, but an apparent decrease in the expression of

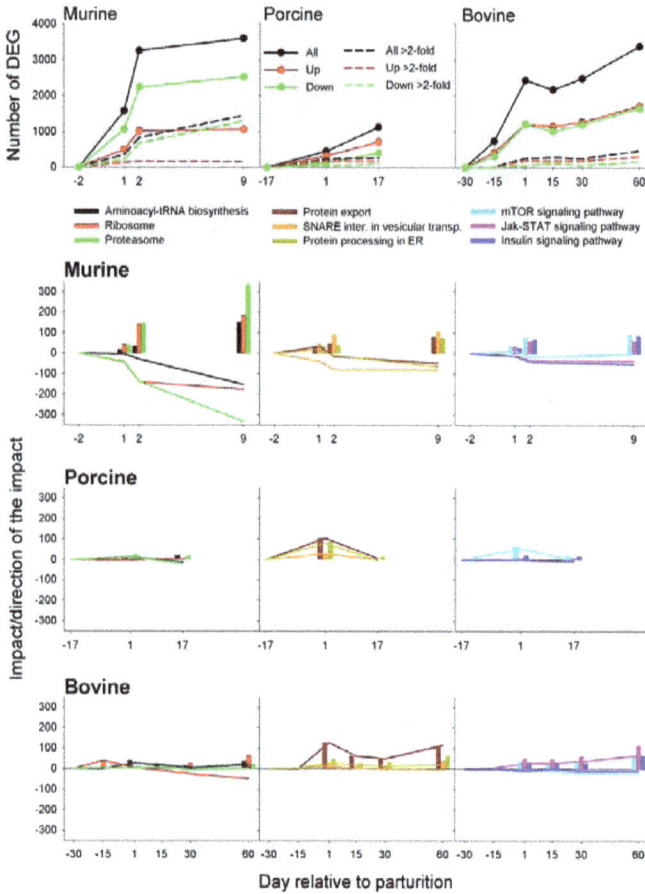

Figure 4. Top panels: number of differentially expressed genes (DEG; FDR ≤0.10, P-value for each time point vs. first time point in pregnancy ≤0.05) without fold change cut-off or with 2-fold cut-off (i.e., >2-fold). Legend denotes all = all DEG, up = up-regulated DEG, and down = down-regulated DEG with or without 2-fold cut-off. Lower panels: impact and direction of the impact in selected Kyoto Encyclopedia of Genes and Genomes (KEGG; see http://www.genome.jp/kegg/) pathways from the Dynamic Impact Approach (DIA) analysis (see [89]) of genes differentially expressed from pregnancy to mid- and late-lactation in the mammary gland of the mouse [91], pig [92], and cow [34]. Original data (i.e., bovine) and data downloaded from the GEO Datasets (at http://www.ncbi.nlm.nih.gov/geo/) as reported in the legend of Figure 1 were statistically analyzed using GeneSpring GX7 with time as fixed effect and with overall false discovery rate (FDR) correction. Day 17 of pregnancy (=-2 day relative to parturition) was used as first time point in pregnancy instead of the 12 day of pregnancy available (i.e., -9 day relative to parturition) from the mouse dataset in order to have all the species at ca. 90% of the pregnancy term. Datasets containing the FDR correction for the overall time effect plus the fold change and the P-value relative to pregnancy (-2d for mouse, -17d for pig, and -30d for bovine) for each time point were uploaded to DIA for analysis with the following criteria: FDR ≤0.10, P-value ≤0.05, and ≥30% of microarray coverage of genes in pathways and the whole annotated microarray as background.

protein synthesis machinery components. In mice the data also suggest a decrease in protein degradation; however, this does not appear to fully account for the increase in overall protein synthesis due to the concomitant reduction of protein synthesis machinery. In addition, the data also suggest that there is a large decrease in expression of many genes in the mouse (Figure 4), but an overall increase in RNA, which is common to all species. As previously suggested [189], the increase in RNA in lactating mammary gland relative to pregnancy is likely due to the dramatic upsurge in expression of a relatively low number of genes coding for high abundant milk proteins, such as caseins and alpha-lactalbumin, and other genes coding for non-specifically secreted proteins, such as fatty acid binding proteins. Therefore, the increase in total RNA is driven by the large increase in expression of a relatively small number of genes. This is also observed in bovine mammary transcriptomics data as evident from Figure 4 (see also [34]).

The decrease, or lack of increase of expression of ribosomal components despite the evident increase in protein synthesis together with the apparent induction of translation during lactation (see Figure 1 and data from other laboratories review above), led us to infer that the decrease of ribosomal components concomitant with a marked increase in mRNA expression of caseins and other milk-specific proteins (Figure 1) is a mechanism for the mammary gland to prioritize at the translational level mRNA coding for proteins related to milk synthesis and secretion rather than non-milk-specific proteins. This suggestion appears to be supported by previous data in mice and goats where competition between secreted milk proteins was observed. Ectopic expression of sheep beta-globulin in mice reduced the expression of other milk proteins (i.e., transcriptional competition) but the amount of total milk protein did not change suggesting that the protein synthesis machinery was already working at full capacity in the lactating mammary tissue [190]. An overall decrease in milk caseins was observed in transgenic goats ectopically expressing high levels of recombinant human butyrylcholinesterase [191]. However, transgenic cows where expression of β- and κ-casein was ectopically enhanced [192] had a significant (ca. 20%) increase in milk protein with a change in casein composition of the milk. In fact, it was observed a 2-fold increase in κ-casein concentration but only ca. 14% increase of β-casein content compared to non-transgenic cows. Those data are surprising considering the contrasting observations from the mouse and goat discussed above. However, caseins and most of the milk proteins are synthesized in polyribosomes attached to the endoplasmic reticulum [193]. It is possible that the decrease in expression observed for some of the ribosomal components encompasses components of the cytosolic protein machinery and not, or with a lower proportion, of ribosomal components of the endoplasmic reticulum [4]. This, in turn will provide a greater competitive advantage for the translation of milk protein genes compared to other genes not translated by the polyribosomes. In the case of the study in transgenic cows discussed above [192] the greater increase in κ-casein but the smaller increase in β-casein suggests that a competition between the two caseins occurs with a clear advantage toward the κ-casein, probably due to a sequence/structural difference between the two caseins.

Based on the above observations, and particularly for the mouse and to a less extent for the bovine, we propose that the protein synthesis machinery in the mammary gland is purposely decreased during lactation in order to increase the competitive advantage of milk-specific proteins, such as caseins and other secreted proteins, and perhaps for other non-secreted proteins with a pivotal role in mammary gland biology, such as the fatty acid binding protein. This hypothesis can account for the extremely large increase in expression of those milk-specific genes in lactating mammary gland, not only due to an increase in transcription but also an increase in half-life driven by the elongated poly-A tail. This goes along with the well-defined teleological role of the mammary gland, i.e., to produce milk for the offspring. In addition, this phenomenon appears to further support the fine regulation of protein synthesis by lactating mammary gland. It appears that mammary gland attempts on one hand to achieve a "set" composition of the milk, while on the other hand to conserve as much energy as possible: in addition to the basic inhibition of translation which is overcome by the stimulation of the insulin-mTOR pathway, the lowered availability of protein translation machinery provides limitations that in turn can be overcome only by a large availability of precursors (e.g., AA) and energy. This hypothesis is in accord with the above reported finding of higher milk protein when animals have higher energy and AA availability. Another observation that can be made based on the above data is that, if there is a decreased availability of ribosomes but there is a concomitant increase in translation, then there must be an apparent increase in translational efficiency. This can be partly explained by the formation of polyribosomes, but may include other unknown factors, as well.

This hypothesis has a technical consequence. If transcripts are under strong competition for the translational machinery, this means that a discrepancy between data related to mRNA abundance and data related to protein abundance may be expected. As for the dilution effect in quantitative PCR of stably expressed genes due to the high increase in lactation-specific genes [189], the increase in abundance of milk protein genes decreases the accessibility to translational machinery by low-abundant transcripts. Thus, we should expect that a significant increase in expression of low-abundant transcripts from pregnancy to lactation, concomitant with a large increase in high-abundant transcripts, will probably end up being translated in the lactating mammary tissue to the same or reduced extent compared to pregnancy, while the increase in abundant transcripts will be translated to a higher rate, i.e., resulting in a large increase in protein abundance. This may be an important point to consider in interpreting transcriptomics data of the mammary gland during pregnancy and lactation.

7. Conclusions

The synthesis and secretion of major proteins found in the milk is a complex phenomenon that still needs to be completely understood. It is clear that those proteins are milk-specific with very minor or no expression in other tissues rather that the lactating (or approaching lactation) mammary gland; furthermore, the expression is present almost exclusively in one type of cell (among more than 200 fully differentiated cells known to compose the mammalian organism): mammary epithelial. The investigations so far have clearly shown that the expression of major milk proteins increases dramatically and in a concerted way during the onset of lactation and remain high until lactation declines (with some exceptions, e.g., *LALBA* in kangaroo). We know that several hormones, the ECM, and the cell-to-cell interactions play a pivotal role in inducing such adaptation with differences between species. We know that nutrition can affect the quantity of milk proteins, with a pivotal role of energy content in the diet. The effect of energy in the diet is mediated by insulin and recent data strongly support a role for the crosstalk between insulin and mTOR in regulating translation. This regulation is chiefly determined by phosphorylation and, with an apparent less importance, by gene transcription. Finally, we have shown that there is an unexpected decrease in expression of ribosomal proteins. This suggests that, at the least in mouse and cow (with a greater emphasis in the former), there is an apparent overall decrease in the protein synthesis machinery despite the known overall increase in protein synthesis. We propose that this is a mechanism for the mammary epithelial cell to increase competitive advantage for translation of milk-related proteins in order to increase the mammary gland's primary function: production of milk.

The control of milk synthesis is still far for being understood. We have not discussed in this chapter the use of bioinformatics to uncover transcriptional networks. This is a semi-*in silico* approach (i.e., use of a mixture of real data with available previously published information) that allows to uncover which transcription factors are involved in controlling the transcriptomics adaptation. The transcription factors or transcriptional networks suggested by this approach to be pivotal in controlling milk protein synthesis need to be considered with caution. In fact, that information is not conclusive but only indicative; however, this approach allows opening new horizons of research not previously considered.

We are living in an exciting moment in science. The advent of 'omics' tools and, particularly, of the system biology approach, open up new possibilities to investigate the complexity of the mammary adaptations to lactation. The integration of all or some of the available 'omics', such as transcriptomics, proteomics, metabolomics, miRNAomics, single nucleotide polymorphisms, epigenomics, and phosphoproteomics, using and integrative systems biology approach, although not feasible at the moment, is probably what we should implement in the future in order to fully understand the biology of organisms [194], including the control of milk protein synthesis in mammary gland of each species of interest.

Author details

Massimo Bionaz, Walter Hurley and Juan Loor
Department of Animal Sciences, University of Illinois, Urbana, Illinois, USA

8. References

[1] Akers RM (2006) Major advances associated with hormone and growth factor regulation of mammary growth and lactation in dairy cows. J Dairy Sci. 89: 1222-1234

[2] Hadsell DL (2004) Genetic manipulation of mammary gland development and lactation. Advances in experimental medicine and biology. 554: 229-251

[3] Loor JJ, Cohick WS (2009) ASAS centennial paper: Lactation biology for the twenty-first century. Journal of animal science. 87: 813-824

[4] Bionaz M, Loor JJ (2011) Gene networks driving bovine mammary protein synthesis during the lactation cycle. Bioinform Biol Insights. 5: 83-98

[5] Goff SA (2011) A unifying theory for general multigenic heterosis: energy efficiency, protein metabolism, and implications for molecular breeding. New Phytol. 189: 923-937

[6] Kapahi P (2010) Protein synthesis and the antagonistic pleiotropy hypothesis of aging. Adv Exp Med Biol. 694: 30-37

[7] Baldwin RL, Smith NE, Taylor J, Sharp M (1980) Manipulating metabolic parameters to improve growth rate and milk secretion. J Anim Sci. 51: 1416-1428

[8] Schingoethe DJ, Byers FM, Schelling GT (1988) Nutrient needs during critical periods of the life cycle.In: DC Church, editors. The Ruminant Animal: Digestive, Physiology, and Nutrition, Illinois: Waveland Press. Inc. pp 421-447

[9] Lescoat P, Sauvant D, Danfaer A (1997) Quantitative aspects of protein fractional synthesis rates in ruminants. Reprod Nutr Dev. 37: 493-515

[10] Hanigan MD, France J, Mabjeesh SJ, McNabb WC, Bequette BJ (2009) High rates of mammary tissue protein turnover in lactating goats are energetically costly. J Nutr. 139: 1118-1127

[11] Kuhn C, Freyer G, Weikard R, Goldammer T, Schwerin M (1999) Detection of QTL for milk production traits in cattle by application of a specifically developed marker map of BTA6. Anim Genet. 30: 333-340

[12] Bequette BJ, Backwell FR, Crompton LA (1998) Current concepts of amino acid and protein metabolism in the mammary gland of the lactating ruminant. J Dairy Sci. 81: 2540-2559

[13] Reynolds CK, Harmon DL, Cecava MJ (1994) Absorption and delivery of nutrients for milk protein synthesis by portal-drained viscera. J Dairy Sci. 77: 2787-2808

[14] Gross J, van Dorland HA, Bruckmaier RM, Schwarz FJ (2011) Performance and metabolic profile of dairy cows during a lactational and deliberately induced negative energy balance with subsequent realimentation. J Dairy Sci. 94: 1820-1830

[15] van Knegsel AT, van den Brand H, Dijkstra J, Tamminga S, Kemp B (2005) Effect of dietary energy source on energy balance, production, metabolic disorders and reproduction in lactating dairy cattle. Reprod Nutr Dev. 45: 665-688

[16] Grieve DG, Korver S, Rijpkema YS, Hof G (1986) Relationship between Milk-Composition and Some Nutritional Parameters in Early Lactation. Livestock Production Science. 14: 239-254

[17] Broster WH (1973) Protein-Energy Interrelationships in Growth and Lactation of Cattle and Sheep. Proceedings of the Nutrition Society. 32: 115-122

[18] Sinclair LA, Lock AL, Early R, Bauman DE (2007) Effects of trans-10, cis-12 conjugated linoleic acid on ovine milk fat synthesis and cheese properties. J Dairy Sci. 90: 3326-3335

[19] Cannas A, Pes A, Mancuso R, Vodret B, Nudda A (1998) Effect of dietary energy and protein concentration on the concentration of milk urea nitrogen in dairy ewes. J Dairy Sci. 81: 499-508

[20] Grigor MR, Allan JE, Carrington JM, Carne A, Geursen A, Young D, Thompson MP, Haynes EB, Coleman RA (1987) Effect of dietary protein and food restriction on milk

production and composition, maternal tissues and enzymes in lactating rats. J Nutr. 117: 1247-1258

[21] Geursen A, Grigor MR (1987) Nutritional regulation of milk protein messenger RNA concentrations in mammary acini isolated from lactating rats. Biochem Int. 15: 873-879

[22] Bauman DE, Harvatine KJ, Lock AL (2011) Nutrigenomics, rumen-derived bioactive fatty acids, and the regulation of milk fat synthesis. Annu Rev Nutr. 31: 299-319

[23] Medeiros SR, Oliveira DE, Aroeira LJ, McGuire MA, Bauman DE, Lanna DP (2010) Effects of dietary supplementation of rumen-protected conjugated linoleic acid to grazing cows in early lactation. J Dairy Sci. 93: 1126-1137

[24] Weerasinghe WM, Wilkinson RG, Lock AL, de Veth MJ, Bauman DE, Sinclair LA (2012) Effect of a supplement containing trans-10,cis-12 conjugated linoleic acid on the performance of dairy ewes fed 2 levels of metabolizable protein and at a restricted energy intake. J Dairy Sci. 95: 109-116

[25] de Veth MJ, Castaneda-Gutierrez E, Dwyer DA, Pfeiffer AM, Putnam DE, Bauman DE (2006) Response to conjugated linoleic acid in dairy cows differing in energy and protein status. J Dairy Sci. 89: 4620-4631

[26] DePeters EJ, Cant JP (1992) Nutritional factors influencing the nitrogen composition of bovine milk: a review. J Dairy Sci. 75: 2043-2070

[27] Hanigan MD, Cant JP, Weakley DC, Beckett JL (1998) An evaluation of postabsorptive protein and amino acid metabolism in the lactating dairy cow. Journal of dairy science. 81: 3385-3401

[28] Rius AG, Appuhamy JA, Cyriac J, Kirovski D, Becvar O, Escobar J, McGilliard ML, Bequette BJ, Akers RM, Hanigan MD (2010) Regulation of protein synthesis in mammary glands of lactating dairy cows by starch and amino acids. J Dairy Sci. 93: 3114-3127

[29] Burgos SA, Dai M, Cant JP (2010) Nutrient availability and lactogenic hormones regulate mammary protein synthesis through the mammalian target of rapamycin signaling pathway. J Dairy Sci. 93: 153-161

[30] Guan X, Pettigrew JE, Ku PK, Ames NK, Bequette BJ, Trottier NL (2004) Dietary protein concentration affects plasma arteriovenous difference of amino acids across the porcine mammary gland. Journal of animal science. 82: 2953-2963

[31] Baumrucker CR (1985) Amino acid transport systems in bovine mammary tissue. J Dairy Sci. 68: 2436-2451

[32] Hanigan MD, Crompton LA, Bequette BJ, Mills JA, France J (2002) Modelling mammary metabolism in the dairy cow to predict milk constituent yield, with emphasis on amino acid metabolism and milk protein production: model evaluation. J Theor Biol. 217: 311-330

[33] Mepham TB (1982) Amino acid utilization by lactating mammary gland. J Dairy Sci. 65: 287-298

[34] Bionaz M, Periasamy K, Rodriguez-Zas SL, Everts RE, Lewin HA, Hurley WL, Loor JJ (2012) Old and New Stories: Revelations from Functional Analysis of the Bovine Mammary Transcriptome during the Lactation Cycle. PLoS One. 7: e33268

[35] Schmidt GH (1966) Effect of insulin on yield and composition of milk of dairy cows. J Dairy Sci. 49: 381-385

[36] Winkelman LA, Overton TR (2010) Is There Opportunity to Boost Milk Protein Production? In: Cornell Nutrition Conference for Feed Manufacturers, (East Syracuse, New York pp 123-132

[37] Herbein JH, Aiello RJ, Eckler LI, Pearson RE, Akers RM (1985) Glucagon, insulin, growth hormone, and glucose concentrations in blood plasma of lactating dairy cows. J Dairy Sci. 68: 320-325

[38] Menzies KK, Lee HJ, Lefevre C, Ormandy CJ, Macmillan KL, Nicholas KR (2010) Insulin, a key regulator of hormone responsive milk protein synthesis during lactogenesis in murine mammary explants. Funct Integr Genomics. 10: 87-95

[39] Menzies KK, Lefevre C, Macmillan KL, Nicholas KR (2009) Insulin regulates milk protein synthesis at multiple levels in the bovine mammary gland. Funct Integr Genomics. 9: 197-217

[40] Te Pas MF, Hoekman AJW, Hulsegge I (2011) From Visual Biological Models Toward Mathematical Models of the Biology of Complex Traits.In: MFW te Pas, H Woelders and A Bannink, editors. Systems Biology and Livestock Science: Wiley-Blackwell. pp 137-159

[41] Liu GE (2011) Recent applications of DNA sequencing technologies in food, nutrition and agriculture. Recent patents on food, nutrition & agriculture. 3: 187-195

[42] Bruggeman FJ, Westerhoff HV (2007) The nature of systems biology. Trends Microbiol. 15: 45-50

[43] Barabasi AL, Oltvai ZN (2004) Network biology: understanding the cell's functional organization. Nature reviews. Genetics. 5: 101-113

[44] Schena M, Shalon D, Davis RW, Brown PO (1995) Quantitative monitoring of gene expression patterns with a complementary DNA microarray. Science. 270: 467-470

[45] Wang Z, Gerstein M, Snyder M (2009) RNA-Seq: a revolutionary tool for transcriptomics. Nat Rev Genet. 10: 57-63

[46] Van Dien S, Schilling CH (2006) Bringing metabolomics data into the forefront of systems biology. Molecular systems biology. 2: 2006 0035

[47] Westerhoff HV, Palsson BO (2004) The evolution of molecular biology into systems biology. Nat Biotechnol. 22: 1249-1252

[48] Palsson B (2009) Metabolic systems biology. FEBS letters. 583: 3900-3904

[49] Loor JJ (2010) Genomics of metabolic adaptations in the peripartal cow. Animal : an international journal of animal bioscience. 4: 1110-1139

[50] Loor JJ, Lin X, Herbein JH (2003) Effects of dietary cis 9, trans 11-18:2, trans 10, cis 12-18:2, or vaccenic acid (trans 11-18:1) during lactation on body composition, tissue fatty acid profiles, and litter growth in mice. The British journal of nutrition. 90: 1039-1048

[51] Morag M (1970) Estimation of milk yield in the rat. Laboratory animals. 4: 259-272

[52] Daly SE, Hartmann PE (1995) Infant demand and milk supply. Part 2: The short-term control of milk synthesis in lactating women. Journal of human lactation : official journal of International Lactation Consultant Association. 11: 27-37

[53] Daly SE, Hartmann PE (1995) Infant demand and milk supply. Part 1: Infant demand and milk production in lactating women. Journal of human lactation : official journal of International Lactation Consultant Association. 11: 21-26

[54] King RH (2000) Factors that influence milk production in well-fed sows. J Anim Sci. 78: 19-25

[55] Hansen AV, Strathe AB, Kebreab E, France J, Theil PK (2012) Predicting milk yield and composition in lactating sows - A Bayesian approach. Journal of animal science.

[56] Bichard M, David PJ (1986) Producing more pigs per sow per year--genetic contributions. Journal of animal science. 63: 1275-1279

[57] Hadsell DL, Parlow AF, Torres D, George J, Olea W (2008) Enhancement of maternal lactation performance during prolonged lactation in the mouse by mouse GH and long-R3-IGF-I is linked to changes in mammary signaling and gene expression. The Journal of endocrinology. 198: 61-70

[58] Capuco AV, Wood DL, Baldwin R, McLeod K, Paape MJ (2001) Mammary cell number, proliferation, and apoptosis during a bovine lactation: relation to milk production and effect of bST. J Dairy Sci. 84: 2177-2187

[59] Moretto VL, Ballen MO, Goncalves TS, Kawashita NH, Stoppiglia LF, Veloso RV, Latorraca MQ, Martins MS, Gomes-da-Silva MH (2011) Low-Protein Diet during Lactation and Maternal Metabolism in Rats. ISRN obstetrics and gynecology. 2011: 876502

[60] Kim SW, Hurley WL, Han IK, Stein HH, Easter RA (1999) Effect of nutrient intake on mammary gland growth in lactating sows. Journal of animal science. 77: 3304-3315

[61] Anderson RR (1975) Mammary gland growth in sheep. Journal of animal science. 41: 118-123

[62] Linzell JL (1972) Milk yield, energy loss in milk and mammary gland weight in different species. Dairy Sci Abstr. 351–360

[63] Mabjeesh SJ, Kyle CE, Macrae JC, Bequette BJ (2000) Lysine metabolism by the mammary gland of lactating goats at two stages of lactation. Journal of dairy science. 83: 996-1003

[64] Guan X, Bequette BJ, Calder G, Ku PK, Ames KN, Trottier NL (2002) Amino acid availability affects amino acid flux and protein metabolism in the porcine mammary gland. The Journal of nutrition. 132: 1224-1234

[65] Millican PE, Vernon RG, Pain VM (1987) Protein metabolism in the mouse during pregnancy and lactation. The Biochemical journal. 248: 251-257

[66] Bequette BJ, Backwell FR (1997) Amino acid supply and metabolism by the ruminant mammary gland. The Proceedings of the Nutrition Society. 56: 593-605

[67] Davies DT, Holt C, Christie WW (1983) The Composition of Milk.In: TB Mepham, editors. Biochemistry of Lactation, Amsterdam: Elsevier Science Publishers B.V. pp 71-117

[68] Oftedal OT (1984) Milk composition, milk yield and energy output at peak lactation: a comparative review. Symposia of the Zoological Society of London. 33-85

[69] Bionaz M, Loor JJ (2008) Gene networks driving bovine milk fat synthesis during the lactation cycle. BMC Genomics. 9: 366

[70] Kolb AF, Huber RC, Lillico SG, Carlisle A, Robinson CJ, Neil C, Petrie L, Sorensen DB, Olsson IA, Whitelaw CB (2011) Milk lacking alpha-casein leads to permanent reduction in body size in mice. PloS one. 6: e21775

[71] Gors S, Kucia M, Langhammer M, Junghans P, Metges CC (2009) Technical note: Milk composition in mice--methodological aspects and effects of mouse strain and lactation day. Journal of dairy science. 92: 632-637

[72] Knight CH, Maltz E, Docherty AH (1986) Milk yield and composition in mice: effects of litter size and lactation number. Comparative biochemistry and physiology. A, Comparative physiology. 84: 127-133

[73] Ley JM, Jenness R (1970) Lactose synthetase activity of alpha-lactalbumins from several species. Archives of biochemistry and biophysics. 138: 464-469

[74] Kumar S, Clarke AR, Hooper ML, Horne DS, Law AJ, Leaver J, Springbett A, Stevenson E, Simons JP (1994) Milk composition and lactation of beta-casein-deficient mice. Proceedings of the National Academy of Sciences of the United States of America. 91: 6138-6142

[75] Jenness R (1979) The composition of human milk. Semin Perinatol. 3: 225-239

[76] Jandal JM (1996) Comparative aspects of goat and sheep milk. Small Ruminant Research. 22: 177-185

[77] Green SW, Renfree MB (1982) Changes in the milk proteins during lactation in the tammar wallaby, Macropus eugenii. Australian journal of biological sciences. 35: 145-152

[78] Messer M, Elliott C (1987) Changes in alpha-lactalbumin, total lactose, UDP-galactose hydrolase and other factors in tammar wallaby (Macropus eugenii) milk during lactation. Australian journal of biological sciences. 40: 37-46

[79] Horne DS, Anema S, Zhu X, Nicholas KR, Singh H (2007) A lactational study of the composition and integrity of casein micelles from the milk of the tammar wallaby (Macropus eugenii). Archives of biochemistry and biophysics. 467: 107-118

[80] Cavaletto M, Giuffrida MG, Conti A (2008) Milk fat globule membrane components--a proteomic approach. Advances in experimental medicine and biology. 606: 129-141

[81] Reinhardt TA, Lippolis JD (2006) Bovine milk fat globule membrane proteome. The Journal of dairy research. 73: 406-416

[82] Reinhardt TA, Lippolis JD (2008) Developmental changes in the milk fat globule membrane proteome during the transition from colostrum to milk. J Dairy Sci. 91: 2307-2318

[83] Affolter M, Grass L, Vanrobaeys F, Casado B, Kussmann M (2010) Qualitative and quantitative profiling of the bovine milk fat globule membrane proteome. Journal of proteomics. 73: 1079-1088

[84] Keenan TW, Mather IH (2006) Intracellular Origin of Milk Fat Globules and the Nature of the Milk Fat Globule Membrane.In: PF Fox and PLH McSweeney, editors. Advanced Dairy Chemistry Vol. 2: Lipids, New York, NY: Springer. pp 137-171

[85] Lonnerdal B (2011) Biological effects of novel bovine milk fractions. Nestle Nutrition workshop series. Paediatric programme. 67: 41-54

[86] Yen CC, Shen CJ, Hsu WH, Chang YH, Lin HT, Chen HL, Chen CM (2011) Lactoferrin: an iron-binding antimicrobial protein against Escherichia coli infection. Biometals : an international journal on the role of metal ions in biology, biochemistry, and medicine. 24: 585-594

[87] Amini AA, Nair LS (2011) Lactoferrin: a biologically active molecule for bone regeneration. Current medicinal chemistry. 18: 1220-1229

[88] Rhoads RE, Grudzien-Nogalska E (2007) Translational regulation of milk protein synthesis at secretory activation. J Mammary Gland Biol Neoplasia. 12: 283-292

[89] Bionaz M, Periasamy K, Rodriguez-Zas SL, Hurley WL, Loor JJ (2012) A Novel Dynamic Impact Approach (DIA) for Functional Analysis of Time-Course Omics Studies: Validation Using the Bovine Mammary Transcriptome. PloS one. 7: e32455

[90] Faucon F, Rebours E, Bevilacqua C, Helbling JC, Aubert J, Makhzami S, Dhorne-Pollet S, Robin S, Martin P (2009) Terminal differentiation of goat mammary tissue during pregnancy requires the expression of genes involved in immune functions. Physiological genomics. 40: 61-82

[91] Rudolph MC, McManaman JL, Phang T, Russell T, Kominsky DJ, Serkova NJ, Stein T, Anderson SM, Neville MC (2007) Metabolic regulation in the lactating mammary gland: a lipid synthesizing machine. Physiological Genomics. 28: 323-336

[92] Shu DP, Chen BL, Hong J, Liu PP, Hou DX, Huang X, Zhang FT, Wei JL, Guan WT (2012) Global transcriptional profiling in porcine mammary glands from late pregnancy to peak lactation. Omics : a journal of integrative biology. 16: 123-137

[93] Lefevre CM, Digby MR, Whitley JC, Strahm Y, Nicholas KR (2007) Lactation transcriptomics in the Australian marsupial, Macropus eugenii: transcript sequencing and quantification. BMC genomics. 8: 417

[94] Wickramasinghe S, Rincon G, Islas-Trejo A, Medrano JF (2012) Transcriptional profiling of bovine milk using RNA sequencing. BMC genomics. 13: 45

[95] Benjamini Y, Hochberg Y (1995) Controlling the False Discovery Rate - a Practical and Powerful Approach to Multiple Testing. Journal of the Royal Statistical Society Series B-Methodological. 57: 289-300

[96] Messer M, Green B (1979) Milk carbohydrates of marsupials. II. Quantitative and qualitative changes in milk carbohydrates during lactation in the tammar wallaby (Macropus eugenii). Australian journal of biological sciences. 32: 519-531

[97] Nicholas K, Simpson K, Wilson M, Trott J, Shaw D (1997) The tammar wallaby: a model to study putative autocrine-induced changes in milk composition. Journal of mammary gland biology and neoplasia. 2: 299-310

[98] Phillippy BO, McCarthy RD (1979) Multi-origins of milk serum albumin in the lactating goat. Biochimica et biophysica acta. 584: 298-303

[99] Shamay A, Homans R, Fuerman Y, Levin I, Barash H, Silanikove N, Mabjeesh SJ (2005) Expression of albumin in nonhepatic tissues and its synthesis by the bovine mammary gland. Journal of dairy science. 88: 569-576

[100] Piccinini R, Binda E, Belotti M, Dapra V, Zecconi A (2007) Evaluation of milk components during whole lactation in healthy quarters. J Dairy Res. 74: 226-232

[101] Vorbach C, Capecchi MR, Penninger JM (2006) Evolution of the mammary gland from the innate immune system? Bioessays. 28: 606-616

[102] Zapico P, Gaya P, De Paz M, Nunez M, Medina M (1991) Influence of breed, animal, and days of lactation on lactoperoxidase system components in goat milk. Journal of dairy science. 74: 783-787

[103] Loor JJ, Moyes KM, Bionaz M (2011) Functional adaptations of the transcriptome to mastitis-causing pathogens: the mammary gland and beyond. Journal of mammary gland biology and neoplasia. 16: 305-322

[104] Elliot JI, Senft B, Erhardt G, Fraser D (1984) Isolation of lactoferrin and its concentration in sows' colostrum and milk during a 21-day lactation. Journal of animal science. 59: 1080-1084

[105] Cheng JB, Wang JQ, Bu DP, Liu GL, Zhang CG, Wei HY, Zhou LY, Wang JZ (2008) Factors affecting the lactoferrin concentration in bovine milk. Journal of dairy science. 91: 970-976

[106] Whetstone HD, Hurley WL, Davis CL (1986) Identification and characterization of a fatty acid binding protein in bovine mammary gland. Comparative biochemistry and physiology. B, Comparative biochemistry. 85: 687-692

[107] Rosen JM, Wyszomierski SL, Hadsell D (1999) Regulation of milk protein gene expression. Annual review of nutrition. 19: 407-436

[108] Neville MC, McFadden TB, Forsyth I (2002) Hormonal regulation of mammary differentiation and milk secretion. J Mammary Gland Biol Neoplasia. 7: 49-66

[109] Forsyth IA (1986) Variation among species in the endocrine control of mammary growth and function: the roles of prolactin, growth hormone, and placental lactogen. Journal of dairy science. 69: 886-903

[110] Amenomori Y, Chen CL, Meites J (1970) Serum prolactin levels in rats during different reproductive states. Endocrinology. 86: 506-510

[111] Jahn GA, Edery M, Belair L, Kelly PA, Djiane J (1991) Prolactin receptor gene expression in rat mammary gland and liver during pregnancy and lactation. Endocrinology. 128: 2976-2984

[112] Mizoguchi Y, Yamaguchi H, Aoki F, Enami J, Sakai S (1997) Corticosterone is required for the prolactin receptor gene expression in the late pregnant mouse mammary gland. Molecular and cellular endocrinology. 132: 177-183

[113] Mepham TB (1987) Physiology of lactation (Milton Keynes ; Philadelphia: Open University Press)

[114] McVey WR, Jr., Williams GL (1991) Mechanical masking of neurosensory pathways at the calf-teat interface: endocrine, reproductive and lactational features of the suckled anestrous cow. Theriogenology. 35: 931-941

[115] Zwald NR, Weigel KA, Chang YM, Welper RD, Clay JS (2005) Genetic evaluation of dairy sires for milking duration using electronically recorded milking times of their daughters. Journal of Dairy Science. 88: 1192-1198

[116] Rijnkels M (2002) Multispecies comparison of the casein gene loci and evolution of casein gene family. Journal of mammary gland biology and neoplasia. 7: 327-345

[117] Zhao FQ, Adachi K, Oka T (2002) Involvement of Oct-1 in transcriptional regulation of beta-casein gene expression in mouse mammary gland. Biochimica et biophysica acta. 1577: 27-37

[118] Groner B (2002) Transcription factor regulation in mammary epithelial cells. Domestic animal endocrinology. 23: 25-32

[119] Reichenstein M, Rauner G, Barash I (2011) Conditional repression of STAT5 expression during lactation reveals its exclusive roles in mammary gland morphology, milk-protein gene expression, and neonate growth. Molecular reproduction and development. 78: 585-596

[120] Hay ED (1993) Extracellular matrix alters epithelial differentiation. Current opinion in cell biology. 5: 1029-1035

[121] Jolivet G, Pantano T, Houdebine LM (2005) Regulation by the extracellular matrix (ECM) of prolactin-induced alpha s1-casein gene expression in rabbit primary mammary cells: role of STAT5, C/EBP, and chromatin structure. Journal of cellular biochemistry. 95: 313-327

[122] Du JY, Chen MC, Hsu TC, Wang JH, Brackenbury L, Lin TH, Wu YY, Yang Z, Streuli CH, Lee YJ (2012) The RhoA-Rok-myosin II pathway is involved in extracellular matrix-mediated regulation of prolactin signaling in mammary epithelial cells. Journal of cellular physiology. 227: 1553-1560

[123] Streuli CH, Bailey N, Bissell MJ (1991) Control of mammary epithelial differentiation: basement membrane induces tissue-specific gene expression in the absence of cell-cell interaction and morphological polarity. The Journal of cell biology. 115: 1383-1395

[124] Kumura H, Tanaka A, Abo Y, Yui S, Shimazaki K, Kobayashi E, Sayama K (2001) Primary culture of porcine mammary epithelial cells as a model system for evaluation of milk protein expression. Bioscience, biotechnology, and biochemistry. 65: 2098-2101

[125] Zavizion B, Gorewit RC, Politis I (1995) Subcloning the MAC-T bovine mammary epithelial cell line: morphology, growth properties, and cytogenetic analysis of clonal cells. Journal of dairy science. 78: 515-527

[126] Zhou Y, Akers RM, Jiang H (2008) Growth hormone can induce expression of four major milk protein genes in transfected MAC-T cells. Journal of dairy science. 91: 100-108

[127] Huynh HT, Robitaille G, Turner JD (1991) Establishment of bovine mammary epithelial cells (MAC-T): an in vitro model for bovine lactation. Experimental cell research. 197: 191-199

[128] Yang J, Kennelly JJ, Baracos VE (2000) The activity of transcription factor Stat5 responds to prolactin, growth hormone, and IGF-I in rat and bovine mammary explant culture. J Anim Sci. 78: 3114-3125

[129] Yang J, Kennelly JJ, Baracos VE (2000) Physiological levels of Stat5 DNA binding activity and protein in bovine mammary gland. J Anim Sci. 78: 3126-3134

[130] Wheeler TT, Broadhurst MK, Sadowski HB, Farr VC, Prosser CG (2001) Stat5 phosphorylation status and DNA-binding activity in the bovine and murine mammary glands. Mol Cell Endocrinol. 176: 39-48

[131] Oakes SR, Naylor MJ, Asselin-Labat ML, Blazek KD, Gardiner-Garden M, Hilton HN, Kazlauskas M, Pritchard MA, Chodosh LA, Pfeffer PL, Lindeman GJ, Visvader JE, Ormandy CJ (2008) The Ets transcription factor Elf5 specifies mammary alveolar cell fate. Genes & development. 22: 581-586

[132] Oakes SR, Hilton HN, Ormandy CJ (2006) The alveolar switch: coordinating the proliferative cues and cell fate decisions that drive the formation of lobuloalveoli from ductal epithelium. Breast cancer research : BCR. 8: 207

[133] Travers MT, Barber MC, Tonner E, Quarrie L, Wilde CJ, Flint DJ (1996) The role of prolactin and growth hormone in the regulation of casein gene expression and mammary cell survival: relationships to milk synthesis and secretion. Endocrinology. 137: 1530-1539

[134] Streuli CH, Edwards GM (1998) Control of normal mammary epithelial phenotype by integrins. J Mammary Gland Biol Neoplasia. 3: 151-163

[135] Ben Chedly H, Boutinaud M, Bernier-Dodier P, Marnet PG, Lacasse P (2010) Disruption of cell junctions induces apoptosis and reduces synthetic activity in lactating goat mammary gland. Journal of dairy science. 93: 2938-2951

[136] Singh K, Molenaar AJ, Swanson KM, Gudex B, Arias JA, Erdman RA, Stelwagen K (2012) Epigenetics: a possible role in acute and transgenerational regulation of dairy cow milk production. Animal : an international journal of animal bioscience. 6: 375-381

[137] Li X, Zhao X (2008) Epigenetic regulation of mammalian stem cells. Stem cells and development. 17: 1043-1052

[138] Stiening CM, Hoying JB, Abdallah MB, Hoying AM, Pandey R, Greer K, Collier RJ (2008) The effects of endocrine and mechanical stimulation on stage I lactogenesis in bovine mammary epithelial cells. Journal of dairy science. 91: 1053-1066

[139] Bionaz M, Loor JJ (2008) Comparative MammOmics™ of milk fat synthesis in Mus musculus vs. Bos taurus. J Dairy Sci. 91: 566-567

[140] Moyes KM, Drackley JK, Morin DE, Bionaz M, Rodriguez-Zas SL, Everts RE, Lewin HA, Loor JJ (2009) Gene network and pathway analysis of bovine mammary tissue challenged with Streptococcus uberis reveals induction of cell proliferation and inhibition of PPARgamma signaling as potential mechanism for the negative relationships between immune response and lipid metabolism. BMC Genomics. 10: 542

[141] Li Y, Limmon GV, Imani F, Teng C (2009) Induction of lactoferrin gene expression by innate immune stimuli in mouse mammary epithelial HC-11 cells. Biochimie. 91: 58-67

[142] Liao Y, Du X, Lonnerdal B (2010) miR-214 regulates lactoferrin expression and pro-apoptotic function in mammary epithelial cells. The Journal of nutrition. 140: 1552-1556

[143] Burdon TG, Demmer J, Clark AJ, Watson CJ (1994) The mammary factor MPBF is a prolactin-induced transcriptional regulator which binds to STAT factor recognition sites. FEBS letters. 350: 177-182

[144] Watson CJ, Miller WR (1995) Elevated levels of members of the STAT family of transcription factors in breast carcinoma nuclear extracts. British journal of cancer. 71: 840-844

[145] LaRocca J, Pietruska J, Hixon M (2011) Akt1 is essential for postnatal mammary gland development, function, and the expression of Btn1a1. PloS one. 6: e24432

[146] Chen CC, Boxer RB, Stairs DB, Portocarrero CP, Horton RH, Alvarez JV, Birnbaum MJ, Chodosh LA (2010) Akt is required for Stat5 activation and mammary differentiation. Breast cancer research : BCR. 12: R72

[147] Maroulakou IG, Oemler W, Naber SP, Klebba I, Kuperwasser C, Tsichlis PN (2008) Distinct roles of the three Akt isoforms in lactogenic differentiation and involution. Journal of cellular physiology. 217: 468-477

[148] Rhoads RP, Kim JW, Leury BJ, Baumgard LH, Segoale N, Frank SJ, Bauman DE, Boisclair YR (2004) Insulin increases the abundance of the growth hormone receptor in liver and adipose tissue of periparturient dairy cows. The Journal of nutrition. 134: 1020-1027

[149] Vernon RG, Clegg RA, Flint DJ (1981) Metabolism of sheep adipose tissue during pregnancy and lactation. Adaptation and regulation. The Biochemical journal. 200: 307-314

[150] Martin-Hidalgo A, Huerta L, Alvarez N, Alegria G, Del Val Toledo M, Herrera E (2005) Expression, activity, and localization of hormone-sensitive lipase in rat mammary gland during pregnancy and lactation. Journal of lipid research. 46: 658-668

[151] Le MN, Kohanski RA, Wang LH, Sadowski HB (2002) Dual mechanism of signal transducer and activator of transcription 5 activation by the insulin receptor. Mol Endocrinol. 16: 2764-2779

[152] Wang X, Proud CG (2006) The mTOR pathway in the control of protein synthesis. Physiology (Bethesda). 21: 362-369

[153] Kimball SR, Jefferson LS (2006) New functions for amino acids: effects on gene transcription and translation. Am J Clin Nutr. 83: 500S-507S

[154] Schieke SM, Phillips D, McCoy JP, Jr., Aponte AM, Shen RF, Balaban RS, Finkel T (2006) The mammalian target of rapamycin (mTOR) pathway regulates mitochondrial oxygen consumption and oxidative capacity. J Biol Chem. 281: 27643-27652

[155] Proud CG (2007) Signalling to translation: how signal transduction pathways control the protein synthetic machinery. Biochem J. 403: 217-234

[156] Feng Z, Zhang H, Levine AJ, Jin S (2005) The coordinate regulation of the p53 and mTOR pathways in cells. Proc Natl Acad Sci U S A. 102: 8204-8209

[157] Inoki K, Ouyang H, Zhu T, Lindvall C, Wang Y, Zhang X, Yang Q, Bennett C, Harada Y, Stankunas K, Wang CY, He X, MacDougald OA, You M, Williams BO, Guan KL (2006) TSC2 integrates Wnt and energy signals via a coordinated phosphorylation by AMPK and GSK3 to regulate cell growth. Cell. 126: 955-968

[158] Sarbassov DD, Ali SM, Sabatini DM (2005) Growing roles for the mTOR pathway. Curr Opin Cell Biol. 17: 596-603

[159] Backer JM (2008) The regulation and function of Class III PI3Ks: novel roles for Vps34. Biochem J. 410: 1-17

[160] MacKenzie MG, Hamilton DL, Murray JT, Taylor PM, Baar K (2009) mVps34 is activated following high-resistance contractions. J Physiol. 587: 253-260

[161] Toerien CA, Cant JP (2007) Abundance and phosphorylation state of translation initiation factors in mammary glands of lactating and nonlactating dairy cows. J Dairy Sci. 90: 2726-2734

[162] Moshel Y, Rhoads RE, Barash I (2006) Role of amino acids in translational mechanisms governing milk protein synthesis in murine and ruminant mammary epithelial cells. J Cell Biochem. 98: 685-700

[163] Prizant RL, Barash I (2008) Negative effects of the amino acids Lys, His, and Thr on S6K1 phosphorylation in mammary epithelial cells. J Cell Biochem. 105: 1038-1047

[164] Toerien CA, Trout DR, Cant JP (2010) Nutritional stimulation of milk protein yield of cows is associated with changes in phosphorylation of mammary eukaryotic initiation factor 2 and ribosomal s6 kinase 1. J Nutr. 140: 285-292

[165] Appuhamy JA, Knoebel NA, Nayananjalie WA, Escobar J, Hanigan MD (2012) Isoleucine and leucine independently regulate mTOR signaling and protein synthesis in MAC-T cells and bovine mammary tissue slices. The Journal of nutrition. 142: 484-491

[166] Hayashi AA, Nones K, Roy NC, McNabb WC, Mackenzie DS, Pacheco D, McCoard S (2009) Initiation and elongation steps of mRNA translation are involved in the increase in milk protein yield caused by growth hormone administration during lactation. J Dairy Sci. 92: 1889-1899

[167] Feuermann Y, Shamay A, Mabjeesh SJ (2008) Leptin up-regulates the lactogenic effect of prolactin in the bovine mammary gland in vitro. J Dairy Sci. 91: 4183-4189

[168] Burgos SA, Cant JP (2010) IGF-1 stimulates protein synthesis by enhanced signaling through mTORC1 in bovine mammary epithelial cells. Domest Anim Endocrinol. 38: 211-221

[169] Appuhamy JA, Bell AL, Nayananjalie WA, Escobar J, Hanigan MD (2011) Essential amino acids regulate both initiation and elongation of mRNA translation independent of insulin in MAC-T cells and bovine mammary tissue slices. The Journal of nutrition. 141: 1209-1215

[170] Bell AW, Bauman DE (1997) Adaptations of glucose metabolism during pregnancy and lactation. J Mammary Gland Biol Neoplasia. 2: 265-278

[171] Manjarin R, Steibel JP, Kirkwood RN, Taylor NP, Trottier NL (2012) Transcript abundance of hormone receptors, mammalian target of rapamycin pathway-related kinases, insulin-like growth factor I, and milk proteins in porcine mammary tissue. Journal of animal science. 90: 221-230

[172] Piantoni P, Bionaz M, Graugnard DE, Daniels KM, Everts RE, Rodriguez-Zas SL, Lewin HA, Hurley HL, Akers M, Loor JJ (2010) Functional and gene network analyses of transcriptional signatures characterizing pre-weaned bovine mammary parenchyma or fat pad uncovered novel inter-tissue signaling networks during development. BMC Genomics. 11: 331

[173] Zhao H, Huang M, Chen Q, Wang Q, Pan Y (2011) Comparative gene expression analysis in mouse models for identifying critical pathways in mammary gland development. Breast cancer research and treatment.

[174] Master SR, Hartman JL, D'Cruz CM, Moody SE, Keiper EA, Ha SI, Cox JD, Belka GK, Chodosh LA (2002) Functional microarray analysis of mammary organogenesis reveals a developmental role in adaptive thermogenesis. Molecular endocrinology. 16: 1185-1203

[175] Lemay DG, Neville MC, Rudolph MC, Pollard KS, German JB (2007) Gene regulatory networks in lactation: identification of global principles using bioinformatics. BMC Syst Biol. 1: 56

[176] Rudolph MC, McManaman JL, Hunter L, Phang T, Neville MC (2003) Functional development of the mammary gland: use of expression profiling and trajectory clustering to reveal changes in gene expression during pregnancy, lactation, and involution. J Mammary Gland Biol Neoplasia. 8: 287-307

[177] Wang C, Li Q (2007) Identification of differentially expressed microRNAs during the development of Chinese murine mammary gland. Journal of genetics and genomics = Yi chuan xue bao. 34: 966-973

[178] Avril-Sassen S, Goldstein LD, Stingl J, Blenkiron C, Le Quesne J, Spiteri I, Karagavriilidou K, Watson CJ, Tavare S, Miska EA, Caldas C (2009) Characterisation of microRNA expression in post-natal mouse mammary gland development. BMC genomics. 10: 548

[179] Silveri L, Tilly G, Vilotte JL, Le Provost F (2006) MicroRNA involvement in mammary gland development and breast cancer. Reproduction, nutrition, development. 46: 549-556

[180] Wang M, Master SR, Chodosh LA (2006) Computational expression deconvolution in a complex mammalian organ. BMC Bioinformatics. 7: 328

[181] Ji F, Hurley WL, Kim SW (2006) Characterization of mammary gland development in pregnant gilts. Journal of animal science. 84: 579-587

[182] Kim SW, Hurley WL, Han IK, Easter RA (1999) Changes in tissue composition associated with mammary gland growth during lactation in sows. Journal of animal science. 77: 2510-2516

[183] Capuco AV, Akers RM, Smith JJ (1997) Mammary growth in Holstein cows during the dry period: quantification of nucleic acids and histology. J Dairy Sci. 80: 477-487

[184] Denamur R (1974) Ribonucleic acids and ribonucleoprotein particles of the mammary gland.In: BL Larson and VR Smith, editors. Lactation: a comprehensive treatise, New York and London: Academic Press. pp 414-491

[185] Nagasawa H, Yanai R (1976) Mammary nucleic acids and pituitary prolactin secretion during prolonged lactation in mice. The Journal of endocrinology. 70: 389-395

[186] Riley LG, Zubair M, Thomson PC, Holt M, Xavier SP, Wynn PC, Sheehy PA (2006) Lactational performance of Quackenbush Swiss line 5 mice. Journal of animal science. 84: 2118-2125

[187] Roh SG, Baik MG, Choi YJ (1994) The effect of lactogenic hormones on protein synthesis and amino acid uptake in rat mammary acinar cell culture at various physiological stages. The International journal of biochemistry. 26: 479-485

[188] Kensinger RS, Collier RJ, Bazer FW, Ducsay CA, Becker HN (1982) Nucleic acid, metabolic and histological changes in gilt mammary tissue during pregnancy and lactogenesis. Journal of animal science. 54: 1297-1308

[189] Bionaz M, Loor JJ (2007) Identification of reference genes for quantitative real-time PCR in the bovine mammary gland during the lactation cycle. Physiol Genomics. 29: 312-319

[190] McClenaghan M, Springbett A, Wallace RM, Wilde CJ, Clark AJ (1995) Secretory proteins compete for production in the mammary gland of transgenic mice. Biochem J. 310 (Pt 2): 637-641

[191] Baldassarre H, Hockley DK, Olaniyan B, Brochu E, Zhao X, Mustafa A, Bordignon V (2008) Milk composition studies in transgenic goats expressing recombinant human butyrylcholinesterase in the mammary gland. Transgenic Res. 17: 863-872

[192] Brophy B, Smolenski G, Wheeler T, Wells D, L'Huillier P, Laible G (2003) Cloned transgenic cattle produce milk with higher levels of beta-casein and kappa-casein. Nat Biotechnol. 21: 157-162

[193] Homareda H, Komine S (1982) Casein synthesis by mouse polysomes and their messenger ribonucleic acid extracts. J Dairy Sci. 65: 915-919

[194] Wheeler MB, Monaco E, Bionaz M, Tanaka T (2010) The Role of Existing and Emerging Biotechnologies for Livestock Production: toward holism In: Acta Scientiae Veterinariae, pp s463-s484

Permissions

The contributors of this book come from diverse backgrounds, making this book a truly international effort. This book will bring forth new frontiers with its revolutionizing research information and detailed analysis of the nascent developments around the world.

We would like to thank Walter L. Hurley, for lending his expertise to make the book truly unique. He has played a crucial role in the development of this book. Without his invaluable contribution this book wouldn't have been possible. He has made vital efforts to compile up to date information on the varied aspects of this subject to make this book a valuable addition to the collection of many professionals and students.

This book was conceptualized with the vision of imparting up-to-date information and advanced data in this field. To ensure the same, a matchless editorial board was set up. Every individual on the board went through rigorous rounds of assessment to prove their worth. After which they invested a large part of their time researching and compiling the most relevant data for our readers. Conferences and sessions were held from time to time between the editorial board and the contributing authors to present the data in the most comprehensible form. The editorial team has worked tirelessly to provide valuable and valid information to help people across the globe.

Every chapter published in this book has been scrutinized by our experts. Their significance has been extensively debated. The topics covered herein carry significant findings which will fuel the growth of the discipline. They may even be implemented as practical applications or may be referred to as a beginning point for another development. Chapters in this book were first published by InTech; hereby published with permission under the Creative Commons Attribution License or equivalent.

The editorial board has been involved in producing this book since its inception. They have spent rigorous hours researching and exploring the diverse topics which have resulted in the successful publishing of this book. They have passed on their knowledge of decades through this book. To expedite this challenging task, the publisher supported the team at every step. A small team of assistant editors was also appointed to further simplify the editing procedure and attain best results for the readers.

Our editorial team has been hand-picked from every corner of the world. Their multi-ethnicity adds dynamic inputs to the discussions which result in innovative outcomes. These outcomes are then further discussed with the researchers and contributors who give their valuable feedback and opinion regarding the same. The feedback is then collaborated with the researches and they are edited in a comprehensive manner to aid the understanding of the subject.

Apart from the editorial board, the designing team has also invested a significant amount of their time in understanding the subject and creating the most relevant covers. They scrutinized every image to scout for the most suitable representation of the subject and create an appropriate cover for the book.

The publishing team has been involved in this book since its early stages. They were actively engaged in every process, be it collecting the data, connecting with the contributors or procuring relevant information. The team has been an ardent support to the editorial, designing and production team. Their endless efforts to recruit the best for this project, has resulted in the accomplishment of this book. They are a veteran in the field of academics and their pool of knowledge is as vast as their experience in printing. Their expertise and guidance has proved useful at every step. Their uncompromising quality standards have made this book an exceptional effort. Their encouragement from time to time has been an inspiration for everyone.

The publisher and the editorial board hope that this book will prove to be a valuable piece of knowledge for researchers, students, practitioners and scholars across the globe.

List of Contributors

Haiyan Sun
China Animal Husbandry Zhihe (Beijing) Biotech Co., Ltd, Beijing, China

Håvard Jenssen
Roskilde University, Dept. of Science, Systems & Models, Roskilde, Denmark

Gabriella Pinto, Marina Cuollo, Sergio Lilla, Lina Chianese and Francesco Addeo
Department of Food Science, University of Naples "Federico II", Parco Gussone, Portici (Naples), Italy

Simonetta Caira
Food Science Institute of the National Research Council (C.N.R.), Via Roma, Avellino, Italy

Teresa Treweek
Graduate School of Medicine, University of Wollongong, Australia

Marcel Jøhnke and Torben E. Petersen
Department of Molecular Biology and Genetics, Aarhus University, Aarhus, Denmark

Bärbel Lieske
Germany

Tomoko Shimamura and Hiroyuki Ukeda
Faculty of Agriculture, Kochi University, Nankoku, Japan

Paolo Polidori and Silvia Vincenzetti
Università di Camerino, Italy

Simonetta Caira, Rosa Pizzano and Gianluca Picariello
Food Science Institute of the National Research Council (C.N.R.),Avellino, Italy

Gabriella Pinto, Marina Cuollo, Lina Chianese and Francesco Addeo
Department of Food Science, University of Naples "Federico II", Portici (Naples), Italy

Barłowska Joanna, Wolanciuk Anna, Litwińczuk Zygmunt and Król Jolanta
University of Life Sciences in Lublin, Poland

Gema Lucia Zambrano-Burbano, Yohanna Melissa Eraso-Cabrera, Carlos Eugenio Solarte-Portilla and Carol Yovanna Rosero-Galindo
University of Nariño, Animal Genetic Improvement Program, Pasto, Colombia

Edyta Molik and Dorota Zięba
Department of Swine and Small Ruminant Breeding, Biotechnology and Genomic Laboratory, Agricultural University in Krakow, Poland,

Genowefa Bonczar and Aneta Żebrowska
Department of Animal Product Processing, Agricultural University in Krakow, Poland,

Tomasz Misztal
Department of Endocrinology, The Kielanowski Institute of Animal Physiology and Nutrition, Polish
Academy of Sciences, Poland

Massimo Bionaz, Walter Hurley and Juan Loor
Department of Animal Sciences, University of Illinois, Urbana, Illinois, USA